The Indies of the Setting Sun

The Indies of the Setting Sun

*How Early Modern Spain Mapped
the Far East as the Transpacific West*

Ricardo Padrón

The University of Chicago Press   CHICAGO AND LONDON

The University of Chicago Press, Chicago 60637
The University of Chicago Press, Ltd., London
© 2020 by The University of Chicago
Published 2020
Printed in the United States of America

30 29 28 27 26 25 24 23 22   3 4 5 6

ISBN-13: 978-0-226-45567-9 (cloth)
ISBN-13: 978-0-226-68962-3 (e-book)
DOI: https://doi.org/10.7208/chicago/9780226689623.001.0001

The University of Chicago gratefully acknowledges the generous support
of the Dean of the College and Graduate School of Arts and Sciences and
the Vice President for Research at the University of Virginia toward the
publication of this book.

Library of Congress Cataloging-in-Publication Data

Names: Padrón, Ricardo, 1967– author.
Title: The Indies of the setting sun : how early modern Spain mapped the
Far East as the Transpacific West / Ricardo Padrón.
Other titles: How early modern Spain mapped the Far East as the
Transpacific West
Description: Chicago : The University of Chicago Press, 2020. | Includes
bibliographical references and index.
Identifiers: LCCN 2019039266 | ISBN 9780226455679 (cloth) |
ISBN 9780226689623 (ebook)
Subjects: LCSH: Pacific Area—Discovery and exploration. | Pacific Area—
Maps—History. | Cartography—Spain—History—16th century. | Pacific
Area—In literature. | Spain—Civilization—1516–1700. | Spain—Relations—
Pacific Area. | Pacific Area—Relations—Spain.
Classification: LCC DU65 .P285 2020 | DDC 950/.3—dc23
LC record available at https://lccn.loc.gov/2019039266

♾ This paper meets the requirements of ANSI/NISO Z39.48-1992
(Permanence of Paper).

Para mi Teresita

Contents

Figures

Introduction

I was working in Spain's Biblioteca Nacional, trying to read everything I could that bore a title like "Description of the Indies" or "Description of America," when one of the librarians brought me an account of early forays into New Mexico that was bound with a description of China, the second edition of Juan González de Mendoza's *Historia de las cosas mas notables, ritos, y costumbres del gran Reino de la China* (History of the most notable things, rites, and customs of the Great Kingdom of China, Madrid, 1585).[1] I had never heard of Mendoza's book and was puzzled as to why anyone would publish these two texts between the same covers. My surprise only grew as I learned that the book had been an early modern best seller, going through forty-five editions in seven languages within fifteen years of its initial publication. This meant that it was not just a sixteenth-century European book about China, but *the* sixteenth-century European book about China, the standard reference on the subject before Jesuit Sinology became dominant in the following century. It was also one of the most influential publications about a secular topic to have come out of Spain during the early modern period. Nevertheless, while it was well known among students of the early modern European encounter with Asia, only a very few scholars in my home fields of early modern Spanish studies and colonial Latin American studies seemed to be aware of its existence. Having never been praised as a masterpiece of Golden Age Spanish writing, or appropriated as an exemplar of early Latin American writing, Mendoza's text had been ignored when the canons of these fields were constituted, and it had eventually fallen off the scholarly radar altogether, as had the vast archive of Spanish writing about East and Southeast Asia, of which Mendoza's book was only one example.

The existing scholarship provided an explanation for the book's strange combination of subject matter. Its author was an Augustinian friar who had been appointed to serve as Philip II's first ambassador to China. Mendoza made for the Middle Kingdom by way of Mexico, where he was to board one of the so-called Manila Galleons that sailed every year from Acapulco to Manila and then travel onward from the Philippines to China, yet for reasons that remain unclear, he turned back after a two-year stint in Mexico City, never making it across the Pacific. The friar nevertheless acquired a reputation as an authority on things Chinese, at least in the eyes of Pope Gregory XIII, who commissioned him to write a history of the country. Cribbing from existing print materials in both Spanish and Portuguese, but also drawing on the accounts of other Spaniards who had actually been to China and whom he had met in Mexico, Mendoza published his book in Rome in 1585. Its second-hand description came accompanied by three of the travel narratives that he had consulted in writing his book, apparently to prop up its credibility. When the second edition of the *Historia del gran reino de la China* came out in Madrid in 1586, one of these travel narratives was revised to include the news of Antonio de Espejo's expedition into New Mexico between 1582 and 1583, which had reached Mexico City while Mendoza was there. It would seem, therefore, that simple biographical accident was the reason that material about New Mexico had appeared between the same covers as a history of China in sixteenth-century Spain. Mendoza had been trying to get to China by way of Mexico, and while he was in the viceregal capital, he had learned about both his intended destination and new discoveries on the northern frontier of the viceroyalty.

It eventually became clear to me, however, that this was not a satisfactory explanation. If it made sense to describe China and New Mexico in the same volume, it was not just because a single individual had experienced these places as stops along a personal route of travel, but because there existed an infrastructure that linked Spain to parts of Asia by way of America and the Pacific. Travel along that route may very well have led other early modern Spaniards to experience what we call East and Southeast Asia as a region that lay to the west, across the South Sea, and to understand that region in relation to the New World and Spain's experience there. That infrastructure, moreover, was itself the product of a long-standing effort to reach the East by sailing west that went back to Columbus and his contemporaries, and that had been institutionalized by the papal encyclicals and international treaties, most famously the 1494 Treaty of Tordesillas, that divided the world between Portugal and Castile. According to these arrangements, as interpreted by the Spanish themselves, everything from the original line of demarcation at the mouth of the Amazon River, more or less, all the way to the mouth of the Ganges River, or at least to the Straits of Malacca, was rightfully Spanish, yet Spaniards could only access this vast expanse by sailing westward. The eastward route around the Cape of Good Hope was closed to them.

This meant that if Spaniards wanted to exercise their legal prerogatives to trade, conquer, and settle in East and Southeast Asia, they had to find a way around the New World that appeared athwart the route west, and a way across the ocean that Balboa discovered when he traversed the Isthmus of Panama in 1513. Spain's efforts to do so began to show promise when the Magellan expedition reached the Spice Islands by way of the Pacific. They began to pay dividends forty years later when, after several failed attempts to exploit Magellan's breakthrough, Miguel López de Legazpi finally established a Spanish colony on the far side of the South Sea. Sailing west to reach the East, therefore, was not just the experience of those who made the voyage on the Manila Galleons, much less that of a single friar who never even boarded them. It was a fundamental part of Spain's entire spatial experience as an imperial power.

This spatial experience, I came to realize, drew upon and in turn nourished a unique geopolitical imaginary. That imaginary ran against two prevailing trends in early modern worldmaking. The first was the mounting tendency to identify the New World as "America," a continent separated from Asia by the vast, empty Pacific Ocean and bound to Europe by the ever-growing network of political, economic, cultural, and demographic ties that spanned the Atlantic Ocean. The second was the much-less-novel tendency that built upon the received geographical tradition and drew upon the spatial experience of Portuguese imperialism to map the region that we call East and Southeast Asia as the extreme end of a broad Orient. These two trends came together in the overall arrangement of the vast majority of sixteenth-century world maps, which oppose the New World in the west to the Old World in the east across an Atlantic Basin that sits very much in the center of things (fig. 1). The Spanish geopolitical imaginary, by contrast, resisted the twin ideas that the New World was entirely separate from Asia and that the Pacific Ocean constituted the geographical and even ontological boundary between these two parts of the world and their inhabitants. Where others saw a vast and indomitable South Sea, the Spanish saw a relatively narrow and navigable oceanic basin. Where the prevailing European imaginary saw separation, the Spanish imaginary saw various forms of continuity and connection. Spanish officialdom eventually institutionalized this way of mapping the world by calling East and Southeast Asia *Las Indias del poniente*, the Indies of the West, or, more poetically, the Indies of the Setting Sun. On the official cartography of the Spanish crown, the Far East became the transpacific West (fig. 2). This book explores some of the many manifestations of this unique geopolitical imaginary, demonstrating just how malleable the world was to sixteenth-century Europeans and thereby reminding us of how malleable it continues to be today.

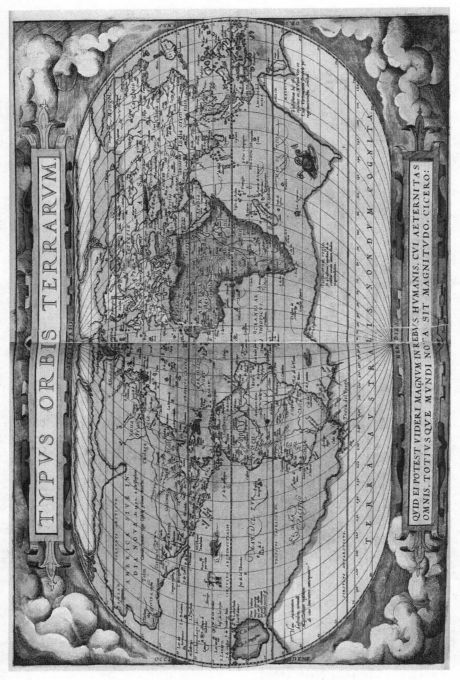

FIGURE 1. One of the sixteenth century's most influential maps of the world. Abraham Ortelius, *Typus Orbis Terrarum* (Antwerp, 1570). 25 x 50 cm. Photograph: Courtesy of the Geography and Map Division, Library of Congress, Washington, DC (call no. G1006.T5 1570; control no. 2003683482).

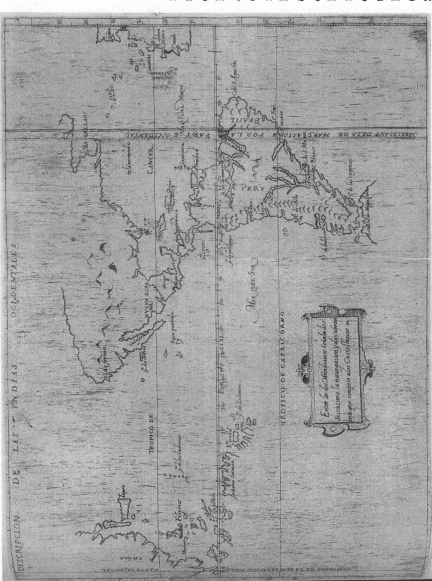

FIGURE 2. The official map of Spain's overseas empire in Antonio de Herrera y Tordesillas, *Descripción de las Yndias Ocidentales* [sic] (Madrid, 1601). 28 x 37 cm. From Herrera y Tordesillas, *Historia general de los hechos de los castellanos . . .* (First Decade) (Madrid, 1601–15). Photograph: Courtesy of the JCB Map Collection, John Carter Brown Library, Brown University (call no. B601 H564h/1-SIZE; file 01808-006).

This Spanish geopolitical imaginary exists in a select number of titles from the notional archive that contains Mendoza's book and what my field used to call the "chronicles of the Indies." The term refers to a corpus of texts written in a variety of genres over the course of the long sixteenth century by Spanish and *mestizo* authors, all dealing in one way or another with the Americas. Some, like Bartolomé de las Casas's *Brevísima relación de la destrucción de las Indias* (A very brief account of the destruction of the Indies, 1552), have long-established reputations, while others, like Felipe Guamán Poma de Ayala's *El primer nueva corónica y buen gobierno* (The first new chronicle and good government, ca. 1616), have received attention only within the last few decades. Over the years, they have been pressed into service as early contributions to the literature of different Latin American republics or of Latin America as a whole, as prototypes of the novel as a literary genre, as ideological instruments of the conquest of the Americas or as sites of resistance to that conquest, as manifestations of colonial and postcolonial subjectivities, and, last but not least, as important sources for the history of early America, both before and after the arrival of Europeans. Yet never, as far as I know, have they been studied as part of a broader category of textual production that *also* includes early modern Spanish writing about those parts of the broader Indies that did not, in the long run, produce a Spanish-speaking nation-state that could claim them as its own, that archive of Spanish writing about East and Southeast Asia that we might call the "chronicles of the Indies of the West."

This book examines selections from both of these collections of "chronicles" and compares them to maps from the same period in ways that I hope are mutually illuminating. It focuses on neglected portions of well-known *crónicas de Indias* that have often been read as contributors to the ongoing invention of America, episodes that reach across the Pacific to include the Indies of the West in the text's overall vision of the Spanish Indies. It compares these texts with maps that have been studied as influential representations of the New World, and it considers how these same maps insert the New World into a broader, transpacific Indies. The book also seeks to uncover the ways in which Spanish texts about East and Southeast Asia, verbal and cartographic, connect the region and its people to the New World, often by looking back upon the history of Spain's effort to reach the East by sailing west. Most of the texts I discuss appeared in print and circulated widely, and they are conventionally thought to have exercised considerable influence upon the ways that Europeans imagined the wider world, but I also attend to manuscript material that informed decision making at the highest levels of Spain's imperial enterprise or that served as sources for print publications. I do not pretend that the corpus of "cartographic literature" that I explore in these pages is the only one that allows us to discuss the issues in question, nor do I intend to make a statement about the relative value of print versus manuscript sources. If print materials prevail over manuscript materials, finished texts over fragmentary documents, it is a result of my training as a scholar of literature rather than as an archival historian, as well as of my own personal experience with the Borgesian library of potential materials. If this book inspires others to find their own way through this material, and perhaps to reach different conclusions about the issues I raise, it will have achieved its purpose.

I present the argument in eight chapters that can be divided into three groups. The first group, chapters 1 and 2, provide the theoretical and methodological framework for everything that follows. Chapter 1, "The Map behind the Curtain," fleshes out what I have said in this brief introduction, placing the argument in the context of existing historiography about the invention of America and current scholarship on the early modern Spanish Pacific. It also makes some preliminary observations about the crucial problem of metageography, arguing that if we are to understand the way the Indies were constructed by the early modern Spanish geopolitical imaginary, we must let go of the entrenched habit of thinking about early modern European efforts to map the globe exclusively in terms of the continents and of the shift from a tripartite to a quadripartite model put in motion by Martin Waldseemüller and his collaborators in 1507. In order to understand how Spanish cartographic literature mapped the Indies, we must first understand that there were a variety of metageographical frameworks available for mapping the world, and that they could be put to use in a variety of ways, sometimes simultaneously. Chapter 2, "South Sea Dreams," develops this assertion by reading and rereading a single nautical chart of the world constructed around 1519 by the Portuguese mapmaker Jorge Reinel and supposedly used by Ferdinand Magellan to make the case for his prospective voyage to the Spice Islands before the court of Charles V. Each reading serves to introduce a different metageography—the architecture of the continents, the theory of climates, and the emerging maritime networks of the Iberian empires—and to demonstrate how it could be used to imagine the world around the time that Balboa discovered the South Sea. It also serves as a model of the general method that I follow in subsequent chapters, involving contextualized close reading of period sources.

The next group, chapters 3 and 4, deals with maps and writing created in the aftermath of the Magellan expedition, between 1523 and 1552. Before Magellan, it was not difficult to imagine geographical and ethnographical continuity between the lands that had been discovered in the

Ocean Sea and the Indies of Marco Polo. The Magellan expedition, however, demonstrated that the distance from the New World to the Spice
Islands was much greater than anyone had expected it to be, and that
an ocean of vast size lay between the two. In fact, many classic histories
of discovery and exploration go so far as to argue that these revelations
served to consolidate the notion that the New World was indeed a continent separate from Asia. Chapter 3, "Pacific Nightmares," therefore,
explores Magellan's supposed discovery of the Pacific as a textual event,
arguing that the Pacific was invented as an impossibly broad, empty
ocean that effectively separated America from Asia in the writing of Antonio de Pigafetta, the author of the most comprehensive and significant
first-hand account of the first circumnavigation of the Earth. Yet it also
explores the Spanish response to Pigafetta's invention, a series of official
and semi-official accounts of the expedition by Maximilian von Sevenborgen (aka Maximilianus Transylvanus, 1523), Pietro Martire d'Anghiera
(aka Peter Martyr, 1530), and Gonzalo Fernández de Oviedo (1530s), as
well as an elaborate copy of Spain's official map of the world constructed
by the crown cosmographer Diogo Ribeiro and presented as evidence of
the rightness of Spain's claim to the Spice Islands (1529). These texts contain the implications of Magellan's discovery, as figured by Pigafetta, by
deploying what I call the rhetoric of smooth sailing and the cartography
of containment, two representational strategies that became constant
features of Spanish Pacific cartography as the century progressed. They
nevertheless reveal anxieties about the very possibility of successfully
navigating the South Sea.

Chapter 4, "Shipwrecked Ambitions," examines what happens in
Spanish cartography and historiography after various attempts to follow
up on Magellan's breakthrough ended in catastrophe. Oviedo's *Historia
general y natural de las Indias* (General and natural history of the Indies,
1530s) and Gómara's *Historia general de las Indias* (General history of the
Indies, 1552) are considered classics of Spanish Americana, as well as contributors to the ongoing invention of America by the European imagination, yet they include extensive accounts of Spain's effort to control the
islands of the western Pacific, from Magellan to the failed expedition of
Ruy López de Villalobos twenty years later. They thus help us understand how the Spanish geopolitical imaginary responded to the defeat of
its transpacific aspirations, and the developing sense that America and
Asia were indeed quite different and separate places. Oviedo appeals to
the shared characteristics of mutually distant tropical locations to map
the transpacific Indies as a natural reality, and to the discourse of heroic
masculinity to claim that Spain can overcome the obstacles of Pacific navigation and Portuguese opposition to convert the transpacific Indies into

a political reality, yet ends up revealing his anxieties about the feasibility of South Sea conquest. These anxieties turn into frustration in Gómara, who expresses doubts about the explanatory power of the theory of climates and prefers to map the world through the architecture of the continents, insisting that the New World is an insular landmass, home to a distinct branch of the human family. Eager, nevertheless, to hold on to Spain's fading dreams of transpacific empire, he insists more adamantly than anyone else that the South Sea is narrow and navigable, and that the Spicery, however different it may be from the New World, is rightfully Spanish.

The third group, chapters 5 through 8, examines cartographic literature produced after the revival of Spain's transpacific ambitions in the conquest of the Philippines by Legazpi and the establishment of the Manila Galleon route. Chapter 5, "Pacific Conquests," turns to the work of Spain's official chronicler and cosmographer of the Indies, Juan López de Velasco, which plays a crucial role in the history of the geographical imaginary at issue in this book. Velasco inherits Gómara's vision of the New World as a place apart, but writing in the aftermath of Legazpi, he turns the tables on the ongoing invention of America and consolidates an official vision of the Spanish Indies as a transpacific expanse. By imagining the Spanish New World through the framework of the theory of climates, as Oviedo had, Velasco reopens the Indies to include the islands in the western Pacific, ultimately forging a tripartite geography composed of the *Indias del mediodía* (Indies of the South, i.e., South America), the *Indias del septentrión* (Indies of the North, i.e., North America), and the *Indias del poniente* (Indies of the West, i.e., East and Southeast Asia). In the process, he also pulls off a powerful discursive and cartographic conquest of the Pacific that allows the Indies of the West to emerge in the Spanish imagination as a new imperial frontier. Velasco nevertheless struggles with those countries that were coming to replace the Spicery as the key destinations on Castile's transpacific horizon, China and Japan. In his major work, the *Geografía y descripción de las Indias* (1574), he tries to squeeze the square peg of civilized China into the round hole of the tropical Indies in order to imagine the Middle Kingdom as a place available for conquest, while in his subsequent *Sumario* (ca. 1580) he imagines China as a commercial partner in a peaceable transpacific space.

The Middle Kingdom emerges front and center in chapter 6, "The Location of China," which analyzes the geopolitical work done by Bernardino de Escalante's *Discurso de la navegación que hacen los Portugueses a oriente y noticia del reino de la China* (Discourse on the navigation that the Portuguese make to the orient and news of the kingdom of China, 1577) and Juan González de Mendoza's *Historia del gran reino de la China*

(History of the great kingdom of China, 1585), the two most important Spanish texts on China from this period, and the first Spanish texts devoted exclusively to East Asia to appear in print. These texts extend and amplify the vision of the Spanish Pacific found in Velasco by proposing a Sinophilic version of Spain's encounter with the empire of the Ming designed to counter the Sinophobic proposals of Spaniards in Manila who wanted to think of China as another Mexico or Peru, a tyrannous regime available for conquest by Spain. Unlike the Sinophobes, Escalante and Mendoza locate China in the northern temperate zone, and near the top of the hierarchy of human societies that was emerging from Europe's encounter with the wider world. Yet while they paint China's government in nearly utopian terms, drawing a stark contrast with the prevailing vision of Amerindian polities, they nevertheless locate China in the Castilian west, at the far end of a hodological space defined by Spain's movement across the Indies and across a narrow and navigable Pacific.

While questions of just and unjust government predominate in my discussion of China, religious matters come to the fore in my discussion of Japan. Chapter 7, "The Kingdom of the Setting Sun," turns to a Franciscan history written in the aftermath of the one of the most dramatic episodes in Europe's early modern encounter with East Asia, the 1597 crucifixion of twenty-six Christians just outside the Japanese city of Nagasaki. It examines Marcelo de Ribadeneira's 1601 history of the Spanish Franciscan enterprise in the Indies of the West, the *Historia de las islas del Archipiélago Filipino y reinos de la Gran China, Tartaria, Cochinchina, Malaca, Siam, Cambodge y Japón* (History of the islands of the Philippine Archipelago and the kingdoms of Greater China, Tartary, Cochinchina, Malacca, Siam, Cambodia, and Japan), as an attempt to map the Far East as the apocalyptic endpoint of Christianity's world-historical pilgrimage from its place of birth in the Mediterranean to the New World and beyond. The book avails itself of the established geopolitical imaginary, and specifically of its persistent treatment of first Zipangu and then Japan as a crucial way station on the way west, but it also departs from other manifestations of that imaginary by failing to contain the Pacific and turning the rhetoric of smooth sailing on its head.

Finally, chapter 8, "The Anxieties of a Paper Empire," turns to the culminating vision of Spain's transpacific dreams in the official historiography of Antonio de Herrera y Tordesillas and Bartolomé Leonardo de Argensola. Herrera's 1601 *Historia general de los hechos de los castellanos en las Islas y Tierra Firme del Mar Océano* (General history of the deeds of the Castilians in the Islands and Mainland of the Ocean Sea) remained unfinished, but nevertheless provides fascinating glimpses into the historian's audacious attempt to map the Spanish Indies as a transpacific expanse, and the history of Spanish imperialism as a transpacific enterprise. I argue that Argensola's 1609 *Conquista de las Malucas* (The conquest of the Moluccas) can be interpreted as a a *de facto* continuation of Herrera's project. It turns the progress of empire away from its American setting, with its nasty war in the south of Chile, and projects it westward across the South Sea, where a recent naval expedition had just succeeded in recovering the Spice Islands from the Dutch. In so doing, it maps an expanded vision of Spain's empire as a space extending from the Strait of Magellan to the Strait of Mozambique, but nevertheless it reveals anxieties about the potential for conflict with the mighty Ming and unwittingly admits to the paucity of Spain's actual capabilities in the region.

A brief conclusion explains why this book ends where it does and sums up what I believe are its principal implications. Before I conclude this Introduction, however, I must say a few words about some of the things this book does not do. This is not a book about Asia in the early modern Spanish imagination, so the reader will find little or no mention of certain authors and texts that would likely enjoy pride of place in a survey of that topic, such as the writings of St. Francis Xavier or the popular *Peregrinação* (Peregrination) of Fernão Mendes Pinto, which enjoyed considerable popularity in Spain during the seventeenth century.[2] Texts like these map East and Southeast Asia as the Far East, following the example of the past and riding on the experience of the Portuguese and the Jesuits in their own day. They do not belong in a book about an ongoing attempt to map the world against the grain of that tradition, at least not as primary objects of analysis. Portuguese and Jesuit maps and writing figure in the following pages only as far as my all-too-mortal capabilities have allowed. Although I recognize that a full, comparative treatment of such materials may very well have enriched this argument, I have found it enough, in the context of this volume, to challenge the conventional separation of the Indies into Spanish America and Hispanoasia. The construction of the boundary between Hispanoasia and Lusoasia is a closely related matter, but it is sufficiently complex to require a book of its own.[3]

This is also not a book about the Philippines in early modern Spanish writing and mapmaking, as can be seen quite plainly from the outline of chapters. Although I devote some attention in the conclusion to one of the most important early books about the Philippines, Antonio de Morga's 1609 *Sucesos de las Islas Filipinas* (Events of the Philippine Islands), I have chosen not to devote a full chapter to any one text about Spain's only real colony in East and Southeast Asia. This is because most of the relevant material, including Morga's secular history and its ecclesiastical analogue, the 1604 *Relación de las islas Filipinas* (Account of the Philippine Islands) of Pedro Chirino, tends to delve deeply into local matters and loses sight of the larger geopolitical frame that is the subject of this book. The Philippines are nevertheless very much present throughout this book, as the central node of the geopolitical imagination of everyone who mapped the Spanish Pacific after 1565, including López de Velasco, Escalante, Mendoza, Ribadeneira, Herrera, and Argensola.

Finally, this book does not engage, as I wish it could, with the cultural production of non-Europeans or with the problematics of encounter. The cartographic projects I examine in this book function on a small scale in the technical sense of the term, that is, they cover a broad swath of territory. As a result, they involve contacts between Europeans and a tremendous variety of non-Europeans, including the indigenous inhabitants of the Americas, the Pacific Islands, insular Southeast Asia, Cambodia, Thailand, Vietnam, southern coastal China, Japan, and New Guinea. No single scholar could possibly assemble the skill set necessary to handle such multiplicity and diversity on her or his own. A cross-cultural approach would require close collaboration across fields that are not in the habit of collaborating, such as Latin American studies and Asian studies, not to mention a much more localized approach to the problems and materials involved. Hopefully, as Spanish Pacific studies continue to emerge as a field of research, the necessary networks of collaboration will also emerge and this book will be rendered obsolete. In the meantime, I hope it serves as a provocation for further exploration, based on maps of the early modern past that look rather different from the ones that we have been using to date.

1

The Map behind
the Curtain

Today, the easternmost tip of Cuba is known as Cape Maisí, but when Christopher Columbus first sighted it during the autumn of 1492, he gave it a much more ominous-sounding name. From the end of October through the middle of November, Columbus had been sailing along the Atlantic coast of Cuba, trying to determine if it was the fabled island of Zipangu or part of the Asian mainland. As he began to favor the second option, he searched for signs that he might be near the city of Quinsay, modern Hangzhou, which Marco Polo had described as the largest, wealthiest, and most fabulous city of the empire of the Great Khan. He even dispatched an embassy inland, equipped with a Latin passport, a letter of introduction from Ferdinand and Isabella, and a royal gift, in the hope of locating one of the Khan's cities and making official contact with his government. When the embassy returned on November 5 having stumbled onto a Taíno settlement that was sizable by local standards but nothing like the metropolis Columbus was looking for, the admiral directed his fleet toward the east and the islands he knew lay in that direction. According to Bartolomé de las Casas, when Columbus traversed the waters that separate Cuba from Haiti, he sighted Cape Maisí and called it "Cape Alpha and Omega." According to the friar, the admiral thought that this promontory was the easternmost tip of the Asian continent and chose to give it a name that marked its significance as the beginning and end of the East (Casas 1986, 1.256–57, 400, 407).

There is reason to doubt the accuracy of this anecdote, since the name does not appear in either Columbus's diary or the letters he wrote reporting his discoveries on his first voyage. As far as I can tell, it makes its first appearance in the extant Columbian corpus in an account of the admiral's voyage to Cuba and Jamaica, written in 1495 (Colón 1992, 290). Accurate or not, however, the anecdote encapsulates one of the major issues that faced anyone who was trying to understand the new discoveries that the Spanish and Portuguese started to make in the ocean west of Europe and Africa starting in 1492. Columbus claimed to have crossed the invisible frontier separating the western Ocean Sea, as it was known to the late medieval imagination, from the Indies as they had been described by Marco Polo, but others doubted him and claimed that the true East still lay beyond the horizon of Columbus's discoveries. As further travel revealed that a previously unknown landmass stretched south of the islands the admiral had discovered, the new lands came to be understood as a New World. In 1507, Martin Waldseemüller and his collaborators in the Rhenish town of St. Dié des Vosges decided that this landmass constituted a fourth part of the world, distinct from and commensurate with the traditional three, Europe, Africa, and Asia. They christened it "America," after Amerigo Vespucci, to whom they attributed its discovery. On Waldseemüller's 1507 map of the world, a hypothetical body of water separates America from the island of Zipangu, helping to constitute the newly discovered fourth part of the world as a massive island separate from Asia.[1] In effect, that body of water plays the role that Columbus had assigned to the waters between Cuba and Hispaniola, now known as the Windward Passage, as the boundary between East and West (fig. 3).[2]

Only a single copy of Waldseemüller's 1507 world map has survived, yet its impact upon the geographical imagination of its day cannot be doubted. Influential cosmographers like Johannes Schöner and Peter Apian copied Waldseemüller's innovation and helped disseminate his ideas through maps of their own, like the world map that Apian included in his wildly popular *Cosmographia* of 1524. It soon became standard practice in much of sixteenth-century Europe, when one flattened the globe into a two-dimensional map of the world, to cut through the ocean between America and Asia, so that the Atlantic stretched down the center, the New World appeared on the left, and the Old World on the right.[3] In this way, America and Asia were not just separated but also opposed to each other as opposite and very different ends of a new and expanded *orbis terrarum* centered on the Atlantic and bound on either side by parts of what we call the Pacific Ocean. This image, of course, is exceedingly familiar to us today, and for many of us educated in the western world, it looks utterly obvious and natural.

Yet to the highly educated elites of Ming China, the sixteenth-century European image of the world looked like complete nonsense. They knew their country as Zhongguo (中国), the Middle Kingdom, the country at the center of the world, and they had their own cartographic tradition that gave graphic form to their Sinocentric cosmology. They could never give credence to a map that consigned China to the edge of the world, as early modern European cartography inevitably did.[4] In response to this exigency, Matteo Ricci, the Jesuit priest at the heart of the early modern

FIGURE 3. America appears named as such for the first time on Martin Waldseemüller's *Universalis cosmographia secundum Ptholomaei traditionem et Americi Vespucii aliorumque lustrationes* (Strasbourg, France[?], 1507). 128 x 233 cm. Photograph: Courtesy of the Geography and Map Division, Library of Congress, Washington, DC (call no. G3200 1507.W3; control no. 2003626426).

Christian mission to China around the year 1600, designed a map of the world based on European cartographic principles but answering to Chinese sensibilities. Eager to court the favor of the Ming elites while at the same time impressing them with European technical achievements, he constructed a map that placed the Pacific, rather than the Atlantic, at the center of the image and thereby moved the Middle Kingdom away from the margins of the world and toward its center, where Ricci's Chinese interlocutors knew it belonged (fig. 4; Zhang 2015, 45–56). This famous anecdote reminds us of something we all know but often forget, that however natural a given map image may seem, it is very often the product of a variety of cultural and ideological biases that not only remain unacknowledged but are actually obscured by the map's claim to objectivity and apparent transparency. Just as it was no accident that the Chinese thought their country should appear at the center of any truthful map of the world, so it was no accident that Europeans placed the Atlantic Ocean at the center of their maps.

The Mexican historian Edmundo O'Gorman helps us understand how they came to do so in his magisterial essay *The Invention of America*. This classic in the literature on the European encounter with the New World is famous for making the case that "America" must be understood as something invented by the European imagination rather than discovered by European explorers, yet that is not all it has to say.[5] O'Gorman emphasizes that the invention of America as a fourth part of the world separate from Asia came hand in hand with the invention of America's indigenous inhabitants as a distinct branch of the human family uniquely needful of the civilizing and evangelizing influence of Europe. His book is not just about physical geography, but also about human geography in a Eurocentric, colonialist register. The process of invention, he argues, not only distinguished America and Americans from Asia and Asians but also tied the destiny of America and its inhabitants to the colonial and imperial designs of Europe (O'Gorman 1961, 144–45).[6] To put it another way, the invention of America was a milestone not just in the history of European geographical thought, but also in the political, military, economic, and social work of forging what we have come to call the Atlantic world, and what O'Gorman himself calls "the new *Mare Nostrum*, the Mediterranean of our day" (O'Gorman 1961, 145; 1986, 139–52).[7] This is the reason that the process of invention produced an image of the world that was centered, not on Europe as we might expect, but on the Atlantic Ocean, what Carla Lois calls the "spinal column of the West" (Lois 2008, 164). It was the apparent availability of America for colonial refashioning that made it different from Asia, where sixteenth- and seventeenth-century Europeans held so little sway, and it was Europe's transatlantic tie with

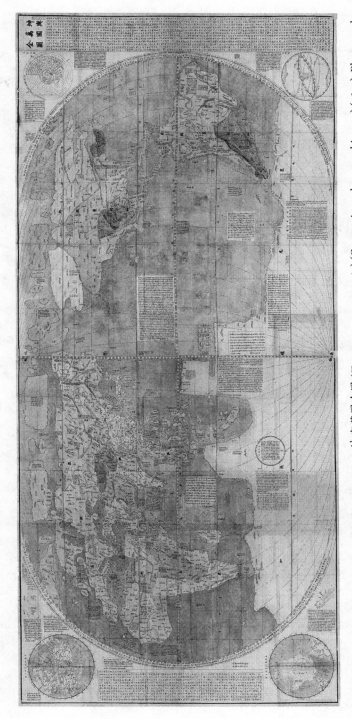

FIGURE 4. The Pacific-centered world of Matteo Ricci, 坤輿萬國全圖 (*Kun yu wan guo quan tu*) (Great universal geographic map) (1602). Photograph: Wikipedia.

its American theater of colonial operations that made for the centrality of the Atlantic Ocean in the European image of the world.

This process was abetted by the very different nature of the European encounter with the Pacific during the sixteenth and seventeenth centuries. That encounter began early in the sixteenth century, when Iberians entered the waters of what we now call the Pacific Ocean from both east and west. Portugal's Antonio Abreu reached the ocean first, when he ventured from Portuguese Malacca to the Spice Islands in 1511, but as far as we can tell, he made no claims to have discovered a new body of water and did nothing to rename the seas he sailed.[8] By contrast, Vasco Núñez de Balboa, who sighted what we call the Pacific from a height in Panama in 1513, claimed to have discovered a new body of water, christened it the "South Sea," and claimed it and all the lands it washed for the crown of Castile and León (Nowell 1947, 7–10).[9] His grandiosity was an attempt to win the favor of the crown, which had become anxious about the broader geopolitical situation. Although Spain and Portugal had divided the world between them in the 1494 Treaty of Tordesillas, no one could map the treaty lines with any accuracy, so there was ample room for disagreement about which kingdom could call the Indies its own. The problem was that the Portuguese had already reached the coveted Spice Islands, while the Spanish found themselves mired in their efforts to get through or around the New World. Balboa's extravagant claim to the entire South Sea basin was meant to put a brake on Portugal's relentless appropriation of East Indian goodies, by asserting Spain's legal claim to a broad swath of territory that included the Spice Islands.

Spain then attempted to give Balboa's claim some legs by launching a series of expeditions across the South Sea. The fleet captained by Ferdinand Magellan sailed first (1519–23), in an effort to finally chart a westward route to the Indies and consolidate Spain's claim to the Spicery. García Jofre de Loaysa (1525–26), Álvaro de Saavedra (1527–29), and Ruy López de Villalobos (1542–44) then tried to establish a permanent Spanish position in the Moluccas or other islands nearby. All of these attempts ended in disaster. Finally, Miguel López de Legazpi (1565–66) established a tenuous foothold in the Philippine Islands, while his companion Andrés de Urdaneta solved the problems of Pacific navigation that had vexed the previous efforts. Shortly afterward, Legazpi initiated the first transpacific commercial route built by and for Europeans, the regular sailings of the so-called Manila Galleons that plied the waters between Manila and Acapulco from 1567 until 1815. The political, economic, and cultural space created by the efforts of Legazpi and Urdaneta, and by the sailings of the Manila Galleons, is what I refer to in this book as the early modern Spanish Pacific.

The Spanish Pacific, however, did not amount to much, at least not by comparison with the Spanish Atlantic. The so-called "Columbian exchange" produced by the European encounter with the New World wrought far-reaching changes to the people and places of Europe, Africa, and America, while the corresponding "Magellanic exchange" had relatively little comparable impact.[10] While millions of Africans were transported across the Atlantic to the New World, migrations across the Pacific remained limited in scope.[11] While the Spanish conquered the Mexica and the Inca to spectacular effect, they could only dream of conquering China and Southeast Asia, and they never controlled anything on the western shores of the Pacific Ocean beyond the Philippine Islands, and for a while, Formosa and the Moluccas. Even in the Philippines, effective Spanish control was limited to the environs of a small number of settlements, primarily Manila. While the Spanish (and other Europeans) learned to sail throughout the Atlantic Ocean, Spanish navigation of the Pacific was limited to the circular route of the Manila Galleons.[12] At the center of the circle lay the Hawaiian Islands, undiscovered and unmolested. Outside it, the Spanish ventured very little. Their encounters with the island people of Oceania were limited to brief and ephemeral contacts made by the expeditions of Álvaro de Mendaña (1567–69, 1595–96) and Pedro Fernándes de Queirós (1605–6), which sailed vast stretches of ocean in a fruitless search for hypothetical islands of gold and silver, and for the fabled *Terra Australis Incognita*.

Of course, this did not mean that sixteenth-century Europeans failed to make contact with East and Southeast Asia, just that the majority of such contact occurred by way of the Portuguese route around the Cape of Good Hope and across the Indian Ocean. Important points of inflection in the ongoing process of developing ties with the region included the Portuguese conquest of Malacca in 1511, the dispatch of an embassy to Beijing in 1516, the start of trade with Japan in 1542, and the establishment of a permanent foothold on the Chinese mainland in 1557, with the creation of Portuguese Macau. The Portuguese brought Catholic missionaries with them, and the Society of Jesus soon stood out as a producer of knowledge about the new world beyond the Cape of Good Hope. Working out of their position in Goa, the Jesuits established missions in the Moluccas (1546), Japan (1549), and Ming China (1582). Other orders joined them. These contacts generated a wealth of new knowledge about the peoples and places of the region, knowledge that began to circulate in Europe after about 1550. The Spanish were thus latecomers to the European encounter with East and Southeast Asia, and even though they made key contributions to the dissemination of new knowledge of the region, they were actually responsible for creating very little of it (Rubiés 2003). Moreover, although they arrived at a time that Portuguese influence in the region was on the wane, they were soon met by the new challenge of Dutch ascendancy and by their own declining capacity to defend, much less extend, their existing empire. In the long run, the transpacific link they created with Asia was nowhere near as significant, for Europe's perception of places like China and Japan, as the link that the Portuguese and others created across the Indian Ocean.

And so, while the Atlantic Ocean emerged as the new Mediterranean of a new and expanded West, the Pacific emerged as something quite different. Early modern Europeans, generally speaking, came to experience the Atlantic as they had long experienced the Mediterranean, and as they were experiencing the Indian Ocean, as a maritime basin domesticated by navigation and therefore capable of integrating the countries and peoples around it into unified and complex, political, social, economic, and even biological space. By contrast, they experienced the Pacific as a vast, largely unknown, and possibly empty expanse that was difficult to cross with safety, and that therefore separated America from Asia, and Americans from Asians. So, when sixteenth-century Europeans flattened the globe into a map of the world, they had good reason to preserve the unity of the Atlantic basin, but they also had good reason to feel comfortable cutting through the Pacific and consigning its parts to opposite edges of the map, as the new Ocean Sea at the edges of the world.[13]

This cartography was naturalized at the expense of many an alternative vision of the world, most notably the cosmologies of Native Americans who found themselves on the business end of sixteenth-century European colonial expansion (Mignolo 2003, 219–314). Yet it also competed with alternatives created by and for Europeans themselves, eventually displacing them altogether. One of these alternatives can be found, of all places, in the endnotes to O'Gorman's *The Invention of America*. There, the Mexican historian admits that the invention of America did not catch on right away, and that Waldseemüller himself, in his 1516 *Carta marina navigatoria Portugallen navigationes*, "went back to the idea" that the landmass he had helped invent as the fourth part of the world was actually an extension of Asia (fig. 5; O'Gorman 1961, 168–69; 1986, 184–85). What O'Gorman does not tell us is that in doing so, Waldseemüller was embracing what was and continued to be the dominant understanding of the geography of the new discoveries throughout the first half of the sixteenth century. Schöner, originally so enthusiastic about Waldseemüller's innovation, joined him in going back on it. On world maps like the one made by Caspar Vopel in 1545, the boundary between East and West is nowhere to be found. Places bearing names derived from Marco Polo appear just over the mountains from "Nova Hispana" (New Spain) and

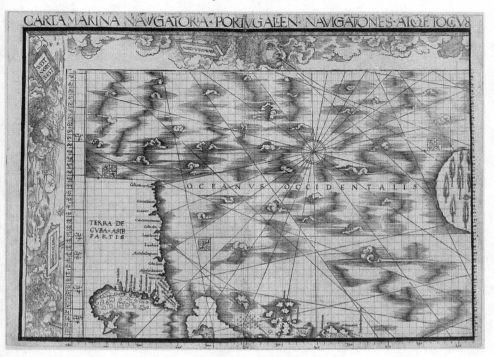

FIGURE 5. North America appears as part of Asia (*Asie Partis*) on this detail from Martin Waldseemüller (cartographer) and Johann Schöner's *Carta marina navigatoria Portvgallen navigationes* (Strasbourg, France, 1516). 46 x 63 cm. Photograph: Courtesy of the Geography and Map Division, Library of Congress, Washington, DC (call no. G1015.S43 1517; control no. 2016586433).

"Temixtitan" (Tenochtitlán). The map gives the impression that one can follow the coast from Mexico to China and stumble along the Spice Islands along the way (fig. 6). According to the historian of cartography George Nunn, it was not until the 1560s that the thesis of American insularity reasserted itself in force and consolidated its grip on the European imagination through the influence of Gerhard Mercator and Abraham Ortelius. Nevertheless, maps that linked America to Asia continued to be made until the 1580s, and the thesis that the two continents were connected, even if they were distinct, was essential to the landmark history of the Indies published by the Jesuit José de Acosta in 1590 (Nunn 1929).[14]

O'Gorman, however, pays no attention to this alternative European cartography of the world. "We are not concerned here with the conflicting opinions," he explains, but only with "the hypothesis that eventually received empirical confirmation" (O'Gorman 1961, 168–69).[15] This

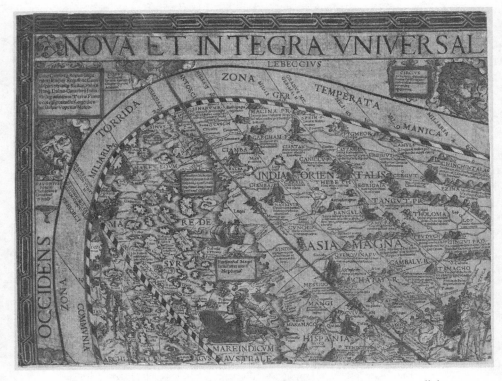

FIGURE 6. On this detail from Alessandro Vavassore's copy of Caspar Vopel's lost 1545 map of the world, the coastline of Mexico is continuous with that of Asia. New Spain (Hispania Nova) appears at the bottom center and "Asia Magna" just above it. From Vavassore's *Nova et Integra Universalisque Orbis Totius . . . Descriptio* (Venice, 1558). Harvard Map Collection digital maps. Cartographic treasures (Hollis no. 990088492540203941). Photograph: Courtesy of the Liechtenstein Map Collection, Houghton Library, Harvard University.

is a strange remark for someone bent upon convincing his reader that "America" is best understood as an ideological invention rather than as a natural object amenable to discovery by simple inspection. In making it, O'Gorman sounds something like the Great and Powerful Oz, commanding that Dorothy and her companions pay no attention to the man behind the curtain. Only in this case, what is behind the curtain is not a man, but a map, a map that suggests the invention of America did not actually happen, at least not for everyone, and certainly not all at once. Waldseemüller's 1507 innovation was indeed revolutionary, but it did not trigger an immediate revolution in the way Europeans mapped the world.

The purpose of this book is to look behind the curtain and unveil the map that O'Gorman does not want us to see. By this I do not mean Waldseemüller's *Carta marina* specifically, but rather a broader geopolitical imaginary that never mapped the New World and Asia as entirely different places, separated from each other geographically and even ontologically by a massive, empty ocean stretching between them. Instead, this imaginary found ways to keep the South Sea small, and the various parts of the Indies either connected to each other or amenable to connection by human activity. This imaginary can be studied across the landscape of sixteenth-century European culture, by examining unduly neglected maps, texts, and objects and by reconsidering materials we thought we already understood.[16] This book, however, focuses on the culture of Spanish imperial expansion. There, the attempt to keep America and Asia together, to imagine the Indies as a geography that spanned the South Sea, was not an alternative cartography of the new discoveries: it was the dominant tradition. Neither was it a matter of idle geographical speculation: it was the cartography of imperial expansion. It fueled the exploration of what we call North America, the continued effort to master Pacific navigation, and the colonization of the Philippines. It bred chimeras of conquest on the Asian mainland and was in turn shaped by the uniquely Hispanic experience of sustained transpacific trade. According to the logic of this tradition, the Far East was not the Far East at all, but the Far West, what the official cartography of the Spanish crown circa 1600 called *Las Indias del poniente*, the Indies of the West, or the Indies of the Setting Sun.

Yet why should we care about this cartography, if the enterprises it made possible amounted to so little, as I have suggested? Why should we bother with a way of mapping the world that eventually lost out to the more familiar model that decisively separates America from Asia and puts each of them where they belong? For one, the Spanish Pacific was nowhere near as inconsequential as I have made it out to be, following the prevailing trend in the historiography. Current scholarship is teaching us to think of the Atlantic world that Europe created over the course of the early modern period as an open system, connected to the rest of the world by way of the Indian Ocean, and the Pacific. The time is ripe, then, to reconsider O'Gorman's argument—which maps the Pacific as a space of separation and distinction and the Atlantic as one of integration, shining the spotlight on one while casting the other into the darkness of historical neglect—and to chart instead a new history of the early modern cartographic imagination, one that recognizes how Europeans—the Spanish in particular—mapped America and Asia into a shared transpacific space. Not only does such a history of cartography respond more capaciously to the lived experience of the early modern Hispanic world, but it also challenges us to imagine how the world might have been mapped differently, had certain imperial projects succeeded, thereby liberating us to imagine how it might be remapped in years to come.

A New Map for Spanish Pacific Studies

"A map is the complete opposite of an illustration; it is the foundational structure of an area of knowledge," writes Frank Lestringant, and many a scholar these days would probably agree (1991, 158). It has become the order of the day in the humanities and social sciences to question the conventional boundaries of established fields of study, under the assumption that there is always something arbitrary and even ideological to the various frameworks, including the metageographies, that are used to configure scholarly fields.[17] They can often lead us to think of potential wholes only as a collection of parts and to lose sight of many things that lie in the interstices of our tidy compartments. This is certainly true of the study of Spanish imperialism and colonialism, which has so often confined its geographical scope to the American continent, when it has not been concerned with the distinct enterprise of Spanish imperialism in Europe and the Mediterranean. Nowadays, such scholarship opens up to the Atlantic basin, and to the ties with Europe and Africa that are so impossible to ignore in any consideration of the history of the Americas, as a matter of course. It also opens up to comparisons with other early modern empires, including those of Britain, France, and Portugal. It even opens up to the Pacific, where the scale of Spain's involvement, as we have seen, was smaller than in the New World, and its impact more difficult to discern.

Scholarship on the Philippines and the Spanish Pacific, however, is nothing new. The study of the Philippines under Spain, in the United States at least, goes back to the acquisition of the islands during the 1898 war, when understanding Philippine history came to be seen as a bulwark of an emerging colonial policy, and of its critique (Bourne 1903; 1907). An early fruit of this early phase of Philippine studies in the United States was the classic work on the Manila Galleons published by William Lytle Schurz in the 1930s, a book that remains useful today despite its many limitations (Schurz 1939).[18] Yet such scholarship, even when it flowed from the pens of established scholars in colonial Latin American or Atlantic studies, like Schurz himself, has not always garnered much attention outside Philippine studies, not even today.[19] In much of the scholarship on colonial America and the early modern Atlantic, the Philippines and the Pacific barely make it out of the footnotes, if they get mentioned at all.

Perhaps more surprisingly, the same can be said about the status of the Spanish Pacific in much of the work done by Pacific studies. Given the enormous size and complexity of the Pacific rim and basin, scholarship in Pacific studies tends to be carried out in separate subfields, each with its own spatial and temporal parameters, its own "Pacific."[20] Work on the European encounter with the Pacific often emphasizes the islands and peoples of Oceania, taking as its point of departure the eighteenth-century voyages of Captain James Cook, the opening salvo of the putatively "fatal impact" of Europeans on the region (Moorehead 2000).[21] While the Spanish Pacific constitutes nothing more than a footnote to colonial Latin American and Atlantic studies, it represents a mere prelude to this strand of Pacific studies. The voyages of Mendaña and Queirós, according to historian Joyce Chaplin, were "overwhelmingly passive and decidedly inconclusive events," typical of the period identified by the title of her essay "The Pacific before Empire, 1500–1800" (Armitage and Bashford 2014, 58).[22] Adam McKeown, meanwhile, recognizes that the Manila Galleons constituted "the first regular trans-Pacific route . . . the last link in establishing trade routes that encircled the globe," and he admits that their travels facilitated important new exchanges among the ocean's "rimlands," but also emphasizes that "the galleons only had a limited effect on reconfiguring the Pacific." That reconfiguration, such scholars argue, did not begin until the 1760s (Armitage and Bashford 2014, 148).

Yet in this nod to the significance of the Manila Galleons, even if only as prelude, McKeown acknowledges the work of another strand of Pacific studies, the strand that is germane to this book. During the late 1970s and the 1980s, the geographer Oskar Spate published a trilogy of volumes entitled *The Pacific since Magellan* (O. H. K. Spate 1979; 1983; 1988). His was not just a history of European exploration, but an account of European efforts to construct a political, social, and economic space out of the natural and human geography of the Pacific rim and basin. The first volume, *The Spanish Lake*, borrows its title from a chapter in Schurz, and it is dedicated almost entirely to the Spanish experience in the Pacific during the sixteenth century. Its publication marks the beginning of a new period in the study of the Spanish Pacific, which took off in earnest during the 1990s, when the economic historian Dennis Flynn and the literary scholar Arturo Giráldez published a series of influential essays that assigned a prominent role to the Manila Galleons in the emergence of the first truly global economy.[23] The lifeblood of this economy was silver, which was produced primarily in Spanish America, and which flowed into the "silver sink" of Ming and Qing China. As the most direct link between the world's major producer of silver and its largest consumer, the transpacific galleon route became a vital artery for its circulation.[24]

By placing a spotlight on the importance of the Manila Galleons to the global circulation of silver, Flynn and Giráldez have provided a powerful answer to the question I posed earlier, about why we should care about the early modern Spanish Pacific, if it looks so inconsequential by comparison with Spanish America and the Atlantic. By their reckoning, China's insatiable demand for silver allowed Spain to pump enormous amounts of the commodity into the global economy without creating as precipitous a drop in its value as would have been the case in the absence of that demand. As a result, they argue, Spain's entire imperial enterprise, dependent as it was on silver remittances from the Americas to secure loans from European bankers, was largely predicated upon the existence of that demand. The Manila Galleons, therefore, were not a marginal phenomenon at all. The trade they facilitated was central to the very existence of the Spanish empire. It was no coincidence that Spain's decline among European powers became irreversible during the same decade, the 1640s, as the fall of the Ming (Flynn and Giráldez 1996).

Nevertheless, the transformations wrought by the Manila Galleons were not just economic in nature, as we learn from the work that has been done in Spanish Pacific studies in the wake of Flynn and Giráldez. The heavy flow of Spanish American silver into China upset the social order and created all sorts of cultural anxieties that may very well have contributed to the collapse of the Ming dynasty.[25] The galleons made it possible for the colonial government of New Spain to exert power over the Philippines, encouraged dreams of reprising the triumphs of Cortés and Pizarro over the empire of the Ming, and nourished plans to make good Catholics out of the Chinese, Japanese, and others (Headley 1995; Ollé 2000; Ollé 2002). They also made it possible for Asians to travel to the Americas, making for the emergence of the first "Chinatowns" in the western hemisphere (Mercene 2007; Seijas 2014). Asian luxury goods made the trip as well, sparking a fascination for all things Asian in the viceroyalties of New Spain and Peru well before the craze reached Europe. Much of the work in the new wave of Spanish Pacific studies that has emerged since the 1990s has been carried out by historians of material culture interested in the impact of silk, porcelain, lacquer ware, ivory carvings, and other luxury commodities on the culture of Spanish America.[26] Spanish American elites consumed Asian luxury goods with marked enthusiasm, and Spanish American artists incorporated Asian techniques and aesthetics into their own work. As we can see, therefore, even though the line that Spaniards flung across the Pacific was not as powerfully transformational as the net they cast over the Atlantic, it nevertheless tied America to Asia in ways that were both significant and without real precedent.

Most importantly for the purposes of this book, however, the link between Spanish America and Asia created by the travels of the Manila Galleons registered in the consciousness of Spanish American elites, and in their sense of collective identity. In Mexico City, to cite the most obvious example, members of the creole elite ate meals flavored with Indonesian spices off Japanese porcelain, directed prayers at religious figures carved from Southeast Asian ivory, adorned their bodies and homes in colorful Chinese silk, and heard mass in churches where the same silk added splashes of bright color to the Baroque display of piety and wealth. All of this made it easy for them to believe that their city was the center of the world, not some outpost on the colonial periphery. As both Serge Gruzinski and Barbara Fuchs have argued, this sense of centrality registers in the 1604 tribute to Mexico City written by the Spanish poet Bernardo Balbuena, *La grandeza mexicana* (The grandeur of Mexico) (Gruzinski 2004, 103–28; Fuchs 2009). In these stanzas, for example, wealth flows into Mexico from both East and West, and the city becomes the fulcrum of global trade:

En ti se junta España con la China.
Italia con Japón, y finalmente
un mundo entero de trato y disciplina.

En ti de los tesoros del Poniente
se goza lo mejor; en ti la nata
de cuanto entre su luz cría el Oriente. (Balbuena 1997, 91)

[In you Spain is joined to China, Italy to Japan, and finally a whole world of trade and knowledge. In you the best of the treasures of the West are enjoyed; in you the cream of everything born under the Oriental sun.]

Novohispanic elites were thus quite conscious of Mexico's location between two great oceans, at the nexus of the trade routes that spanned them. The Spanish Pacific was not just the lived space of exchange that connected Mexico City to Manila. It existed at the level of consciousness, in a variety of representations that, like Balbuena's poem, served to construct the identity of the viceroyalty as a crucial node in networks of communication and trade that spanned the globe.

Yet the transpacific space that opens up in Balbuena's imagination, as it does in that of other Spanish writers from around the year 1600, was not created *ex nihilo* when the first Manila Galleon arrived in Acapulco. The historian John Headley reminds us that it had important precedents that went back to the beginnings of Spain's encounter with the Indies. As

Headley puts it, "During the last third of the sixteenth century, the no-tion of *el Nuevo Mundo* was being transformed by Spain's experience in the Pacific; and the more comprehensive view of the New World, *latent all along in the papal bulls of 1493–94,* came at last to prevail" (1995, 644; emphasis added). The Spanish Indies, Headley reminds us, had origi-nally been conceptualized as all the lands that fell within the hemisphere assigned to Spain by the papal donation and the Treaty of Tordesillas, not just the lands that came to be known as the New World. Headley would have us believe, however, that this "comprehensive view of the New World," as he calls it, had remained "latent" in the Spanish impe-rial experience and so presumably had not informed the way Spaniards imagined the New World until the conquest of the Philippines made it relevant. To put it another way, he seems to believe that America was ef-fectively invented in the Spanish imagination, in more or less the manner we have described, and was only reconnected with Asia after the Manila Galleons started to sail.

This, I argue, was not the case at all. From the very start of the en-counter with the New World, Spanish ideas about the geography of the new discoveries often remained open to the possibility of an Asian con-nection, when they did not actually invent some sense of connectedness or continuity between America and Asia. The transpacific imaginary so in evidence in Balbuena's poem, Headley's "comprehensive view of the New World," was always present in the Spanish imagination, in one form or another, throughout the sixteenth century, inspiring and informing at-tempts to press the project of empire westward, all the way to the Indies of the Setting Sun. The late sixteenth-century transformation to which Headley refers did not reattach America to Asia across a Pacific Ocean that had successfully severed the two. It consolidated a cartography that had always sought the sort of continuity that had finally been realized in practice by the sailings of the galleons.

To see things this way, we have to learn to suspend judgment about when "America" finally came into view as the fourth part of the world, separated from Asia by a vast ocean that effectively marked the onto-logical boundary between East and West. We have to come to terms with the fact that *The Invention of America*, along with the rest of the historiography of discovery and exploration that tries to track when, pre-cisely, America became America, is much too invested in establishing the historical and geographical foundations of certain fields of study to provide reliable guidance to a scholarly project that tries to question the conventionalized boundaries of those very fields.[27] The map of America, to echo Frank Lestringant, is indeed "the foundational structure of a field of knowledge," the field of American studies. If we are to fully open our understanding of colonial America to the Pacific and Asia, we need another map. We need the map, or maps, that actually informed the Spanish imperial project, the maps that O'Gorman consigns to the footnotes. These are the maps that can inform the emerging field of Spanish Pacific studies, by helping it understand the geopolitical imaginary that nourished and was in turn nourished by those exchanges that the economic and art historians have helped us understand.

Cartography and the Spanish Geopolitical Imaginary

Now I must confess that when I refer to the maps in O'Gorman's footnotes, I mean this figuratively, not literally. I do not really mean maps like Waldseemüller's *Carta marina*, which clearly depict the newly discovered landmass as a peninsular extension of Asia, but rather a body of maps that resist the invention of America in a variety of ways that can be spotted only when we emphasize what maps do rather than what they say. This statement obviously requires clarification. First, when I use words like "maps" and "cartography" I am referring to representations of various kinds that configure geographies, including written texts deeply invested in the configuration of space, such as travel narrative and official historiography, as well as cartographic images of the sort that these words immediately bring to mind. In previously published work, I have referred to this inclusive category as "cartographic literature" (Padrón 2004, 12). References to "mapmakers," in turn, should be interpreted in a similarly inclusive manner, unless the context suggests otherwise, as references to the producers of cartographic literature in any genre. Second, when I refer to what maps do rather than what they say, at least in the context of this argument, I am referring to the work they do in fostering a sense of connectedness between the landmasses we call Asia and America, something that can be accomplished in a variety of ways that can be discovered only through historically contextualized close reading, not through simple inspection of the ways various maps depict landmasses and the connections among them, or lack thereof. What follows is my attempt to elaborate on this final point.

As we have seen, we can identify two cartographic traditions that, over the course of the sixteenth century, took very different approaches to the geography of the discoveries in the Ocean Sea. The more famous of the two, the one at the heart of O'Gorman's *The Invention of America* and of the positivistic historiography of America's "discovery" against which he sets himself, maps the discoveries as an enormous island of continental status, a fourth part of the world christened "America" by Waldseemüller and his collaborators. This tradition, with its theory of American insularity, is usually celebrated as the more progressive of the two, insofar as it boldly breaks with the tripartite model of the *orbis terrarum* bequeathed to modernity by the bookish culture of antiquity and the Middle Ages. The second tradition, by contrast, depicts the new discoveries as an enormous peninsular extension of the old *orbis terrarum* in the manner of the *Carta marina*. If it recognizes any of the geography in question as a "New World," it limits that recognition to what we call South America, and it identifies the Atlantic coast of North America as part of continental Asia. This tradition, with its theory of Amerasian continuity, is less well known than the first, and less admired by those familiar with it. It has been thought to demonstrate the profound hold that the bookish culture of the past had on the minds of many European humanists, who refused to redefine the *orbis terrarum* in order to make room for a previously unknown fourth, insular part of the world.[28]

The geopolitical imaginary whose history I sketch in this book drew on both of these cartographic traditions but cannot be identified with either of them. For one, the vast majority of the extant maps that portray America and Asia as a single continuous landmass were made in France, Italy, and Germany, by people who were not directly involved in the enterprise of Spanish territorial expansion. The few surviving examples of Spanish maps that embrace the theory of Amerasian continuity were made in Flanders or Italy, not in Iberia.[29] The surviving copies of the Spanish crown's official map of the world, meanwhile, remain agnostic regarding the geographical relationship between the New World and Asia, leaving the relevant part of the image blank, even though most of them were made to insist on the propriety of Spanish claims to sovereignty over the Spice Islands in Southeast Asia and might have been able to have used the theory of Amerasian continuity to advantage (fig. 7). Yet while the theory makes only a few appearances on maps of Spanish origin and remains absent from the most important corpus of surviving Spanish maps, it nevertheless informed the thinking of many a Spaniard in the New World. The whole arc of exploration along the shores of New Spain and into what is now the southwestern United States during the twenty years following the siege of Tenochtitlán is incomprehensible without it. Therefore, we can affirm that the theory of Amerasian continuity played a significant role in the history of the Spanish exploration and the geopolitical imaginary that informed it, but must acknowledge that Spanish maps tended to keep it at arm's length.

Something similar can be said about the relationship between the Spanish geopolitical imaginary and the theory of American insularity. The blank spaces on the crown's official maps cut both ways, remaining agnostic about the possibility that the New World was separate from

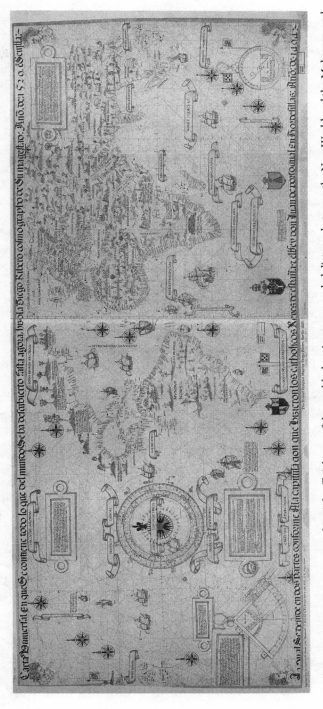

FIGURE 7. A presentation copy of Spain's official map of the world, showing an expanded distance between the New World and the Moluccas, and a blank space between Asia and our North America. From Diogo Ribeiro's *Carta universal en que se contiene todo lo que del mundo se ha descubierto fasta agora* (London, 1887). 58 x 140 cm. Photograph: Courtesy of the Geography and Map Division, Library of Congress, Washington, DC (call no. G3200 1529.R5 1887 MLC; control no. 85690293). This is a facsimile of the highly faded original at the Vatican Library.

Asia just as they remained agnostic about the possibility that the two landmasses might be connected. More importantly, however, the word "America," appears only rarely in the Spanish of the sixteenth century, and then only late in the game. In the Spanish Royal Academy's *Corpus del nuevo diccionario histórico del español* (*CNDH*), it occurs only fifty-five times across thirteen documents written between the years 1500 and 1600, and only two of these were written prior to 1572. They include the *Historia de las Indias* of Bartolomé de las Casas, which explains that "America" is what foreigners call the New World, and that this is how it is called on maps imported from abroad. They also include passages from a Spanish translation of Peter Apian's *Cosmography*, a popular treatise originally written in Latin.[30] "America," moreover, rarely appears on Spanish maps as a name for the New World. A notable exception can be found in the massive, elaborately engraved 1562 map of the New World published in Antwerp with authorization from the crown, which bears the title *Americae sive quartae orbis partis nova et exacctissima descriptio* (A new and most accurate description of America, also known as the fourth part of the world; fig. 8). The title, however, might very well have been chosen by the Flemish engraver Hieronymous Cock, rather than the Castilian cosmographer Diego Gutiérrez. The map itself remains ambiguous about the insular status of the New World by filling the part of the map image that matters to us most with decorative materials and cutting it off with the map frame itself.

Spanish speakers overwhelmingly tended to refer to the newly discovered lands in the Ocean Sea as *las Indias* (the Indies), although they sometimes used variations of *las Indias occidentales* (the West Indies). References to "las Indias" are abundant in the *CNDH*, where the term occurs 6,602 times in 691 documents written over the course of the sixteenth century, providing a stark contrast to the scarcity of "América."[31] In the Spanish linguistic context, the continental landmass that others called "America" became either *el nuevo mundo* or more commonly and more simply, *Tierra firme* (the mainland). Of course, the term *las Indias* sometimes functioned as a synonym for "America," as it does in the work of Francisco López de Gómara, which I examine in a later chapter. Gómara's use of the term as a synonym for America, however, proves to be the exception rather than the rule. For the most part, *las Indias* referred to something broader, in some ways more ambiguous, but always more inclusive than what others called America. This is the geopolitical object that is at issue in this book, the Spanish Indies, which over the course of the sixteenth century could accommodate *either* the theory of Amerasian continuity *or* that of American insularity, or remain agnostic about the matter altogether.

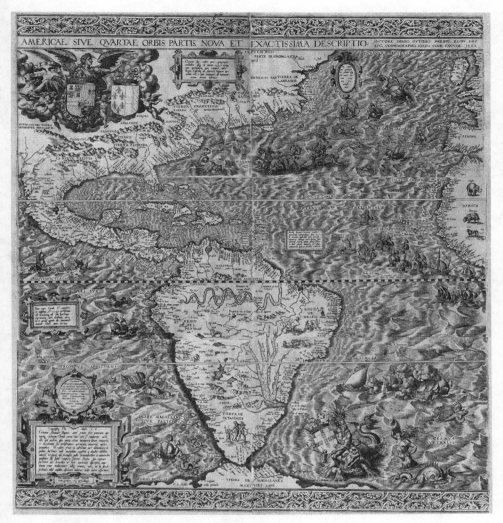

FIGURE 8. The only printed sheet map of the New World that we know was made using Spanish sources during the sixteenth century. It is also one of the only maps of Spanish origin to call the New World "America." Diego Gutiérrez and Hieronymus Cock's *Americae Sive Qvartae Orbis Partis Nova et Exactissima Descriptio* (Antwerp, 1562). 100 x 102 cm. Photograph: Courtesy of the Lessing J. Rosenwald Collection, Geography and Map Division, Library of Congress, Washington, DC (call no. G3290 1562.G7; control no. map49000970).

On one level, the geography of the Indies is easily defined. *Las Indias* refers to the half of the world assigned to Spain by a series of papal encyclicals and the 1494 Treaty of Tordesillas, as delimited by the so-called line of demarcation that split the world between Spain and Portugal. The trouble was that no one could agree about the location of the line, not only because there was no way to measure longitude with any accuracy, but also because there was no agreement about such basic issues as the true circumference of the Earth and the overland distance from Portugal to Cathay. Thanks in part to Ptolemy's prestigious *Geography*, fifteenth- and sixteenth-century Europeans tended to underestimate the former and overestimate the latter. This trend allowed the Spanish to argue that when one extended the line of demarcation that ran through the New World more or less at the mouth of the Amazon River into a full meridian that split the globe in two, its other half, the "antimeridian," sliced through Asia near the mouth of the Ganges River.[32] This is how the cartography of the demarcation works out, for example, in the 1518 Spanish geography of the world by Martín Fernández de Enciso, the first to include a description of the New World based on Spanish sources (Fernández de Enciso 1948, 25). A graphic representation of the idea appears on a 1524 map of the world on a polar projection constructed to demonstrate the validity of the Spanish claim to sovereignty over the Spice Islands. When one draws the line of demarcation as a full meridian on this map, it assigns much of continental Asia to Spain (fig. 9). Other sixteenth-century Spanish maps shift the antimeridian eastward to Malacca, but none so much as budges it any farther east. Until the eighteenth century, when changes in European cartography and geography definitively falsified the geodesical and geographical assumptions that made this cartography possible, Spanish officialdom mapped the Castilian demarcation as a hemisphere stretching from the Marañon to Malacca, a transpacific space that included most of the New World along with East and Southeast Asia. It continued to do so, moreover, as European knowledge of the Pacific and East Asia grew and changed, and as prevailing ideas about the status of the New World began to favor the theory of American insularity.

Spanish mapmakers, however, rarely relied on the line of demarcation alone to convince their readers of the reasonableness of Spain's claim to transpacific empire, particularly when they made maps rather than globes. The reason for this becomes readily apparent when we look at a map made for the Hapsburg monarch in 1630, during the period known as the "Union of the Crowns" (1580–1640), when Spain and Portugal shared the same king (fig. 10). The map was made by the Portuguese cosmographer João Teixeira in an effort to defend Portugal's territorial

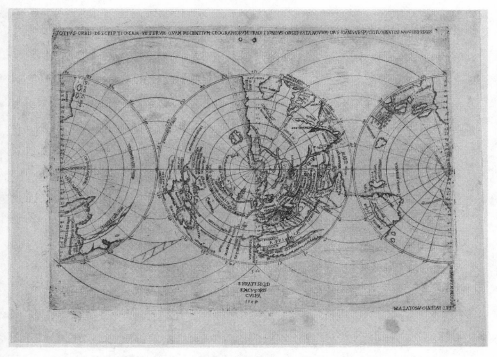

FIGURE 9. The northern hemisphere appears in the middle. The southern hemisphere is divided in two halves appearing on either side. Juan Vespucci, *Totius orbis descriptio iam veterum quam recentium geographorum* (Italy [?]), 1524). Harvard Map Collection digital maps. Cartographic treasures (Hollis no. 990088492040203941). Photograph: Courtesy of the Liechtenstein Map Collection, Houghton Library, Harvard University.

prerogatives in Asia against Castilian encroachment. On this map, as on all flat maps, the meridian that divides the globe into a Spanish and a Portuguese hemisphere becomes two vertical lines, the original line of demarcation running through South America and the so-called antimeridian running through East Asia. Teixeira places the antimeridian where the Portuguese always placed it, farther east than it appears on Spanish maps, in such a way that the Spice Islands, China, the Philippines, and much else fall within the "Comquista de Portugal" (Portuguese Conquest), the Portuguese hemisphere, but leave Japan, New Guinea, and the Marianas Islands to Spain.

The map's anti-Castilian posture, however, is not limited to the Lusophilic placement of the antimeridian. In making his map, Teixeira has joined countless other European mapmakers in choosing to place the

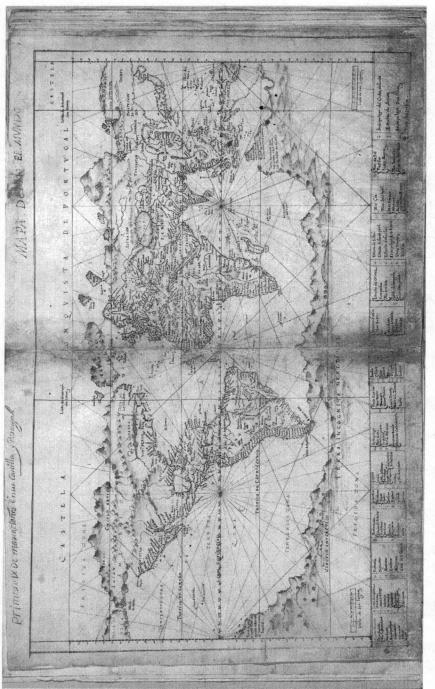

FIGURE 10. On this world map, part of the Spanish hemisphere appears on the left, and another on the right. From João Teixeira Albernaz, Jeronimo de Attayde, and Francisco de Seixas y Lovera's *Taboas geraes de toda a navegação* (1630). Photograph: Courtesy of the Geography and Map Division, Library of Congress, Washington, DC (call no. G1015.T4 1630; control no. 78653638).

Atlantic in the middle, the New World on the left, and the Old World on the right. Authorized and even naturalized by prevailing trends in European cartography, this design has a convenient political effect: it dismembers the Spanish demarcation in ways that are advantageous to Portugal. Most of the Spanish hemisphere appears on the left, where we find an insular New World clearly labeled "America," while the rest appears on the right, as a fragment of Asia arbitrarily cut off from the rest of the continent and assigned to the Castilian demarcation, but apparently beyond Spain's effective reach. So, while the placement of the antimeridian concedes certain parts of East Asia and the western Pacific to the sovereignty of Spain, the overall design of the map mocks that kingdom's pretension to rule in any part of the region.

Clearly, anyone who mapped the world for the crown of Castile and León could not afford to dismember the Indies this way, so they had no choice but to put East and Southeast Asia on the left-hand side of the map, at the outer edge of the Spanish west, despite the fact, or maybe because of the fact, that this meant isolating places like the Spice Islands, China, and Japan from the rest of Asia. Every surviving planisphere produced by the cosmographers at Seville's Casa de la contratación, the nerve center of official Spanish cartography, maps the world this way, although some actually depict East and Southeast Asia at the right-hand edge of the map as well (see fig. 7). The general map of the Indies produced for use by members of the Council of Indies does something similar. Although it depicts only the Spanish demarcation, not the whole world, it keeps the Indies together the same way the world maps do, by placing East and Southeast Asia on the left, across the South Sea from the New World (fig. 11). So does the print map of the Indies that was derived from this confidential map, in order to equip the crown's official history of the Indies with an official map of Spain's overseas empire (see fig. 2). The makers of these maps knew that in order to make Spain's claim to authority over East and Southeast Asia look convincing, the Spanish Indies had to appear as a continuous, unbroken expanse stretching from the line of demarcation in the New World westward to the antimeridian in Asia. That line, not the Pacific Ocean, had to serve as the boundary between East and West, the Alpha and Omega of the Orient and Occident.

All of this we already know from the work of historians of cartography who have provided invaluable insights into Spain's efforts to map the New World (Martín Merás 1992; Cerezo Martínez 1994). Yet there is something else that we find in the work of Spanish cartographic literature, something beyond the tendency to locate the antimeridian so as to maximize Spain's territorial claim in Asia, and the accompanying tendency to project the Far East into the far west, something that has not yet been

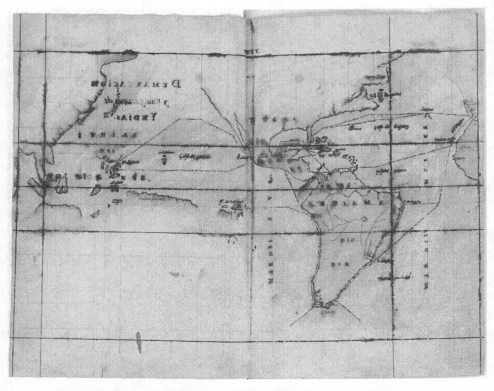

FIGURE 11. Like other Spanish maps made under official auspices, this one places East and Southeast Asia on the left side, in the Castilian west. Juan López de Velasco, *Demarcacion y nauegaciones de Yndias* (Madrid [?], 1575). 20.8 x 32.5 cm. Photograph: Courtesy of the JCB Map Collection, John Carter Brown Library, Brown University (call no. Codex Sp 7/1-SIZE; file 17000-1).

studied. That something is an ongoing attempt, or perhaps a persistent desire, to lend internal coherence to the disparate territories corralled by the lines of demarcation into the hemisphere that constituted the Spanish Indies. It was never enough to map the lines of demarcation in politically convenient ways or to project a part of Asia into the west rather than the east. It was also necessary, or at least commonplace, to imagine some sort of internal structure or logic that held together the resulting assemblage. This was necessary so that the map of Spain's overseas empire would not look like an arbitrary collage, but like a coherent territory that could and should be governed from Spain.

The work was made difficult by the speed at which knowledge of the New World and Asia was changing. As reports poured into Europe from east and west, ideas about geography and ethnography struggled to keep up. As we have seen, the New World and its strange inhabitants became America, the fourth part of the world, home to a distinctly barbarous branch of the human family. Medieval knowledge of the Indies, Cathay, and Zipangu was replaced by modern knowledge that reconfigured these places into the Moluccas, China, and Japan. Initial beliefs that the New World and Cathay might be connected, if they were not identical, yielded to beliefs that they were in fact different and separated from each other by an ocean of indeterminate breadth and unknown boundaries. In order to preserve any sense that the Indies exhibited some sort of internal coherence, the Spanish geopolitical imaginary had to adapt to these changes, assimilating or rejecting new knowledge, adapting it to frameworks that were at once politically expedient and epistemologically valid.

It could only do so because it had a variety of frameworks to work with, not just what geographers Martin Lewis and Kären Wigen call the "architecture of the continents," that familiar metageography that divides the globe into its major constitutive landmasses and treats each of them as a basic unit for the consideration of other matters, such as human geography (Lewis and Wigen 1997, 21–46). In other words, the Spanish geopolitical imagiary does not just divide the globe into major physical parts, like America, Asia, Africa, and Europe, but also divides the human family into the principal branches that correspond to the parts, Americans, Asians, Africans, and Europeans. It also mapped the world and its people according to the theory of climates, which divided the globe into frigid, torrid, and temperate zones, and which insisted that there existed crucial differences between the inhabitants of temperate climates and hot ones. As Nicolás Wey Gómez (2008) has demonstrated, the theory of climates may have even mattered more than the architecture of the continents, at least during the early stages of early modern expansion. In addition to the theory of climates, there was also a growing sense of mastery over the Ocean Sea that accompanied an important transformation in the architecture of the continents and the developing experience of oceanic navigation. As Europeans built maritime routes that spanned the Atlantic, Indian, and Pacific Oceans, that general sense of mastery developed into an entirely novel way of imagining the globe in and through the network of maritime networks that came to embrace it.

The Spanish geopolitical imagination drew upon all these frameworks over the course of the sixteenth century, picking up one when another failed, or combining them in a variety of ways, in order to imagine various forms of internal coherence, continuity, or connectedness among the lands and peoples that lay in its half of the world. As prevailing ideas about America and Asia changed, so did the cartographic literature of Spain's imperial project, constantly adapting to new realities, constantly fighting the centrifugal forces that tended to pull America and Asia apart. If we are to understand how it did so, however, we must first understand the pertinent frameworks a little better, and identify ways to spot them on maps and in writing.

2

South Sea Dreams

The sense of connectedness between the New World and Asia that Spanish cartographic literature constructed over the course of the long sixteenth century was not always a matter of continental geography. During that long century, some mapmakers involved in the Spanish imperial enterprise embraced the theory of Amerasian continuity, while others advanced the theory of American insularity, and still others refused to take a position about the whole question of whether or not the New World was physically connected to Asia in the North Pacific. All of them, however, were able to suggest that the new discoveries in the Ocean Sea toward the west and the world that Marco Polo had discovered toward the east were parts of a single whole. For some, its unity was an arbitrary effect of the lines of demarcation. The whole in question was nothing more than the hemisphere assigned to the crown of Castile by the Treaty of Tordesillas, whatever people and places it happened to include or exclude. Most mapmakers, however, assigned some sort of internal coherence to the physical and human geography embraced by the lines. They could do so thanks to the availability of other conceptual frameworks besides the architecture of the continents, the metageography upon which modern historiography has been so focused. There were other ways to map the globe during the long sixteenth century, other ways to imagine significant connections among people and places that did not rely on physical connections among the major landmasses.

One of these frameworks was the theory of climates, which mapped the globe as a series of latitudinal zones rather than as an assemblage of landmasses. As Nicolás Wey Gómez argues, the theory of climates underwent important changes during the fifteenth century, as European cosmography developed in tandem with European knowledge of the world between the Tropics of Cancer and Capricorn. The revised theory provided important ideological underpinnings to Columbus's Enterprise of the Indies, and indeed to the entire project of early modern territorial expansion (Wey Gómez 2008; 2013). Its development came hand in hand, however, with changing ideas about the globe itself, particularly the relative prevalence of land over water on its surface. Ideas about the Ocean Sea changed dramatically, again as a function of both theoretical and experiential developments. Renaissance geography developed something that medieval geography had lacked, an architecture of the oceans that both responded to and supported an increasing sense of mastery over the element of Water. As the sixteenth century progressed, that sense of mastery over the oceans, now plural rather than singular, developed into an entirely novel sort of metageography, one that mapped the globe primarily in terms of the network of sea lanes that had been constructed by Europeans.

The purpose of this chapter is twofold. The first is to review some of the key characteristics of these various frameworks for mapping the globe, in order to equip the reader with the background he or she will need in order to follow the argument in subsequent chapters. The second, however, is to demonstrate what it means, in the context of this book, to read cartographic literature with what Baxandall (1980) calls the "period eye" that we can reconstruct by becoming acquainted with the period's various metageographies. In order to do so, I do not just rehearse what we know about these frameworks, but use them to read a portolan-style map of the world created sometime around 1519. My reading of this map is meant to reconstruct how some of the movers and shakers in the Spanish imperial project might have imagined the world between 1513, when Balboa discovered the South Sea, and 1523, when the survivors of the Magellan expedition returned with news of the first-ever European crossing of the Pacific. It is also meant to demonstrate that any attempt to capture a period "image of the world" cannot proceed by simply inspecting a map as if it were a transparent representation of geographical knowledge, but rather requires us to interpret the map on the basis of a plausible reconstruction of the set of beliefs and expectations available to the map's makers and readers. In this way, the chapter provides not only necessary background knowledge but also a model for the method of historically contextualized close reading that I apply in the book as a whole.[1]

The map I have chosen is known to historians of cartography as the anonymous planisphere of Munich, after the city where it was held until its disappearance at the end of World War II; it is sometimes referred to as the Kunstmann IV, after the German historian and geographer who first published the map in facsimile. Sadly, it is known to us now only through a color copy made by the German designer and military officer Otto Progel in 1843 and now in the collection of France's Bibliothèque Nationale. I refer to it here simply as the Reinel chart, after the man who is thought to have made it, sometime around 1519 (fig. 12).² Jorge Reinel was a prominent Portuguese mapmaker who produced numerous charts for the crown of Portugal, but in this case, he seems to have been working for some powerful Spanish personage, perhaps even the Emperor Charles V himself. By focusing on a chart by a Portuguese mapmaker working on behalf of Spanish interests, I hope to make it clear that by "Spanish mapmaker," here and elsewhere, I mean a mapmaker involved with the project of Spanish territorial expansion, regardless of his or her country of origin. Some of the most famous "Spanish maps" of the sixteenth century were actually made by natives of Portugal, like Reinel, in the context of the ongoing dispute between the two Iberian kingdoms over the position of the lines of demarcation, while still others were made by mapmakers from other parts of Europe. One of the reasons we can be certain that, in this case at least, Reinel was working for a powerful Spaniard is that the chart takes a stand favorable to Spain when it comes to the coveted Spice Islands, in modern Indonesia. The original line of demarcation cuts right down the center of the map, so everything on the right is Portuguese and everything on the left is Castilian. The left- and right-hand edges of the map, therefore, constitute the antimeridian. Flags and various legends dot the map's surface, supporting its work of political territorialization. The "Ilhas de maluqua," or Malucca Islands, appear in the far left, within the Castilian demarcation, but unmarked by a Castilian standard.

Kunstmann published his facsimile of the Reinel chart along with twelve other early maps in an atlas that bore the telling title *Atlas zur entdeckungsgeschichte Amerikas* (Atlas of the history of the discovery of America).³ Clearly, he believed that the chart's most important feature was its treatment of the emerging geography of the American continent. Hence the editor's decision to include only that part of the map that depicts the various disconnected coastlines that we know as parts of America. These include a large landmass that the chart calls "Brasil," which stretches from roughly 35° S, around Cape St. Augustine, and along the shores of what we recognize as Venezuela, Colombia, and Central America, all the way to the Yucatan Peninsula.⁴ A Latin inscription identifies this landmass as the fourth part of the world, but does not call it America. North of it we find the Ātilhas de Castela (Castilian Antilles), as well as bits of what look to us like the North American coastline, including the Teradimini (Florida), Bacalnaos (Newfoundland), and DoLavrador (Labrador).⁵ As my description suggests, it is easy for us to draw upon latter-day geographical knowledge to connect the dots and glimpse on the Reinel chart a shadowy image of America slowly coming into view, leaving the Spice Islands isolated by water and distance on the edge of the map. In this chapter, however, I resist the temptation to read the Reinel

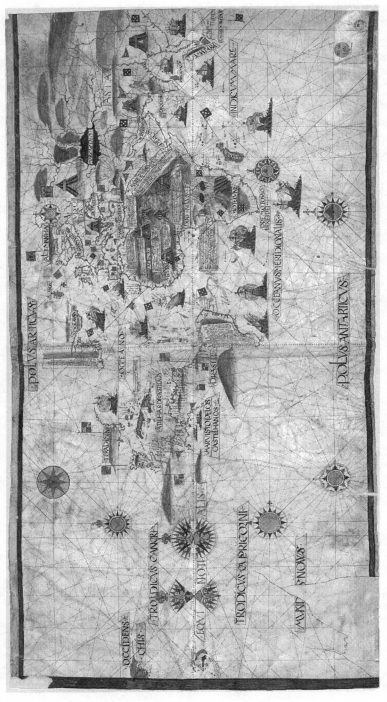

FIGURE 12. This nautical chart of the world (Carte du monde) by a Portuguese mapmaker locates the Spice Islands in the Castilian half of the world, near the left-hand edge of the map. This is a copy of Jorge Reinel's lost 1519 original made in 1843 by Otto Progel, published in Munich. Manuscript facsimile on parchment. 65 x 124 cm. Photograph: Bibliothèque nationale de France (registry c 05627; no. FRBNF40750142).

chart so anachronistically and instead try to read it in light of the various metageographies that were available to the communities that made and read it. What we will discover is not a preliminary image of America starting to come into view as a continent separate from the Spicery and the rest of Asia, but a complex assemblage of competing ideas, expectations, and desires, which in various ways struggle to make everything in the Castilian demarcation hang together, around a South Sea that was largely the stuff of cosmological dreams.

Empty Space on the Reinel Chart

First, we must understand a little better what it is that we are seeing, and not seeing, on the Reinel chart. A great deal of it is empty space. It was not at all uncommon during the sixteenth century for mapmakers to fill the space of the unknown, not only with illustrations and cartouches of various sorts but also with speculative geographies. Extrapolating from what they knew on the basis of available theoretical models, or relying on sources that we know to have been spurious but which they had good reason to trust, mapmakers inferred the existence of coastlines and landmasses that had not been fully charted, if at all, and included them on their maps, along with hypothetical islands and chimerical cities (Fernández-Armesto 2007a, 740–43). Most famously, they inferred the existence of an enormous landmass in the southern hemisphere, a *Terra Australis Incognita*, and made it a standard fixture on many a map of the world.[6] One prestigious Spanish cosmographer, Alonso de Santa Cruz, mocked this tendency to speculate, and he singled out the Frenchman Oronce Finé for singular ridicule (Santa Cruz 2003, 394v). Finé's cordiform world maps from the 1530s were responsible for introducing the idea of the *Terra Australis* to European mapmaking, although there were important precedents for the idea in the work of earlier mapmakers (Broc 1986, 168–72; Hiatt 2008, 184–210; fig. 13). Given all the reckless speculation that in his view marred the work done on the other side of the Pyrennees, Santa Cruz insisted that the only maps of the world that could be trusted were those made by the royal cosmographers of Spain and Portugal, who might fill empty spaces with cartouches and illustrations, but who generally refrained from including speculative landmasses and coastlines, depicting only what they believed had been reliably charted by modern exploration (Santa Cruz 2003, 394v). The Reinel chart reflects this practice, remaining empty below 40°S latitude, and throughout much of the Castilian half of the world, and depicting various landmasses as cartographic fragments, clearly admitting to the incomplete state of geographical knowledge.

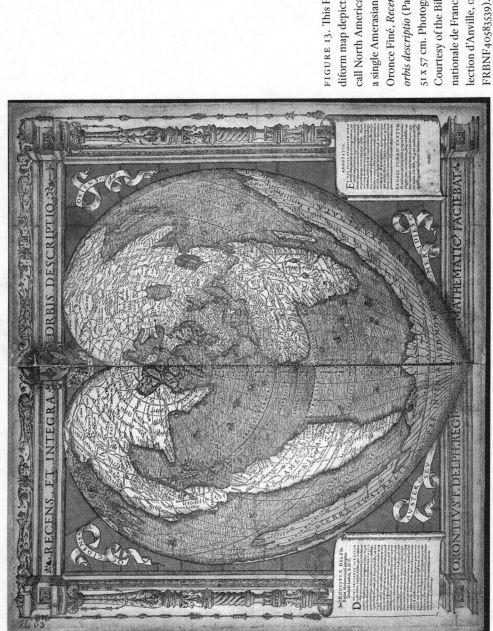

FIGURE 13. This French cordiform map depicts what we call North America as part of a single Amerasian landmass. Oronce Finé, *Recens et integra orbis descriptio* (Paris, 1534). 51 x 57 cm. Photograph: Courtesy of the Bibliothèque nationale de France (Collection d'Anville, 00063; no. FRBNF40583539).

Empty spaces and fragmentary geographies invite us to fill in the blanks and round out the partial wholes, just as they did for many an early modern reader. We tend to do this on the basis of our own knowledge of the world, seeing Reinel's fragmentary coastlines in the western hemisphere as bits and pieces of North and South America, gradually coming into view. In fact, we have been trained to do this by historians like Kunstmann who have inserted this map and others like it into Whiggish accounts of the discovery of America. Yet what would an early modern reader have imagined in Reinel's empty spaces? In order to answer this question, it helps to destabilize an assumption that we might be tempted to make about Reinel's blank spaces. Our constant exposure to modern maps of the world has given us not only a readily available image of the continents that we can use to round out Reinel's fragmentary landmasses, but an image of the negative space, so to speak, constituted by the world's oceans. Since the ocean covers roughly 70 percent of the Earth's surface, and modern maps tend to depict it as empty space colored in blue, we are accustomed to having a large amount of empty space in our mental image of global geography, and we tend to associate such emptiness with water. This can lead us to mistake much of the empty space on the Reinel chart for something it is not, a depiction of the world's oceans. If we are to imagine what an early modern reader might project into the empty spaces of the Reinel chart, we would do well to first free ourselves of this anachronism.

The geographer Philip Steinberg can help us here. He argues that during the sixteenth century, blank cartographic space did not signify oceanic space, as it would starting in the middle of the seventeenth century, and as it does for us today. Hence the perfect nautical chart in Lewis Carroll's "The Hunting of the Snark," which is nothing more than a blank piece of paper (Carroll 1939). Before that, mapmakers tended to mark oceanic space as such in one way or another. Not only did they name seas and ocean basins, but they also populated large bodies of water with telling images, such as ships or marine animals. (Steinberg 2001, 99–106).[7] Such illustrations were not merely decorative, nor were they necessarily fanciful. They played an important role in the map's signifying work. There were other ways, of course, of marking oceanic space as such. The engraver Hieronymous Cock fills every inch of the ocean on his elaborately engraved 1562 map of America with choppy waves (fig. 8). Others might include islands, whether known or hypothetical, since to draw an island was to affirm that the space surrounding it is in fact aquatic. Empty space that remained unmarked in these or other ways was nothing more than blank cartographic space, neither land nor sea, but the space of the unknown.

Where, then, are the oceans on the Reinel chart, and where is the empty space of the unknown? The North Atlantic, which Reinel calls simply the "Occeano," presents the most straightforward case. Four ships sail its waters, and the surrounding coastlines of Europe, Africa, Brasil, Teradimini, Bacalnaos, and DoLavrador define its approximate boundaries.[8] The basin features a number of prominent islands on both the Castilian and the Portuguese sides of the central line of demarcation. Reinel's "Occeanus Meridionalis," our South Atlantic, and his "Indicum Mare," our Indian Ocean, present more difficult cases, insofar as they demonstrate the inherent ambiguity involved in using illustrations to mark oceanic space as such. The map tells us that the spaces south of the Cape of Good Hope, India, and China are ocean by depicting clearly drawn continental coastlines, islands, and ships that sail the waters of the Occeanus, the Occeanus Meridionalis, the Indicum Mare, and an unnamed body of water that we know as the South China Sea. But unlike the Occeano, both the Occeano Meridionalis and the Indicum Mare lack a southern boundary. Presumably, the farther south we move, beyond the coastlines, the toponyms, the islands, and the illustrations, the weaker their signifying power becomes, until at some indeterminate point, we are no longer in oceanic space, but in empty cartographic space. Is there land somewhere south of Africa or India? If so, where does the ocean end and the land begin? According to the Reinel chart, we simply do not know. The practice of marking oceanic space through the inclusion of illustrations is inherently ambiguous. It identifies certain spaces as aquatic, but it admits to ignorance regarding the dimensions or boundaries of the body of water in question.

The enormous space that we are tempted to identify as the Pacific Ocean or the South Sea offers the most interesting example. The only maritime toponym anywhere west of Brasil is the name "Mar visto pelhos castelhanos" (The sea sighted by the Castilians), which appears in a yellow banderole just below the coast of what we call Panama, as well as two images of what appear to be large ocean-going canoes. With this inscription, the chart acknowledges the discovery made by Vasco Núñez de Balboa, who marched across the Isthmus of Panama in 1513 in search of a large body of water whose existence had been recounted by indigenous informants. Balboa called it the South Sea, and he apparently assumed that it was the same body of water that washed the shores of the Spicery. With time, the name came to be applied to the entire expanse that we know as the Pacific Ocean. Yet Reinel places the name and the canoes close to the coastline, indicating that he does not know how far that sea extends or what its boundaries might be. The modern eye travels westward, over an empty expanse that it easily mistakes for the Pacific Ocean, but the period eye correctly identifies that expanse as blank cartographic space, not as oceanic space. It recognizes that the map makes no claims about the relationship between the sea spotted by the Castilians and the unnamed maritime space to the west, whose existence is marked by the coastline of the "Occidens Chis" and the "Ilhas de maluqua."[9] It makes no claim about what might lie between the two seas, whether empty water, an undiscovered continent, or islands in numbers large or small. By leaving that expanse unmarked, the map marks it as the space of the unknown, inviting the reader to speculate if he or she wishes about what might lie there waiting to be found. As we shall see in the next section, where I address the first of the two errors mentioned above, that period reader is unlikely to have imagined the empty ocean that the modern reader is tempted to see.

The Reinel Chart and the Terraqueous Globe

While we tend to project oceans into the empty spaces of Iberian planispheres, even if only unconsciously, early modern readers were more likely to imagine land. Were we discussing French, German, or Flemish cartography, in which the *Terra Australis Incognita* became a fixture of world maps as of the 1530s, this assertion would be self-evident, but it is also true of the Iberian milieu, where both intellectuals interested in cosmography and geography and practical people engaged in commerce and colonization, were well aware of the theoretical developments that underpinned such speculation, and were perfectly capable of bringing those ideas to bear on the empty spaces of their reticent planispheres. Those developments involved a revolution in prevailing ideas about the relationship of land to water on the surface of the globe, a revolution that had deep implications for the architecture of the continents, and for the way Europeans mapped it into global space. By reading the Reinel chart with these developments in mind, we can easily see how a period reader would tend to imagine land where we know there to be vast stretches of water.

These developments are best understood if we begin with prevailing medieval ideas about the world's geography, as captured by the two cartographic traditions that are so often invoked to illustrate it, the T-O–style *mappamundi* and the Macrobian zonal map.[10] The T-O map depicts the world's geography along highly traditional lines inherited from the Babylonians, Greeks, and Romans, as a tripartite disk surrounded by a river-like Ocean Sea, with East at the top and, sometimes, the city of Jerusalem at the center (fig. 14). The Macrobian map depicts that same island-world as part of the larger sphere of Earth and Water that was thought to sit at the center of the classical cosmos. It divides that sphere into five climatic zones and often includes another island similar to the *orbis terrarum* in the southern hemisphere (fig. 15). The existence of these two separate strands of cartography speaks volumes about how medieval intellectuals understood "the world," *oikoumene* in Greek and *orbis terrarum* in Latin. The T-O maps could ignore the rest of the sphere because only the tripartite island, the *orbis terrarum*, counted as "the world," the place created by God for habitation by his creatures. The rest was not just unknown, but also irrelevant, and in any case, inaccessible, thanks to the vast size of the Ocean Sea and the impassability of the hot, arid Torrid Zone. The Macrobian map, by contrast, does not map the world, but rather the larger sphere of Earth and Water where the *orbis terrarum* is found, and facilitates reflection about the existence of other worlds that might serve as homes to alternative creations.[11]

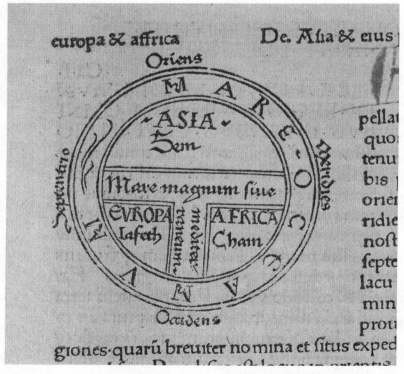

FIGURE 14. T-O style *mappamundi* from an early printed edition of Isidore of Seville's *Etymologia*. The *orbis terrarum* is figured as a tripartite disk surrounded by the Ocean. Isidore of Seville's *Medieval Depiction of Earth as Wheel Encircled by Ocean*, from *Etymologiae* (Augsburg, 1472).

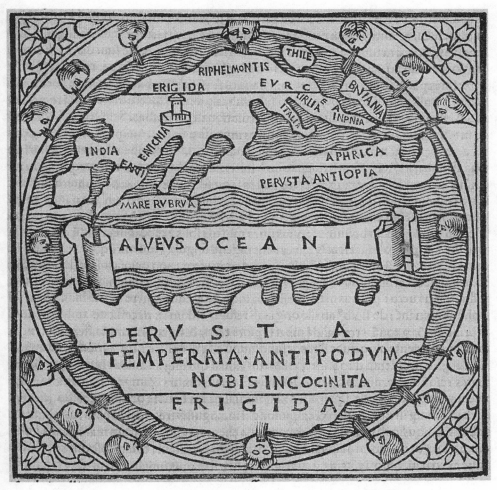

FIGURE 15. Macrobean-style map of the globe, which divides it into five climatic zones and speculates about the existence of an unknown world in the southern hemisphere. From Ambrosius Aurelius Theodosius Macrobius et al., *Macrobii Interpretatio in Somnium Scipionis a Cicerone Confictum . . .*, edited by Niccolò Angeli. (Venice, 1521). Photograph: Courtesy of the Tracy W. McGregor Library of American History, Albert and Shirley Small Special Collections Library, University of Virginia (call no. E 1521.M33).

By the fifteenth century, the distinction embodied by the separateness of these two cartographic traditions began to collapse. The whys and wherefores of this change have been studied before, so we need not rehearse them in detail (O'Gorman 1961, 1986; Randles 2000; Besse 2003; and Wey Gómez 2008). Suffice it to say that that the transformation involved a shift in the preferred answer to a crucial question that Aristotle had left unanswered in *De caelo* (IV. 3), one of the foundational texts of classical and medieval cosmography. According to Aristotle, the four elements, Earth, Water, Air, and Fire, naturally existed as four nested, concentric spheres at the center of the cosmos. The Philosopher, however, provided no explanation for what everyone could plainly see, the fact that the Earth was not entirely submerged. So medieval scholastics came up with an answer of their own, invoking a primordial moment when Providence displaced the Earth from its natural position at the center of the cosmos so that part of it broke through the surface of the sphere of Water. The result was a model in which the spheres of Earth and Water were no longer concentric. Jaime Pérez de Valencia, a Spanish bishop with a strong cosmographical bent, helps us imagine this by comparing the Earth to an apple bobbing in a basin (Gil 1995, 262). The *orbis terrarum* was the part of the apple that stuck out above the surface of the water, a relatively small island surrounded by a massive Ocean Sea. No other such islands could possibly exist.

Over the course of the Middle Ages, Al-Farghānī, Albertus Magnus, Roger Bacon, and Pierre d'Ailly criticized this model and proposed an alternative, but their arguments did not acquire much traction until the fifteenth century, when the recently rediscovered *Geography* of Claudius Ptolemy took hold of the European cosmographical imagination and the Portuguese began to sail down the coast of Africa (Randles 2000, 28–29). Ptolemy mocked the insular *oikoumenē* of the inherited tradition and its encircling Ocean Sea, insisting that the known world extended across a full 180° of longitude and 79° of latitude, and suggesting that it could extend even farther than that (Ptolemy, Berggren, and Jones 2001, 177; Also Randles 2000, 16). What he was describing was not the relatively small, insular, and isolated *orbis terrarum* of the medieval tradition, but something much larger, altogether different. The geographical discoveries of the Portuguese added fuel to the fire. They demonstrated that the coast of Africa extended farther south than anyone had imagined and therefore that the *orbis terrarum* could not be as small, relative to the larger sphere of Earth and Water, as the scholastics had supposed.

Under the influence of Ptolemy and the new discoveries, Pérez de Valencia, the bishop who gave us the apple metaphor, joined the ranks of those intellectuals who began to favor a different answer to the question of how dry land could exist in a cosmos ordered on Aristotelian principles. In fact, he was one of the first to articulate in writing the difference between the scholastic model and the new Ptolemaic model, in his 1484 *Commentaria in psalmos* (Commentary on the Psalms; Randles 2000, 41). The new cosmological model explained the existence of dry land, not by imagining a primordial moment in which God rearranged the spheres, but by appealing to the irregularity of the surface of the Earth, something readily observable in the existence of mountains and valleys. As the French cosmographer Jean Fernel put it, the sphere of Earth was like "a wooden globe in which there are many hollows in which water can gather" (cited by Randles 2000, 68). Maybe he would have come up with a better metaphor had he had a modern golf ball in hand. Settling into those hollows, the sphere of Water became the world's lakes, seas, and oceans. Rather than speak of two concentric spheres, therefore, one of Earth and another of Water, the new cosmography spoke of a single sphere of Earth and Water conjoined, the terraqueous globe.

Unlike the old model, the new one placed no limits on the potential size of the *orbis terrarum*, or on the amount of land, relative to water, that could exist on the surface of the globe. This means that the architecture of the continents, which had always been a closed system composed of three major parts, Europe, Asia, and Africa, had the potential to become an open system. It became possible to suggest that there were portions of the three parts of the world that had not been charted and remained unknown, perhaps islands or stretches of continental land, or that there were other worlds much like the *orbis terrarum* waiting to be discovered. One could even speculate that land, insular and continental, actually predominated over water on the surface of the globe. Pérez de Valencia cites Scripture (2 Esdras 6:42) to that effect, putting the ratio at seven parts land to one part water. Once again, the geographical discoveries of the Portuguese and the Spanish were pressed into service as evidence in support of the new cosmological paradigm. By 1506, the Portuguese cosmographer Duarte Pacheco Pereira had concluded that the various coastlines that had been charted in the Ocean Sea were really parts of a single continuous coastline running southward from 70°N to 28°S, with no end in sight. This demonstrated, he argued, that the *orbis terrarum* wrapped around most of the globe, revealing that "o mar não cerca a terra, como Homero e outros autores disseram, mas ante a terra por sua grandeza tem cercadas e inclusas todalas áuguas dentro na sua concavi-dade e centro" (The sea does not surround the land, as Homer and other authors said it did, but rather the earth, in its greatness, has enclosed and encloses all of the waters in its concavity and center; Carvalho 1991, 539).

Columbus, of course, was well ahead of Pacheco Pereira in mapping an *orbis terrarum* so large relative to the size of the terraqueous globe that it was possible to sail across the Ocean Sea from Europe to Asia. So had his contemporaries, the Florentine mathematician Paolo Toscanelli and the German merchant Martin Behaim. On Behaim's 1492 globe, the first of its kind, the *orbis terrarum* stretches around the globe a full 234°

of longitude and reaches south beyond the Tropic of Capricorn, leaving
an ocean only 130° wide between Europe and the Indies (fig. 16).[12] The
cosmological model that underpinned Behaim's globe, moreover, per-
sisted after 1492. Many informed Europeans continued to suspect that
land was plentiful, perhaps even predominant, on the surface of the terraqueous globe, and the discovery of what seemed to be a new world
in the southern hemisphere only confirmed this suspicion, as it did for
Pacheco Pereira. So, not only would a period reader have seen blank cartographic space where a modern reader might see empty ocean, but he
or she would have also imagined that there was more land than water in
that blank space. While there might be significant bodies of water among
the islands and continental landmasses awaiting discovery, none of them
could possibly be as vast and empty as the Pacific Ocean.

The cartographic literature of the decade prior to the making of the
Reinel map suggests various ways that the period reader might have imagined the lands that were awaiting discovery. Some of that literature adamantly denied that the newly discovered lands in the Ocean Sea were the
Indies described by the ancients or Marco Polo. One of the most forceful
cases was made by Bishop Rodrigo Fernández de Santaella, who published a 1503 edition of Marco Polo and a 1518 translation of the book into
Spanish that were meant to demonstrate that the newly discovered lands
were something geographically distinct, and substantially different from
anything previously known. According to the bishop, the Castilian "An-
tilia" belonged to a newly charted West, while the Indies belonged in the
traditional East (Gil 1987, 174–77; see also Gruzinski 2014, 44–45). The
argument was taken up again by Juan López de Palacios Rubios, who was
responsible for the first systematic attempt to justify Spain's sovereignty
over its new colonial possessions. Writing in 1513, Palacios Rubios cites
Santaella, noting that "the Islands of the Ocean Sea" have not yielded the
spices, cities, precious stones, or elephants that we read about in Marco
Polo and therefore should not be confused with them (Palacios Rubios
1954, 6–8). Clearly these authors were moving in the same direction as
Waldseemüller, laying the foundations for what would become the hegemonic image of the world by the last third of the century.

We can imagine that like-minded readers of the Reinel chart might
have been prone to round out its depiction of the newly discovered lands
along the lines suggested by the theory of American insularity. They might
have imagined a solution similar to the one proposed by Waldseemüller
in 1507, believing that *both* Brasil and Terradimini, if not Bacalnaos and
DoLavrador, were newly discovered lands that had nothing to do with
Asia or the Indies (see fig. 3). They might have found subtle support for
the idea in the arc of compass roses that figure so prominently on the left

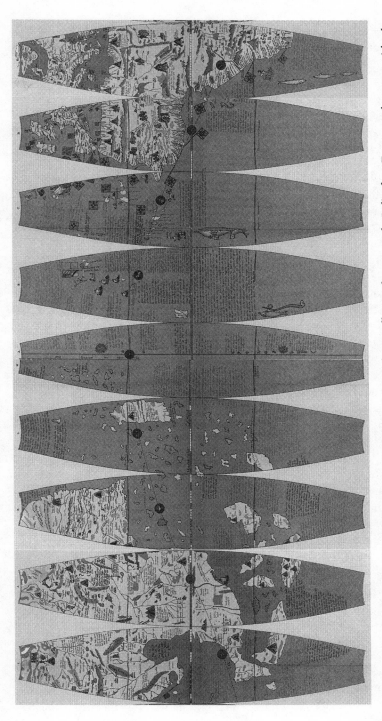

FIGURE 16. Detail from facsimile gores of Martin Behaim's 1492 globe. Only 130° of longitude separate the Indies from Europe, and numerous islands, known and hypothetical, promise to provide the necessary way stations for the trip across. From Ernst Georg Ravenstein, *Martin Behaim, His Life and His Globe* (London, 1908). 76 x 61 cm. Photograph: Courtesy of the Hathi Trust Digital Library.

side of the map, and that might be understood as constituting a notional boundary between the newly discovered lands in the west and the Asian east. Alternatively, such readers might have imagined the invisible geography of Reinel's map along the lines suggested by Johannes Ruysch in 1508 (fig. 17). It is easy to imagine how the fragmentary coastline of the Occidens Chis might stretch eastward to dovetail with the coastline of the Terradimini, and perhaps even Bacalnaos and DoLavrador, leaving a maritime passage open to the south, separating this easternmost part of Asia from the Yucatan Peninsula, the northernmost tip of an insular fourth part of the world. In either case, the Castilian hemisphere would not exhibit any internal geographical coherence. Instead, it would appear to be a motley collection of newly discovered lands in the West and bits and pieces of the East, lumped together by the arbitrary geometry of the lines of demarcation.

Yet the Reinel map does not necessarily suit readers whose geographical ideas coincided with those of Santaella or Palacios Rubios. The bishop's mental cartography distinguishes between an absolute East and an absolute West, and so it does not seem to have made the global turn characteristic of Toscanelli, Behaim, Columbus, and so many others. It seems to forget that on a round Earth, west and east eventually meet, attenuating whatever distinctions one might try to make between the two as absolute locations. While the Reinel chart is a flat map and not a globe, it nevertheless takes this global turn, insofar as it places the easternmost tip of Asia, the Occidens Chis, and a putatively oriental archipelago, the Moluccas, on the western edge of the map surface, thereby confounding East and West. While Santaella assumes that the boundary between East and West is stable and readily discernible, Reinel invites us to wonder where it might be. Rubios, by contrast, seems to think with a globe in mind, insofar as he understands that the way west eventually leads to the East, but he maintains the distinction between a Castilian West made up of newly discovered lands and a Portuguese East made up of previously known ones by arranging the two regions as stops along the one-dimensional space of a notional itinerary of travel. The two-dimensional space of the Reinel map has the potential to complicate this way of thinking, particularly if one suspects that any of the newly discovered lands might actually be *pars Asiae*. If that is the case, then the Atilhas de Castela and Brasil lie south rather than east of Asia. Once again, the Alpha-Omega point where the West becomes the East proves difficult to identify and the absolute distinction between the two difficult to sustain.

I argue that the Reinel chart actually lends subtle support to the theory of Amerasian continuity, particularly through the placement and prominence of its banderoles. The right-hand edge of the map truncates

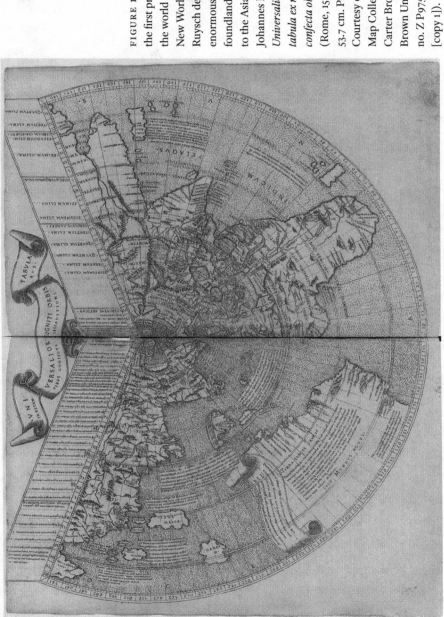

FIGURE 17. In one of the first printed maps of the world to include the New World, Johannes Ruysch depicts it as an enormous island. Newfoundland is attached to the Asian continent. Johannes Ruysch, *Universalior cogniti orbis tabula ex recentibus confecta observationibus* (Rome, 1508). 39.8 x 53.7 cm. Photograph: Courtesy of the JCB Map Collection, John Carter Brown Library, Brown University (call no. Z P975 1507/2-SIZE [copy 1]).

the continent of Asia, indicating that it extends eastward, onto the left-hand edge of the map. Only a fragmentary coastline appears, one that soon trails off into the unknown, but only to hand the reader's eye over to the series of banderoles that provide the map's names for China and the Tropic of Cancer, and finally to the hills that mark the unknown horizon of the Brasilian west. This series of banderoles stretching from the Chinese coast to the Brasilian hills pierces the arc of compass roses that seem to isolate the newly discovered lands from China and the Moluccas, suggesting continuity instead of discontinuity. They become an ersatz Amerasian coastline, rising like a specter from the chartmaker's attempt to avoid such speculation.

A reader seduced by the continuity of the banderoles rather than by the discontinuity suggested by the arc of compass roses might fill the blank spaces of the Reinel chart with a geography akin to the one that we find on one of the so-called Zorzi maps. Alessandro Zorzi was a Venetian collector who helped Bartolomeo Columbus translate his brother's account of his fourth voyage, and who sketched a map in the margins of his translation that is widely believed to capture the admiral's ideas about the geography of the new discoveries. The map identifies the New World as the landmass we call South America and connects it via the Isthmus of Panama to mainland Asia.[13] The map was unknown in the Spain of 1519, but the basic geographical idea that it expresses was quite popular (Nunn 1929, 23; Nowell 1947, 6). It is at work in the writing of Pietro Martire d'Angheira, known in English as Peter Martyr, the Milanese humanist and emigrant to Spain who was responsible for the first history of Castilian activity in the Ocean Sea. Writing in the wake of Balboa's discovery of the South Sea, in 1514, Martyr marveled at the fact that only a narrow isthmus of land separated the North Sea from the South (Anglería 1989, 196, and see also 243). He might just as well have noted that only a narrow isthmus connected the New World in the south to the landmass in the north. His further remarks about the aftermath of Balboa's discovery leave no room for doubt that he thought this landmass was the continent of Asia. Reporting in 1516 on the discovery of islands rich in pearls just off the coast of Panama, Martyr anticipates that Spanish ships will soon reach the Spicery:

Varias esse in australi pelago insulas, Sancti Michaelis et Ditis insulae sinui occidentales, in quarum pluribus aiunt procreari colique arbores, fructus gignentes eosdem quos terra ollocutea gignit. Est Collocuti una et Cochini et Camemori Portugalensium nundinaria aromatum terra, inde arguunt non longe abesse aromatum omnium initialem tellurem (Anghiera 2005, 1.446).

[A number of islands lie in the South Sea to the west of the gulf of San Miguel and the Isla Rica, and on most of these islands are grown and cultivated trees that bear the same sort of fruit that we find in Calicut. Calicut, Cochin, and Camemor constitute the market where the Portuguese procure spices. Thus it is thought that the country where spices grow is not far from there.][14]

Clearly, Martyr believes that the ocean Balboa had discovered was the same one that washed the shores of India. The humanist's *australi pelago* is the same body of water that Reinel calls the *Indicum mare*, and that the Portuguese historian João de Barros, writing about his countrymen's voyages to India thirty years later, would refer to as the *mar do sul* (South Sea) (Barros 1552, 1:28r). This would mean that, in Martyr's imagination, a continuous coastline extended from the spot where Balboa waded into the South Sea westward to Cathay and Mangi. The South Sea was "south" not by way of a simple contrast with the North Sea, as Las Casas would argue roughly twenty years later, but because it lay south of the great bulk of the *orbis terrarum*, the Amerasian landmass that stretched from the New World to the Indies of Marco Polo and the Portuguese.[15]

So, just as a reader of the Reinel chart could fill in its blanks along the lines suggested by the theory of American insularity, so he or she could also imagine its invisible geographies along the lines suggested by the theory of Amerasian continuity. In fact, given the popularity of the theory in early sixteenth-century Spain and its embryonic colonies, most readers of the Reinel chart would probably have interpreted its possibilities in this way. Of course, they might also have projected countless other possibilities onto the blank spaces of the map, involving a southern landmass or countless islands, perhaps. Yet they might also have read the map in an entirely different way, one that did not attach so much importance to the architecture of the continents. The history of exploration and discovery places undue emphasis on the fortunes of this particular metageographical framework, forgetting that it was not the only one available to early modern culture. As Nicolás Wey Gómez has demonstrated, the early modern West also mapped the globe in terms of the theory of climates, with its emphasis on latitudinal position rather than physical geography.[16] In the next section, I explore what it might mean to read the Reinel map through this largely forgotten metageography.

The Tropics on the Reinel Chart

Like Bishop Santaella, the conquistador and historian Gonzalo Fernández de Oviedo believed that the lands discovered by Christopher Columbus were not part of the Indies as they were known to the classical and medieval tradition, but something new and different. He makes this point many times in his voluminous writings, but one passage from his *De la historia natural de las Indias* (Of the natural history of the Indies), better known as the *Sumario de la natural historia* or simply the *Sumario* (Summary of the natural history of the Indies, or Summary) is particularly revealing. Composed at the request of Charles V and published in 1526, the *Sumario* describes the flora, fauna, and human societies of the Caribbean, including the mainland of Central and South America, as unprecedented novelties that do not answer to the period's bookish knowledge of the Indies as derived from ancient authors. Chapter 11 of the text describes what Oviedo calls the "tiger" of the Castilian Indies, the animal we know as the jaguar, drawing contrasts between its characteristic torpor and the liveliness of the Indian tiger as described by Pliny. This contrast and others like it lead him to affirm that the places where this animal lives constitute a "tierra que hasta nuestros tiempos era incógnita, y de quien ninguna mención hacia la Cosmografía del Tolomeo ni otra, hasta que el almirante don Cristóbal Colón nos la enseñó" (a land until our times unknown, of which no mention is made in the *Cosmography* of Ptolemy or in any other, until the admiral Christopher Columbus revealed it to us; Fernández de Oviedo 1950, 146). By making such distinctions, it has been argued, the *Sumario* effectively summons a "New World" into being, "America" in all but name (O'Gorman 1972, 1.46–49; Ballesteros Gaibrois 1981, 171–74).

Recently, however, Nicolás Wey Gómez has interpreted this passage anew, in ways that dislodge Oviedo from the historical teleology of the invention of America narrative, and its exclusive concern with continental geography. Just as Oviedo makes explicit comparisons between different tigers, Wey Gómez points out, so he also draws a tacit but nevertheless clear contrast between the big cats of far-off places and the small cats of Spain. He does something similar with bats and the yucca plant. By way of these contrasts with the mother country, the two Indies that seemed so different from each other begin to look more alike. It becomes clear that even though the newly discovered Indies are not the same as those of the classical tradition, by Oviedo's reckoning they are both tropical locations that are to be distinguished from temperate ones. The metageographical framework that governs his distinctions is not the architecture of continents, but the theory of climates, which divided the world into five latitudinal zones rather than three geographical "parts." (Wey Gómez 2013, 626).

I would add that this sense that the two Indies are similar to each other in their shared tropicality comes across in the way the *Sumario* refers to the region it describes. Only once does the text refer to it as a "new part of the world" in a way that recalls Waldseemüller, and never does the text use the names "America" or "New World" (Fernández de Oviedo 1950, 119). It invariably calls the mainland of the Indies "Tierra-Firme," and says nothing about its relationship (or lack thereof) to continental Asia. Only two passages hint at the distinction that later became conventional between the "East Indies" and the "West Indies," but they are clearly concerned with affirming Castilian sovereignty over these lands rather than addressing differences on the level of the natural order.[17] Sometimes the *Sumario* calls the region "estas Indias" (these Indies), but most of the time it just refers to it as "las Indias" (the Indies), or even "aquellas partes" (those parts). In the end, while the *Sumario* does indeed emphasize that its Indies are not the same Indies that one can read about in Pliny and other classical authors, it also draws attention to the region's "Indianness," that is, its tropicality, something it shares with Pliny's Indies and not with the countries of the temperate zone.

Wey Gómez's interpretation of Oviedo builds upon his earlier work demonstrating that the theory of climates was just as important, if not more so, than the architecture of the continents in fifteenth-century world-making (Wey Gómez 2008). This theory finds graphic representation in the Macrobian map we saw above, which divides the globe into two frigid zones, two temperate zones, and one torrid zone (fig. 15). Each zone felt the influence of heavenly bodies, the sun in particular, in different ways, with profound consequences for the nature and even the very possibility of life within it. According to the version of the theory that the ancients bequeathed to the Middle Ages, the Torrid Zone, which lay between the tropics of Cancer and Capricorn, at 23.5° north and south latitude, was too hot and arid to allow for life, while the Frigid Zones, which began at 66° north and south and stretched to the poles, were too cold and icy for life to exist. Within the temperate zones, the only zones that were inhabitable, the nature of life varied according to latitude. As Surekha Davies explains, environmental factors that changed according to one's distance from the equator, such as temperature and the influence of the heavenly bodies, helped determine the so-called "complexion" of human bodies, the balance of the four humors that constituted them. Variations in complexion manifested themselves as variations in physical appearance, personality, mental aptitude, and character. Cold northerly climes tended to produce light-skinned, rugged people who were brave but thick-headed, while hot southerly ones made for dark-skinned, delicate people who were clever but submissive. At either extreme, the environment produced people incapable of forming civil societies, and even creatures whose humanity was questionable at best (Davies 2016, 25–30).

According to Wey Gómez, the cosmological paradigm shift of the fifteenth century that we reviewed in the previous section brought significant changes to the way the Torrid Zone was imagined. The same thinkers who argued, on theoretical grounds, that the *orbis terrarum* must be larger than it was commonly made out to be also argued that humid conditions mitigated the heat in much of the Torrid Zone, just as they did along the Nile and in India, making for a climate that was not only amenable to life, but that tended to produce it in abundance, along with the kinds of riches that could be found only where the sun really shone, like spices and gold. Portuguese expansion along the coast of Africa confirmed these expectations. Sailors who were afraid they would spontaneously combust as they sailed south into the Torrid Zone found themselves very much alive, sailing along a lush coast rich in abundant vegetation, inhabited by human populations, and rich in coveted commodities like spices (malagueta pepper) and gold. Slowly, the classical Torrid Zone began to morph into the modern tropics, rich in all sorts of natural resources, and home to people whose character and mental aptitude did not render them fit to form civil societies conducive to human flourishing, if any at all (Wey Gómez 2008, 229–334). This was the perfect destination for colonial powers on the make. So, as Columbus sailed west to reach the East, he also veered south to reach the tropics, where he knew he would find the Indies, the particular part of Asia in which he was most interested (335–434).

Building on Wey Gómez and other scholars who have drawn our attention to the importance of climate in early modern understandings of human geography, Surekha Davies teaches us to always look at early modern maps and cartographic literature with latitude in mind, thereby mimicking the sort of "spatial thinking" in which early modern Europeans so often engaged. According to this way of thinking, one should expect to find similarities among people and places found at the same latitude, no matter what island or continental landmass they inhabit, and contrasts among people and places at different latitudes, even if they inhabit the same landmass (Davies 2016, 44–45, 250). This sort of spatial thinking often ran afoul of empirical realities, such as the unexpectedly light skin tone of tropical Brazilians, or the existence of complex, urbanized cultures in Mexico and the Andes (217–56). It also implied, quite inconveniently for Europeans who were settling in their new tropical possessions, that over time their descendants would become just like the native inhabitants, their skin turning darker and their mental capacities weaker.[18] Nevertheless, the theory of climates and its attendant notions of human geography proved either too prestigious or too useful to be abandoned altogether. Although some questioned and even undermined it, others found ways to use it selectively or to interpret empirical observations in ways that made them fit into the theory's ideological framework. Bartolomé de las Casas, for example, did a little of both in his effort to turn the theory of climates against efforts to brand the native inhabitants of the New World as Aristotelian natural slaves. He mapped the New World as a single, large climatic zone stretching from the forty-fifth parallel north of the equator to the same parallel south of it, and claimed that its environment was so favorable that its inhabitants were fully rational and capable of self-governance (Casas 1967). We shall see it at work throughout most of this book, even in works written after 1570 or so, when the theory of American insularity became the preferred way of mapping the new discoveries, and the architecutre of the continents gained new prominence as the preferred metageography of European culture.

For now, let us ask what it might mean to read the Reinel map with the theory of climates in mind. Up until now, we have been most concerned with the various fragmentary landmasses that appear in the left-hand side of the map, and we have wondered how a period reader might round out or connect the various pieces of land. We have been guided by an assumption that underpins almost all of the historiography of the so-called Age of Exploration, that the so-called "parts of the world" provided the basic ground for medieval and early modern notions of human geography. Just as European cosmography divided the *orbis terrarum* into three parts, Europe, Asia, and Africa, so it divided the human family into the three branches that descended from the three sons of Noah, Japhet, Shem, and Ham, and populated each of those three parts, becoming Europeans, Asians, and Africans. It is from the perspective of this metageography that the novelty and insularity of the lands discovered in the Ocean Sea posed a problem. The existence of a previously undiscovered landmass that was home to human beings who went unmentioned in the bookish tradition and apparently had never heard the gospel message challenged not only the traditional architecture of the continents, but the biblical anthropology that went with it. This was particularly true if that landmass was not physically connected to the *orbis terrarum,* making it difficult to explain how those people had reached the new land in the aftermath of the Deluge.

To read the Reinel map with the theory of climates in mind is to set aside the unsolved puzzle of the world's physical geography and take the edge off the challenge it posed to notions of human geography. From the perspective afforded by the theory of climates, human difference does not map onto the parts of the world, but onto the latitudinal zones. It does not matter if people live in Africa, Asia, or the New World. If they all live in the Torrid Zone, then they should all be people of the tropical sort, clever but timid, lacking in any real ability to govern themselves rationally. By the same token, if they all live in one of the temperate zones, then they should all resemble Europeans in their ability to cultivate the arts of civilization and live a fully political life. From the perspective provided by this framework, it does not matter if Reinel's Brasil is part of Asia or not, at least as far as empire is concerned. What really matters is its location in the Torrid Zone, which the map reader can easily appreciate thanks to the prominent banderoles marking the Tropics of Cancer and Capricorn. There it lies, along with other resource-rich places home to servile people, like sub-Saharan Africa, India, Southeast Asia, and the Moluccas. The banderoles labeling the tropics and the eye-catching red letters that spell out "Equinoctialis" lead the eye from the Moluccas to Brasil and back again, inviting the reader to consider the possibility of reaching the Spice Islands by sailing west, but they also mark out the blank space in the Castilian tropics, an area that, according to the new cosmology, should be rich in undiscovered lands yielding gold, silver, spices, and slaves.

The Reinel Chart and the Ocean Sea

The Reinel chart proclaims in a variety of ways that the culture in which this map was made and consumed was confident in its ability to reach those distant lands by sea. This sort of confidence had appeared in European culture relatively recently, over the course of the fourteenth, fifteenth, and sixteenth centuries. Before then, speaking in very general terms, the open ocean was regarded as a menacing expanse at the edge of the world, ultimately unfit for human beings, a space where they ventured only at great risk. Medieval *mappaemundi* of the T-O variety reproduced Greco-Roman notions of the Ocean Sea as the ultimate boundary of the *orbis terrarum*, a space of unbounded chaos that stood in contrast to the bounded orderliness of the world itself (Romm 1992, 20–26; Mack 2013, 72–104).[19] The Greeks and Romans also bequeathed a common rhetorical trope, the "diatribe against navigation," which figured seafaring as a foolhardy violation of the natural order of things undertaken in the vain pursuit of wealth and power (Boruchoff 2012, 136–38). In English, the expression "all at sea" implies that someone is lost beyond hope (Mack 2013, 72). In Spanish, meanwhile, the verb *engolfar* and its reflexive form, *engolfarse*, which means "to sail out to sea, beyond sight of land," also means "to get involved in difficult and arduous affairs, the kind in which one can never touch bottom."[20] To sail the open sea, therefore, was to run into trouble in the deep end, where the water was over one's head, and risk losing oneself in ways both figurative and frighteningly literal.

All of this began to change as the societies of the medieval Mediterranean learned from Muslims and Jews, like the chart-making Cresques family of Majorca, how to use the compass and the astrolabe, how to build sturdier, more agile vessels, and how to make and use nautical charts. According to Fernández-Armesto, these crucial technology transfers joined forces with the fondness for adventure typical of late medieval Iberian cultures that fed on chivalric romance to produce a unique sense of maritime daring, a marked willingness to *engolfarse* (Fernández-Armesto 2007b, 144–48). The Portuguese and later the Spanish made for distant islands, the Canaries in particular, expanding the horizons of the late medieval world to include large stretches of open ocean. Ever-increasing experience with the pattern of winds and currents in the Atlantic made it possible to imagine that one could safely sail with the wind out into the Ocean Sea, in the hope of reaching its further shore (Fernández-Armesto 2007b, 149–51). Everyone knew that seafaring continued to be highly unpleasant at best and deadly at worst (Pérez-Mallaína Bueno 1998, 129–90). Diatribes against navigation never stopped flowing from early modern pens.[21] Nevertheless, confidence grew around the art of navigation and, as David Boruchoff explains, the navigational instrument that was at the heart of this transformation, the compass, was elevated to symbolic status. Since it had allowed the moderns to succeed where the ancients had failed, in sailing the Ocean Sea to previously unknown lands, the compass joined gunpowder and the printing press in a novel canon of "great inventions" that served as icons of the period's modernity (Boruchoff 2012, 142–48). The diatribes against navigation now found a clear rejoinder in celebrations of seafaring as one of the quintessential accomplishments of modern culture.

This transformation is very much in evidence in the Reinel chart, in a variety of ways, not the least of which is its genre. We find ourselves before a "plane chart," a kind of map that emerged from Portuguese efforts to adapt the portolan chart of the medieval tradition to the needs of Atlantic navigation and the emerging practice of using astronomical observations to determine the position of a ship and of a given geographical location. As such, the Reinel preserves some of the key characteristics of the portolan chart, notably the web of "rhumb lines" that crisscross its surface, with compass roses at key points of intersection, as well as the tendency to populate coastlines with densely packed rows of toponyms set at right angles to the coast itself, but the Reinel chart abandons parchment for paper and adds a coordinate grid system. It also adopts a global rather than a regional scale, which was not unknown in the portolan tradition, but was nevertheless rare. The map itself is therefore very much a product of the new culture of oceanic navigation, which brought together the established practices of seafaring in the Mediterranean and along the shores of the Atlantic with emerging technologies, an expanded vision of the world, and the cartographic innovations of the new Ptolemaic cosmography. Yet there are also more specific ways in which the Reinel chart speaks of the newfound sense of mastery over the Ocean Sea. One of these has to do with its treatment of oceanic space, which echoes a revolution in hydrography that developed as a consequence of the new cosmography. The other has to do with its iconographic program, particularly its use of ships to "decorate" maritime spaces. As we saw above, these images serve to mark oceanic space as such, but they also have a lot to say about the ongoing domestication of the ocean.

The *Geography* of Ptolemy rejects the insular island-world of the poets and pours the watery element into the concavities, large and small, of the pockmarked surface of the sphere of the Earth, converting them into lakes, seas, and ocean basins (Ptolemy, Berggren, and Jones 2001, 117; also Randles 2000, 16). Pacheco Pereira echoes Ptolemy in the passage cited above, "O mar não cerca a terra, como Homero e outros autores disseram, mas ante a terra por sua grandeza tem cercadas e inclusas todalas áuguas dentro na sua concavidade e centro" (The sea does not surround the land, as Homer and other authors said it did, but rather the earth, in its greatness, has enclosed and encloses all of the waters in its concavity and center) (Carvalho 1991, 539). Not only does this passage affirm that the *orbis terrarum* is much larger than the scholastics thought it to be, but it also draws out the consequence of that belief for the hydrography of the Ocean Sea. The ocean is no longer an open expanse that surrounds the *orbis terrarum*, as it does on medieval *mappaemundi*, but a basin of water contained by its opposing shores, Europe and Africa in the east and what we call America in the west. What had been for so long imagined as the hostile waters at the edge of the world becomes in Pacheco's cosmography a new version of the old *mare nostrum*, a "medio terrano" or "Mediterranean" (Carvalho 1991, 542). Meanwhile, the other maritime space with which the cosmographer was familiar, the Indian Ocean, becomes nothing more than a large *lagoa* (lake) (690). The new cosmography, therefore, brought with it a new hydrography.

This new hydrography found cartographic expression in any number of sixteenth-century maps. As Martin Lewis explains, the cartography of the period generally abandoned the "unitary" approach characteristic of earlier times, by which the entire watery element was mapped as a single surrounding Ocean Sea, in favor of a "basin perspective," by which the oceans were mapped as a series of "discrete units of sea space" (Lewis 1999, 199). According to Lewis, the process reached its culmination in Abraham Ortelius's 1570 "Typus orbis terrarum," the world map that accompanied his influential atlas, the *Theatrum orbis terrarum*. The map depicts the world's oceans as a series of four interconnected ocean basins, the Mar del Nort (North Sea, or North Atlantic), the Oceanus Ethiopicus (Ethiopian Ocean, or South Atlantic), the Mar di India (Sea of India, or Indian Ocean), and the Mar del Zur (South Sea, or Pacific Ocean). "Sea" and "ocean" are treated as fully interchangeable terms, not only in the "Typus," but in the *Theatrum* as a whole (200). The massive *Terra Australis*, meanwhile, serves in large measure to equip three of these four basins with a clearly defined, if hypothetical, southern shore (see fig. 1). The "Typus orbis terrarum" thus gives full expression to what I call the "cartography of containment," which depict the world's oceans as reasonably well-bounded basins, contained by the major landmasses, or "continents."

This cartography of containment is not as easy to detect on the Reinel chart, with all its blank spaces, particularly in the southern latitudes. We can nevertheless detect its presence in the use of different names to identify different oceanic basins and in its treatment of what we call the North Atlantic. "Occeano," which would have named the entire Ocean Sea on a medieval *mappamundi*, becomes here the name of a particular oceanic basin, defined by the shores of Europe, Africa, Brasil, the Terradimini, and DoLavrador. This basin is connected by a discrete passage to Occeanus Meridonalis, which is in turn connected by the route around the Cape of Good Hope to the Indicum Mare. The words for "ocean" and "sea" are used interchangeably to refer to these various basins, as they would be in the Ortelius atlas more than sixty years later. The two seas in the southern hemisphere lack geographical boundaries to the south, but we must remember that the period reader would have likely imagined that such boundaries were only waiting to be found. While our eyes, conditioned as they are by modern maps of the world, see the boundless expanse of the South Atlantic and the Indian Ocean, their eyes saw blank cartographic space that was more likely to contain undiscovered land than endless ocean.

It is the ships, however, that speak most loudly of the new sense of mastery over the sea, particularly when we consider them in light of the extremes that early modern rhetoric could reach in this regard. Just as early modern writers raised the compass to iconic status, so they also celebrated singular voyages of exploration, like those of Columbus, Da Gama, and Magellan as milestones in the history of navigation, yet they also bragged that travel across the oceans had become a commonplace occurrence. Around midcentury, for example, Francisco López de Gómara boasted that the route across the North Sea to Spain's possessions in the New World, "está ya tan andado y sabido, que cada día van allá nuestros españoles a ojos (como dicen) cerrados" (is now so practiced and well known that every day our Spaniards go there as if with their eyes closed; López de Gómara 1999, chap. 6).[22] José de Acosta made a similar point toward century's end:

> A la piedra imán se debe la navegación de las Indias, tan cierta y tan breve, que el día que hoy vemos muchos hombres, que han hecho viaje de Lisboa a Goa, y de Sevilla a Méjico y a Panamá; y en estotro mar del sur hasta la China y hasta el estrecho de Magallanes: y esto con tanta facilidad como se va el labrador de su aldea a la villa. Ya hemos visto hombres que han hecho quince viajes, y aun dieciocho a las Indias: de otros hemos oído, que pasan de veinte veces las que han ido y vuelto, pasando ese mar océano, en el cual cierto no hallan rastro de los que han caminado por él, ni topan caminante a quien preguntar el camino. (Acosta 1999, 1. cap. 17)[23]

> [It is to the lodestone that we owe Indies navigation, so sure and brief, that today we see many men who have made the trip from Lisbon to Goa, and from Seville to Mexico and Panama; and in this other South Sea to China and even to the Strait of Magellan, and this with the same ease that a worker has in going from his village to town. We have already seen men who have made fifteen voyages, and even eighteen, to the Indies. We have heard of others who have come and gone more than twenty times, crossing the ocean sea, upon which one certainly finds no trace of those who have traveled upon it, nor does one encounter a traveler of whom one can ask the way.]

This was sheer hyperbole. Both writers knew perfectly well, Acosta from personal experience, that travel by ship was miserable when it was not fatal, yet they choose in these passages to set that aside in order to marvel at the fact that what had once looked like a foolhardy ambition beyond the limit of human capability had become a routine activity. Although relatively few people actually traveled long distances by ship, travel across the ocean had become a part of life. Gómara and Acosta capture this extraordinary transformation in these remarks, which I offer as instances of new rhetorical trope to complement encomia of the compass needle, and which I call the "rhetoric of smooth sailing."

As Fernández-Armesto notes, voyages across unpreceented stretches of open ocean had become more or less routine by the 1430s, so Gómara and Acosta are voicing a sensibility that was probably already well in place by the time they were writing, and certainly by the time Reinel drew his chart (Fernández-Armesto 2007b, 127). Hence, it is possible to identify in his images of ships sailing the open ocean an iconographic analogue to the rhetoric of smooth sailing. According to Michel de Certeau, images of ships like the ones we find on Reinel's chart recall the process by which charts like Reinel's were created, the voyages of reconnaissance that charted previously unknown shores, but the Reinel ships are not the vessels of singular explorers making new discoveries (1984, 119–20). Like ship images on many maps, they depict specific types of vessels, rendered with a sense of drama and movement, yet none of them is clearly identified as a particular ship, the way Magellan's *Victoria* is singled out on some of the maps made after the first circumnavigation of the globe (Unger 2010, 11–12). Their placement, moreover, is by no means arbitrary. Not only do they serve to mark oceanic space as such, as we saw above, but they also trace, in a very general way, the maritime routes that had been constructed by the Iberian powers over the course of the previous generations. Four ships mark the circular route that served to sail the Occeano, while a fifth sails from the Occeano to the Occeanus Meridionalis, indicating that the two basins are connected by a navigable passage. Six ships mark the routes of Portuguese shipping from the Atlantic to India, indicating that the Indicum Mare is indeed accessible via the Cape of Good Hope, marking the Arabian Sea and the Bay of Bengal as choice destinations and gesturing eastward toward Malacca.[24] None of these vessels is foundering. None is under attack by ocean creatures. Like the rhetoric of smooth sailing, Reinel's iconographic program sets aside the risks and dangers of navigation in order to emphasize its routine nature. These are ships that cross the oceans all the time. Perhaps the pilots that guide them even do so with their eyes closed.

The Reinel chart, therefore, is not just a map of the world's landmasses as they were known to its Iberian maker in 1519, or of the distribution of those landmasses among the world's climatic zones. It is also very much a map of the world's oceans that testifies to a growing sense of mastery over the watery element itself. The old Ocean Sea, that boundless, menacing expanse that surrounded the world and defined its insularity in more ways than one, is nowhere to be found. In its place is a new architecture of the oceans, albeit still under construction, as a series of well-bounded basins. The Occeano, that partially enclosed basin that announces the shape of things to come once the other oceans have been charted, does not constitute a space hostile to humanity, irreducible to human uses, but a lake of sorts, to borrow Pacheco Pereira's word, a new *mare nostrum* of an emerging Atlantic world. The ships that sail across its waters speak of the mastery that Europeans have achieved over this ocean basin, just as other ships speak of mastery over the Occeanus Meridionalis and the Indicum Mare. Together, Reinel's cartography of containment and his iconography of smooth sailing speak of a world where oceans have the potential to connect rather than separate, and where those connections have actually been forged by European seafaring.

Conclusion

In sum, by emphasizing the discovery or invention of America as the crucial event in the history of early modern geography and cartography, legions of historians have taught us to look at early modern maps exclusively in terms of a single metageography, the architecture of the continents, and to trace its transformation by the new cosmography and the discoveries made by exploration in the Atlantic. Recent scholarship, however, has drawn our attention to the importance that the theory of climates held for the early modern geographical imagination, inviting us to consider the fact that the architecture of the continents was not the only framework it had available for imagining the world. In this chapter, I have suggested that the analysis of cartographic literature from this period must be undertaken with this metageographical plurality in mind. Not only do we need to think carefully about the way any text maps the world's major landmasses, but we also need to look for the ways it handles questions of latitude and climate, oceanic architecture, and the emerging network of global maritime routes.

In the chapters that follow, we shall see how these various metageographies are invoked to construct and reconstruct transpacific space in reaction to developing geographical knowledge and changing circumstances on the ground. In general, we shall see the theory of climates yield pride of place to the architecture of the continents, reaching a crucial point of inflection with Gómara, who favors the continents over the climates as the best way of mapping global human diversity in his *Historia general de las Indias* of 1552. In doing so, Gómara was a herald of things to come. If we take Ortelius's famous title-page engraving for his *Theatrum orbis terrarum* (1570) as a privileged indicator, it would seem that the architecture of the continents emerged as the privileged framework of European world-making by the last quarter of the sixteenth century, and with it, the idea that that the discoveries in the Atlantic were best understood as America, a fourth part of the world separate from Asia in a variety of significant ways (fig. 18). Nevertheless, we shall also see how the tendency toward distinction and separation along continental lines embodied by this shift met its match in a continued search for connectivity through appeal to the theory of climates, which was never abandoned as

FIGURE 18. The frontispiece from Abraham Ortelius's popular atlas depicts the parts of the world as allegorical female figures. From Abraham Ortelius, *Theatrum orbis terrarum* (Antwerp, 1570). Photograph: Courtesy of the Geography and Map Division, Library of Congress, Washington, DC (call no. G1006.T5 1570; control no. 2003683482).

an explanatory framework, the cartography of containment, and various forms of the rhetoric of smooth sailing, whose usefulness only grew as Spanish shipping expanded in scope and regularity.

Attending to metageographical plurality in the cartographic literature of early modern Spain allows us to discover various aspirations and anxieties regarding the possibilities for secular and spiritual empire in the Pacific Rim. In the case of the Reinel chart, for example, the network of maritime routes embodied in the chart's iconography proves to be most revealing. I have interpreted that iconography as an attempt to represent a particular historical achievement, the conquest of the Ocean Sea through the establishment of regular maritime routes across the Atlantic and Indian Oceans, yet my reading has said nothing about four of the vessels that appear on the map. Two of them appear just southeast of Brasil, and one of those seems to be heading toward the southern tip of the coastline. This particular ship speaks of the project that this map was supposedly designed to support, the plan proposed by Ferdinand Magellan and Ruy Faleiro to Charles V to reach the Moluccas by sailing through a maritime passage in the southern latitudes of the New World. By gesturing toward this possibility, the map's iconography triggers speculation about the blank space south and west of Brasil. Given everything we have seen in this chapter, such speculation would lead to the expectation that significant amounts of land were awaiting discovery in that space, and that the new discoveries in the tropics would be rich in precious metals, spices, and submissive human populations.

The other two ships that I have not mentioned, by contrast, would raise some significant questions for which there would be no ready answers. In the waters east of Malacca, we find a vessel with five masts that Unger believes to be a Chinese junk, or rather, a rough approximation of one. Its presence here speaks of the sophisticated ship-building tradition of Ming China with which the Portuguese had recently become acquainted (2010, 83). Meanwhile, in the "Mar visto pelhos castelhanos" we find two vessels propelled by oars or paddles rather than sails. I suggest that these are ocean-going indigenous canoes answering to descriptions like those we find in Martyr, where the canoe is defined as a single-hulled vessel of varying size that could accommodate as many as eighty rowers, used on both inland waterways and offshore waters (Anghiera 2005, 1.44; 1989, 12). These images testify to the seafaring capabilities of non-European peoples and identify the bodies of water they sail as non-European maritime worlds, beyond the outer limits of established European networks. They also invite one to ask whether or not the waters at the right-hand edge of the map, where the junk sails, could possibly be of a piece with the "Mar visto pelhos castelanos," where the canoes ply the waters. Is it all one ocean, or is it a series of various lake-like basins? Are there undiscovered lands separating the two, and if so, are they insular or continental? Is it possible, moreover, to incorporate those lands into the European network that was starting to make its way around Brasil?

Although we do not know for certain what Magellan and Faleiro said to Charles V and his courtiers, we can imagine that the answers they would have given to all these questions would have been a hearty yes. We know they expected to find rich islands in the space between Brasil and the Moluccas, because of the privileges they requested from the emperor. In a proposal that Sancho Panza would certainly have admired, each of them was supposed to be able to pick an island of their own from among those they discovered, beyond the first six, to govern, to exploit, and to pass on to his descendants (Magellan and Faleiro 1954). Neither man ever reaped his reward. Faleiro never left Spain, and Magellan died in the islands we now call the Philippines. More importantly, however, the gaunt survivors who limped off the leaky *Victoria* when it returned to Seville in 1523 put the lie to these South Sea dreams, triggering a crisis in Spain's efforts to keep the New World connected to Asia, a crisis that would mark every subsequent effort to map the Castilian Indies as a coherent whole. The story of how official Spanish historiography and cartography responded to this crisis is the subject of the next chapter.

3

Pacific Nightmares

On November 28, 1520, the fleet commanded by Ferdinand Magellan exited the strait that now bears his name and entered the waters of what he himself christened the Pacific Sea. Having discovered the maritime passage through the New World, Magellan believed he had met the most difficult challenge facing his expedition, and that the Spice Islands, his ultimate objective, could not be more than a few weeks away, across a relatively narrow ocean dotted with islands potentially rich in spices, gold, silver, and precious stones. Yet rather than island-hop his way to the Spicery in short order, he found himself sailing for months across vast tracks of featureless ocean, encountering only two desolate, uninhabited islets along the way. The ships did not make any significant landfall until March 6, 1521, when they raised an island in the Marianas group, most likely Guam, the crew starving, thirsty, and wracked by scurvy. Eighteen months later, the expedition's sole surviving vessel, the *Victoria*, limped into Seville carrying a cargo of cloves and a small group of emaciated survivors, the expedition's commander lost in combat with the indigenous inhabitants of what would later come to be known as the Philippines. It brought the surprising news that the ocean between the New World and the Spicery was much broader and emptier than anyone had expected it to be.

The historiography of exploration often identifies this revelation as Magellan's "discovery of the Pacific," and argues that it had a significant impact on European ideas about the world's geography (Bourne 1904, 132; Basch et al. 1962, 248; Parry 1974, 258; Morison 1971, 2.466). Simply put, the discovery of the Pacific made it more difficult to believe that what we call North America was part of Asia, and thus it helped consolidate the invention of America. In making this argument, historians of exploration and discovery often point to the surviving manuscript planispheres produced during the 1520s at Seville's Casa de la contratación, utilizing privileged information drawn from the logbooks of the Magellan expedition, which the crown did not allow to circulate outside official circles.[1] A large empty space opens up west of the New World on these charts, testifying to the surprising distance that the fleet had to sail to reach the Spicery from the Strait of Magellan, and presumably, to the uncharted dimensions of the Pacific Ocean. The distance ranges from 120° of longitude between Peru and the Spicery on the anonymous planisphere of 1523, to 130° on the famous 1529 planisphere by Diogo Ribeiro, but it is invariably greater than the distance that had appeared on any previous map, even those made in Seville (see fig. 7).[2] Historian J. H. Parry lauds this particular planisphere for its apparent "completeness," and he offers it as an example of how the discovery of the Pacific brought "received ideas" about the world's geography, most of them stemming from Ptolemy, crashing to the ground (1974, 260).[3]

The maps and globes constructed on the other side of the Pyrenees during the years following the return of the *Victoria*, meanwhile, look quite different. As the historian of cartography Edward Stevenson points out, the second quarter of the sixteenth century witnessed a noticeable shift away from the theory of American insularity in favor of Amerasian continuity, even among cosmographers like the German Johannes Schöner, who had previously shown enthusiasm for the geography of the 1507 Waldseemüller map (Stevenson 1971, 1.106–7; Nunn 1929, 11). The same maps and globes that depicted what we call North America as part of East Asia, moreover, also tended to fill the southern hemisphere with the *Terra Australis Incognita*, following the pioneering example of the French cosmographer Oronce Finé, whose influential cordiform maps of the world, constructed during the 1530s, favored both of these geographical hypotheses (see fig. 13). On maps and globes like those of Finé, Caspar Vopel (1545), and Giacamo Gastaldi (1565), the South Sea acquired boundaries that had not been reconnoitered by the men who first crossed it, and that are nowhere to be seen on the Seville planispheres. (See figs. 6 and 19). On Finé's map, it is even quite narrow. While Ribeiro stretches the distance between Peru and the Spicery to 130° of longitude, Finé puts it at a mere 60° of longitude, less than half Ribeiro's value. As a result, Finé's South Sea looks nothing like the Pacific Ocean as we know it today, an enormous watery expanse covering two-thirds of the globe, but rather like a well-bounded oceanic basin of manageable dimensions. The cartography of containment has come into play to fashion the South Sea out of the Pacific Ocean.[4]

Historians have explained this apparent divergence between the two post-Magellanic cartographic cultures in a variety of ways. Some have lauded the empiricism of Iberian nautical charts over the bookish conservatism of much of the period's print cartography, which was

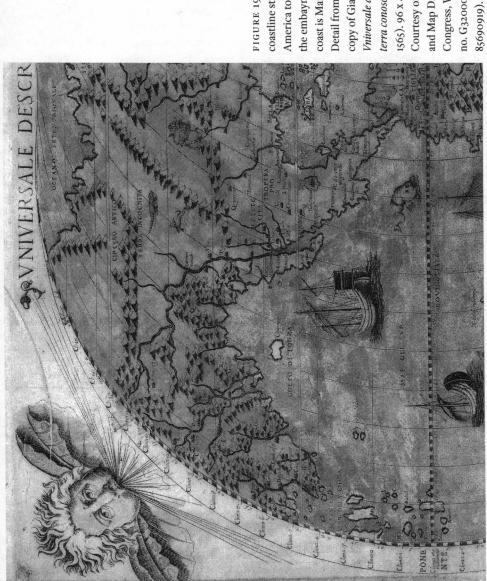

FIGURE 19. A continuous coastline stretches from America to Asia. The island in the embayment formed by the coast is Marco Polo's Zipangu. Detail from Paolo Forani's copy of Giacomo Gastaldi, *Vniuersale descrittione di tvtta la terra conosciuta fin qvi* (Venice, 1565). 96 x 40 cm. Photograph; Courtesy of the Geography and Map Division, Library of Congress, Washington, DC (call no. G3000 1565.F6; control no. 85690919).

predicated upon the new cosmography and its insistence that land had to predominate over water on the surface of the Earth. This meant that the Ocean Sea had to consist of an interconnected series of relatively small, well-contained, and bounded oceanic basins, leaving no room for a vast, largely empty ocean that spanned the distances noted in the expedition logbooks and that found its way onto the charts of the Seville cosmographers. Others have insisted that the makers of print maps outside of Iberia were just as interested in mapping the world according to information garnered by modern expeditions, but simply had less access to current information, thanks to the often sluggish and uneven pace at which information moved through early modern knowledge networks. Mapmakers like Finé, for example, knew nothing of the logbooks from the Magellan expedition, so they had to rely on the figures that appeared in the flawed 1525 Paris edition of Antonio Pigafetta's account of the voyage, which dramatically underestimated the breadth of the South Sea. Add to this the remarkable coincidence that news of the first circumnavigation of the globe began to circulate at the same time as Hernán Cortés's accounts of the conquest of Mexico. The rich cities of the Mexica confirmed many a suspicion that what we call North America was really East Asia, and that the world of Marco Polo was just over the next mountain range from where the Spanish were building their new city atop the ruins of Tenochtitlán. Of course, these arguments are by no means incompatible. It is possible that the prestige of the reigning intellectual paradigms conspired with the peculiarities of knowledge flows to favor the production of maps that looked like Finé's, maps that contained Pigafetta's Pacific within the confines of a South Sea shaped by the prejudices of the new cosmography.

Yet it is possible to overdraw the contrast between the manuscript charts of Iberian Europe and the print maps of Trans-Pyrenean Europe. While the Seville planispheres did indeed incorporate an unprecedented distance between the New World and the Spicery into their images of the world, they also made sure not to stretch that distance so far as to place the Moluccas outside the Spanish hemisphere and in Portuguese territory. In fact, the surviving maps are all presentation copies of the official map of the world maintained by the crown's cosmographers at the Casa de la contratación, designed to convince their readers that Spain was in the right in laying claim to the Spicery (Sandman 2007, 1113–16; Cerezo Martínez 1994, 173–74; Martín Merás 1993, 87; See also Chaplin 2012, 40). As such, they might be understood as examples of "cartographic chicanery," the deliberate misrepresentation of geographical knowledge for political purposes.[5] By this account, the Seville charts do not mark the dawn of a new, empirically driven scientific modernity, but rather exemplify the subordination of knowledge to power in the context of early modern empire building. Their apparent commitment to cartographic empiricism, which is most evident in their agnosticism regarding hypothetical geographies like the *Terra Australis*, thus functions rhetorically to underwrite the authority of their highly politicized territorial claims.

The charts were not the only tools that Spain used to advance its case. Narrative was weaponized as well. Charles V is said to have ordered the destruction of the journal of Sebastián del Cano, the man who captained the *Victoria* on the final leg of the voyage, because it said too much about how long it had taken Magellan's fleet to get from the Strait of Magellan to the Moluccas (Chaplin 2012, 37). The crown may have also been behind the production and circulation of an account of the expedition by the emperor's young Latin secretary, Maximiliaen von Sevenborgen, who went by the pen name Maximilianus Transylvanus. His brief narrative of the first circumnavigation of the Earth, *De Moluccis Insulis* (Of the Molucca Islands), became the first account of the Magellan expedition to appear in print, in Cologne in 1523. The crown's official account, however, fell to the experienced court historian Peter Martyr, who sent his chronicle of the expedition to the pope. Although the text was destroyed in the 1527 sack of Rome, a version of it appeared in the author's *De Orbe Novo Decades* (Decades of the New World), published for the first time in Alcalá de Henares in 1530. Like the Seville planispheres, these accounts acknowledge that the distance from the New World to the Spicery was unexpectedly long, and they even celebrate the discovery of a new and previously unknown ocean, but arrange things so as not to jeopardize Spanish claims to the Spicery. Like the apparent empiricism of the charts, the claim these narratives make to the authority of the traveler serve a rhetorical purpose, that of convincing the reader of the truth of the text's geographical claims.

Once again, however, we must be careful not to overstate the argument. To explain away the Spanish attempt to keep the Moluccas in the Castilian hemisphere as mere chicanery, whether cartographic or narrative, is to assume that the mapmakers involved knew the truth, but followed the directives of their superiors in producing charts and narratives that claimed the opposite. Yet as Alison Sandman (2001, 2007) and Felipe Fernández-Armesto (2007) remind us, early modern mapmaking did not have the luxury of knowing the truth with any clarity or certainty. The information provided by the pilots was gathered using highly inaccurate methods and techniques, when it was gathered at all. The cosmographers were aware of the limitations in the material they used, and they often looked down on the pilots as ignorant amateurs. Numbers that did not suit them could easily be dismissed as errors. Then there were the problems involved in converting the information from sources like log books to charts, problems having to do with the inadequacy of available cartographic projections, the puzzle of magnetic variation, and the difficulties surrounding the measure of longitude. In sum, the knowledge that sailed into Seville was often highly problematic, and it never flowed onto charts with ease, speed, transparency, or efficiency.

In mapping the Pacific, three issues stood out above all others. First, and most famously, no technique existed for the accurate measure of longitude, which meant that one could only estimate coordinate position by converting distance traveled into degrees. The Spanish and Portuguese, however, used different conversion factors. Both conversion factors were politically convenient, but both were also intellectually defensible, because they were based on different values for the circumference of the Earth, provided by different ancient authorities.[6] This second limitation implied that the Portuguese and the Spanish would never agree about how the world should be mapped. The third limitation had to do with the method that was used to determine the distance traveled in the first place, dead reckoning. Fair winds and following seas add an appreciable degree of error to estimates of distance determined by dead reckoning, the only method that was available to early modern pilots.[7] On the Pacific crossing, however, ships experienced favorable conditions over an enormous longitudinal distance, with the result that the degree of error introduced was correspondingly high, and the degree of correction required, debatable. This meant that it was particularly easy to underestimate the distance that one had traveled across the Pacific or to question someone else's estimate of that distance.

It is impossible, therefore, to point to differences between figures in a logbook and longitudes on a chart and cry fraud. Under the circumstances of early modern cartographic production just described, geographical truth was not easy to come by and often remained open for interminable debate. The space of the early modern map was not a transparent representation of an empirically determined reality: it was just as Seth Kimmel (2010) describes it, radically indeterminate and irreducibly inaccurate. The will of a king who needed the world to look one way rather than another, therefore, did not impose a clear fiction over a known truth, but served to cut the Gordian knot among a variety of plausible solutions to a series of intractable problems. This is why mapmakers in the service of Spain and Portugal could map the position of the Moluccas differently, and argue interminably over whose map was correct. Although they would never have seen them this way, their maps constituted world-making fictions much like those examined by Ayesha Ramachandran (2015), at once creative and unstable in their very creativity. Much the same, I argue, can be said about narratives of the Magellan voyage and their verbal mappings of the South Sea.

In this chapter I attempt to let go of the dichotomies that often structure our understanding of the maps and narratives produced in the wake of the *Victoria*'s return, those contrasts between Iberian empiricism and Trans-Pyrenean bookishness, between deliberate propaganda and geographical truth. Instead, I examine the way that post-Magellanic mapmakers used different measures of the same ingredients, including the principles of Ptolemaic cosmography, whatever numerical data they had at their disposal and considered trustworthy, and, yes, the demands of their powerful employers, to map the South Sea during the decade after Magellan. My emphasis, however, does not fall on the Seville planispheres primarily, but rather on the narratives of the first European crossing of the Pacific written by Pigafetta, von Sevenborgen, Martyr, and Oviedo. By privileging narrative over cartography, I hope to avoid the quantitative trap into which we often fall when we discuss maps in isolation from other texts. We fasten on miles, leagues, or degrees, and forget that distance, as Mary Helms so effectively demonstrates, has many other measures, because it is constructed by culture, not simply measured with a ruler (1988, 6). In many of the societies Helms examines myth constructed notions of near and far, while in early modern Europe travel writing often did the job. Writing of various sorts could render peoples and places as proximate or distant, familiar or exotic, in a variety of ways. By examining the cultural construction of transpacific distance in my four accounts of Magellan's Pacific crossing, I unpack some of the anxieties and contradictions, political, ideological, and cosmographical, that mark Spanish world-making projects in the wake of the *Victoria*'s return. I then turn to a representative Seville chart in order to read it as an equally fraught attempt to both acknowledge and deny some of the realities that emerged from that epochal voyage.

I begin with the account of Antonio Pigafetta, the Venetian gentleman who joined the fleet as a supernumerary and survived to compose the most lengthy and thorough eyewitness account of the voyage, which was first printed in Paris in 1525, and was later translated into a variety of European languages.[8] The numerical information that appears in this edition maps the Moluccas well inside the Castilian hemisphere, but the treatment of the Pacific crossing tells a different story. Pigafetta's version of the event comes closer than any other period narrative to inventing the Pacific Ocean as a vast, empty, practically boundless space that has the potential to call into question Ptolemaic frameworks of the world's geography, to suggesting that the New World might indeed by something quite separate and different from Asia, and to mapping the South Sea as a new version of the old Ocean Sea, inherently hostile to human use and even life.

I then turn to three of the narratives that emerged from Spanish circles, those of von Sevenborgen, Martyr, and Oviedo. As one might expect, all of these accounts take Spain's side in the controversy over the Moluccas. None of them do so, however, simply by providing a convenient longitude for the Spice Islands. Instead, they tell the story of the Pacific crossing in ways that are meant to manage the reader's understanding of transpacific space. In various ways, they silence or mitigate the impact of the key elements in Pigafetta's version of the story, and they even undermine his authority altogether through dismemberment and selective citation. In this way, they contain the Pacific within the fundamentally Ptolemaic framework of the South Sea, inviting the reader to experience the waters between the New World and the Spicery as a bounded oceanic basin amenable to mastery by European navigation. They tend to remain agnostic about the continental status of the New World, and they forge a sense of spatial continuity between it and East Asia by appealing to the theory of climates. It can be difficult to tell, sometimes, whether they are engaging in deliberate propaganda or simply struggling with the challenges of figuring the world at a time of rapid epistemic change. Nevertheless, their projects are fraught with contradictions that reveal unspoken anxieties, whether about the credibility of their propaganda efforts or about the truth value of their South Sea cartographies.

The Sublime Ocean: Pigafetta's Pacific

The definitive version of the Pigafetta's Italian-language narrative remained unknown until 1800, when it was discovered among the manuscripts of Milan's Ambrosiana Library. Until then, the only version of the text that was generally available was an abbreviated French edition that bore the title *Le voyage et navigation faict par les Espaignolz és Isles de Mollucques* (The voyage and explorations of the Spaniards among the Moluccas).[9] As an extensive eyewitness account of the Magellan expedition, the text attracted immediate attention, and it seems to have had considerable impact on the way cosmographers mapped the Pacific. According to Thomas Suárez, Oronce Finé drew upon the Paris Pigafetta to construct his cordiform maps of the world and may very well have determined that the South Sea was relatively narrow on the basis of its flawed longitudes (2004, 38–45). That text puts Tidore, one of the principal islands of the Moluccas, at only 171° west of "la ligne de la partie" (the line of departure), presumably, the longitude of Seville. The Ambrosiana manuscript, by contrast, puts Tidore 161° west of the line of demarcation, pushing the Moluccas farther west by a distance roughly equivalent to the breadth of the Atlantic Ocean, although still mapping them within the Castilian hemisphere (Pigafetta 1999, 306; 2007, 131). The reasons for the difference between the two versions of the text need not concern us. What matters is that the larger distance is more consistent with the way that Pigafetta maps the South Sea in and through his narrative. In his version of the event, the first crossing of the Pacific Ocean by Europeans becomes a nightmare that the travelers only barely survive. Distance is not measured in numbers or leagues, but in the toll it takes on human bodies and souls. The ocean, in turn, becomes a boundless expanse bereft of land and hostile to human life.

All of this is consistent with the way Pigafetta invites the reader to experience the world as a whole. His account follows the fleet from Spain to Patagonia by way of the Canary Islands, the coast of Sierra Leone, and Brazil. Pigafetta mentions headlands sighted, marvels witnessed, and storms endured along the way, but dwells primarily on the expedition's experiences in Verizin (Brazil) and Port St. Julian (Patagonia), the two places where the ships lingered the longest. On these occasions, the text even shifts into descriptive mode, providing detailed accounts of Brazil's cannibals and Patagonia's savage giants. In the nineteenth of the Paris Pigafetta's 114 chapters, the text follows the fleet out of the Strait of Magellan and across the ocean that Magellan christens the Pacific Sea. It then lingers in the islands of insular Southeast Asia, paying special attention to the fleet's experiences in and around Cebu, in the modern Philippines, and Tidore, in modern Indonesia. Once again, description comes to the fore, as does the narrative of the expedition's dealings with the locals. Finally, the text follows the *Victoria* back to Spain by way of the Indian and Atlantic Oceans.

The text provides enough locational information so that the reader can construct a rough cartography of the voyage, yet unless one draws upon either period or modern maps to fill in the many blanks left by Pigafetta's text, the result would not look anything like the Ptolemaic terraqueous globe, with its enormous landmasses and neatly contained oceans. The narrator never does anything to organize the various stopping places along the way into a continental architecture. Verizin, for example, is said to be "very vast and larger than Spain, Portugal, France and Italy combined, very vast indeed," but it is never identified as the New World, America, or the fourth part of the world, nor is anything ever said about its relationship with Asia (5). The reader is thus invited to experience Brazil as a massive island of sorts, much larger than Cebu, Tidore, or any

of the other islands that the fleet encounters, but no less singular and isolated than they. The same could be said of Port St. Julian, the other major
stopping place on the South American continent.

The result is an oceanic world of the kind that Denis Cosgrove identifies with early modern *isolarios*, atlases of island maps. These texts typically devote an entire page to a single island and do very little if anything
to locate the island in larger spaces. They thus figure the world as a series
of discrete geographical fragments floating in an undifferentiated ocean
(2001, 79–101). Pigafetta may very well have had some experience with
such texts. As Theodore Cachey has argued, the Ambrosiana manuscript
itself represents an *isolario* of a kind, particularly when we consider its
only illustrations, the rough charts of islands and of the Strait of Magellan, which Pigafetta drew to accompany his text. The islands are named,
but no coordinates are given. Like the islands of an island book, they
appear on Pigafetta's pages in all of their glorious insularity, that is, their
singularity and isolation.[10] Of course, the analogy between the *isolarios*
and Pigafetta breaks down when we consider the fact that the locational
information that is missing from the maps appears in the narrative. Unlike the authors of the *isolarios* that Cosgrove has in mind, Pigafetta locates all of the places where the fleet stops along the route of travel, and
at times, in the global coordinate grid. Those stops, however, preserve
their insularity, their sense of isolation from each other. What emerges
from Pigafetta's pages, therefore, is not the terraqueous globe of Ptolemaic cosmography, but its unsettling other, an impossible waterworld
dotted by fragmentary landscapes.

That waterworld blows open in the chapter that follows Magellan's
fleet across the Pacific, the *locus classicus* of a qualitative, narrative cartography of the Pacific that every subsequent Spanish map of the transpacific Indies will have to confront. It begins as follows:

> They sailed out from this strait into the Pacific Sea on the 28th of No-
> vember in the year 1520, and they were three months and twenty days
> without eating anything (i.e. fresh food), and they ate biscuit, and when
> there was no more of that they ate the crumbs which were full of mag-
> gots and smelled strongly of mouse urine. They drank yellow water, al-
> ready several days putrid. And they ate some of the hides that were very
> much toughened by the sun, rain and winds. And they softened them in
> the sea for four or five days, and then they put them in a pot over the fire
> and ate them and also much sawdust. A mouse would bring half a ducat
> or a ducat. The gums of some of the men swelled over their upper and
> lower teeth, so that they could not eat and so died. And nineteen men
> died from that sickness, and the giant together with an Indian from the

land of Brazil, and twenty-five or thirty were so sick that they could not help with arm or limb. (Pigafetta 1969, 24–25)[11]

Like all the other major episodes in Pigafetta's narrative, this passage begins by locating the expedition and the reader, in time (November 20, 1520) and space (the mouth of the Strait of Magellan), only to then open up an enormous stretch of undifferentiated time, three months and twenty days, sailing across trackless sea with no sight of land. All sense of specific location in either time or space dissolves away, as the text turns to the sufferings of the crew. The narrative of their ordeal begins *in medias res*, without relating anything about the events that led up to the crucial moment when the hardtack ran out. It then relates the desperate measures taken by the crew to survive as a series of disconnected but graphic snapshots of suffering and desperation. In the absence of any real narrative development, these snapshots expand to fill the entire three months and twenty days of the crossing, and the whole distance the ships traverse. They become icons akin to an image of a cannibal on a map. Just as a drawing of this kind on a map of Brazil marks the whole country as a land of man-eaters, rather than document a particular instance of anthropophagy, so these graphic verbal images of the suffering of Magellan's men mark the entire South Sea as a space hostile to the presence of human beings.[12]

This is no celebration of the conquest of the sea by modern navigation. It is an adventure story featuring a foolhardy voyage across a broad maritime space where human beings simply do not belong. The chapter develops this perspective as it goes on:

> And others (but only a small number) were by the grace of God spared any sickness. In these three months and twenty days they went four thousand leagues in an open stretch of the Pacific Sea. And it is indeed pacific, for in all this time without sighting land, there were neither storms nor tempests. And they saw but two uninhabited islands, and for this reason called them the Unfortunate Isles, and the distance between them is two hundred leagues. They found no anchorage there, and there were many fish called *tiburons* there. The first island lies in 15° and the second in 9° of south latitude. Daily they sailed fifty, sixty or seventy leagues. And if God had not given them fine weather, they would all have perished of hunger in this vast sea. And they were certain that such a voyage would never again be made (24–25).

Now we get some numbers, the total distance sailed, the latitudes of the two islands encountered along the way, and the healthy daily distances covered by the ships. A subsequent chapter (22) provides an additional detail, that the fleet reached the equator at 120° of longitude west of the "line of departure" (26). None of this, however, allows us to map the route in anything but the most general way. The numbers are not there to locate us in cartographic space, but to help us imagine the South Sea as a sublimely vast and empty expanse, entirely uncontained by any landmasses, large or small.[13] Pigafetta makes sure to report that they found no other land in the far reaches of the southern hemisphere, other than those immediately surrounding the strait (29–30).[14] To engulf oneself in the South Sea, we discover, is to risk almost certain loss of self. God, Pigafetta piously claims, held death at bay by providing fair winds and following seas, but only out of his mysterious mercy, not apparently out of any eagerness to support the expedition's objectives. This is not a voyage that can lay the foundations of transpacific empire, because it should never have been attempted and will never be repeated.

The two landfalls made along the way only mock the men's expectations and prolong their suffering. After sailing between the wealthy islands of Zipangu and Sumbdit Pradit without sighting them, the fleet raises two barren islands, which Magellan christens the Unfortunate Isles, a parody of the paradisiacal Fortunate Isles of Greco-Roman mythology and certainly a death warrant of the hopes that Magellan and his men had of finding rich islands along the way to the Spicery (27). The next landfall comes when the fleet reaches what we know as the Marianas Islands in chapter 23:

> On the sixth of March about seventy leagues from the said course, in 12° of north latitude and 146° longitude, they discovered a small island. . . . And the General wanted to anchor by the large one (probably Guam) in order to rest, but he could not because of the people of this island, who boarded the ships and stole one thing after another, to such an extent that our men could not protect their belongings. (28; see also 1999, 202–3; 2007, 27.)

The Unfortunate Isles were a disappointment. Guam is a cruel joke. By locating us once again in time and space with a date, a latitude, and a longitude, the text seems to announce that the hellish crossing is over and that a new phase of the journey has begun. Readers of Columbus might expect the new island to host welcoming and generous natives, only to be shocked by the apparent rapaciousness of these islanders. Magellan certainly seems to have been. He christens the islands *Las islas de los Ladrones* (The isles of Thieves) and responds aggressively to the thievery of the locals: "And the Captain was exceedingly angry; he landed with forty armed men and burned forty or fifty houses. . . . And they left immediately following the same course; and before they landed, the sick crewmen begged them that if they killed any men, they should bring back their entrails, and this would soon cure them" (28).

This is a stereotypical scene of encounter turned upside down. Instead of experiencing hospitality from innocent natives, the men suffer theft and mete out violence. Instead of filling their bellies with fruit and pork, they hope to devour human flesh and become themselves the cannibals that Europeans so often feared and despised. In the South Sea, it seems, those who do not die run the risk of becoming monsters.

With the departure of the ships from these islands, Pigafetta remarks, "From the signs they made, these thieves thought that they were no other people in the world besides themselves" (31; see also 1999, 205; 2007, 28). Did the islanders really believe this, or was Pigafetta projecting onto the locals the sense of radical isolation that Magellan and his men must have felt after more than three months of sailing an ocean that defied every expectation of what an ocean should be? As a projection of their own loneliness, the remark seems to encapsulate what the Pacific is like on the pages of Pigafetta's account. It is nothing like the basins of Ptolemaic cosmography, but more like the old Ocean Sea, impossibly large and therefore untamable. It is broader than anyone had thought possible, apparently unbounded to the north and south, and bereft of the islands that it should have contained. As such, it threatens to break out of the confines imposed upon the waters of the Earth by Ptolemaic cosmography, and it utterly defies mastery by European navigation.

That broad expanse, moreover, seems to separate two very different regions along the route of travel. The inhabitants of Verizin and Port St. Julian are cannibals and savages who go naked or wear animal skins, live in huts, speak unintelligible tongues, and devour charred meat using their bare hands when they are not eating other human beings (5–9, 11–18). The inhabitants of the islands on the far side of the ocean, by contrast, are "hommes de raison" (a rational people) who live in "cities," have recognizable forms of government and religion, eat off porcelain plates, flavor their foods with spices, write and read books, and speak a language that Enrique, Magellan's slave, can understand (32–42). Brunei, with its large capital and wealthy monarch, is especially magnificent (90–100). So even though Pigafetta says nothing about continental geographies and how they should be mapped, he provides perfect grist for the mill of anyone interested in conceptualizing America as a place fundamentally different and separate from East and Southeast Asia. Not only does Pigafetta, therefore, invent the Pacific, but he also provides support to the ongoing invention of America.

Pigafetta's world, however, would not do, at least not in Spanish circles. It begs to be brought under the controlling gaze of an imperial cartography eager to figure the Moluccas as Spanish territory, readily accessible from the Atlantic; of a Ptolemaic cartography that has no room for a broad, empty, boundless, watery expanse; and of a zonal cartography that cannot accept such a marked difference in the level of civility between tropical peoples on one side of the ocean and those on the other. There were reasons to spare, therefore, to doubt Pigafetta's tale, and those reasons were both political and epistemic. It would be up to the official historians of the Spanish crown to remap Pigafetta's Pacific in terms that served Spanish interests and Ptolemaic expectations about global geography.

The Imperial Pacific: Martyr, von Sevenborgen, and Oviedo

None of the writers examined in this section, Peter Martyr, Maxilimian von Sevenborgen (aka Maximilianus Transylvanus), and Gonzalo Fernández de Oviedo, ever sailed the Pacific, although Oviedo had plenty of experience crossing the Atlantic. All of them, however, had close ties to the court of Charles V and thus privileged access to knowledge produced by the Magellan expedition, including accounts by the survivors that have not come down to us, as well as, presumably, the charts produced at the Casa de la contratación. This allowed them to claim considerable authority as historians of the expedition and as interpreters of the Pacific crossing for a broad European audience. All of them, in various ways, map the South Sea in ways that contain and domesticate Pigafetta's Pacific, squeezing it into existing metageographical frameworks and thereby defending Spain's imperial claims and ambitions. They do so by telling imperial versions of the story of Magellan's Pacific crossing. I take them up in the order in which they appeared in print.

De Moluccis Insulis, by Maximilian von Sevenborgen (aka Maximilianus Transylvanus), was born as a personal letter to a relative, the cardinal-archbishop of Salzburg, and perhaps as an exercise in Latin composition. The secretary's letter matured into something else when it was printed in Cologne, then in Paris and Rome, all in 1523, only months after the return of the *Victoria* and two years before the publication of *Le voyage et navigation faict par les Espaignolz*. As the first printed account of the expedition, *De Moluccis Insulis* clearly caught the eye of European readers, and eventually achieved canonical status when it made its way into the second volume of Ramusio's *Navigationi e viaggi*, alongside an Italian translation of Pigafetta's narrative derived from the Paris edition (1559). *De Moluccis Insulis* is so favorable to Spain's interests that one has to wonder if it was not meant for publication from the start, that it was not in fact meant to serve as a piece of propaganda, camouflaged as a personal letter published far from Spain. Its version of the Pacific crossing could not be more different from that of Pigafettta. Von Sevenborgen presents it as an episode in the ongoing conquest of the Ocean Sea, and therefore, a successful effort to lay the foundations of a nascent Spanish transpacific empire.

Near the beginning of the text, von Sevenborgen tells how Spain and Portugal divided the world between them in the 1490s, and how some people subsequently came to wonder if the Portuguese had not overstepped the boundary into Castilian territory when they conquered the city of Malacca. According to von Sevenborgen, Magellan and his collaborator, Ruy Faleiro, settled the matter when they successfully argued before Charles V that the Spice Islands and China lay in Spain's half of the world (A iv r–A iv v).[15] The few numbers the pamphlet provides regarding distances and longitudes confirm this. The text puts Port St. Julian and the Strait of Magellan both at 56° west of the Canaries, and Inaguana, or Guam, at 158° west of Cádiz (A vi r, A viii r, A viii v). The first number is not far off the true value of roughly 60°, but the second falls far short of the 207° of longitude that separates Cádiz from Guam moving west around the globe. Although the pamphlet does not provide any locational information for the Moluccas themselves, it is on track for a longitude that places them well within the Castilian demarcation. The controversy with Portugal never comes up again. At the end of the account, where we might expect it to reiterate the rightness of Castilian claims against Portuguese protests, it marvels at the fact that the world has been circumnavigated by praising the survivors of the expedition as the clear betters of Jason and his Argonauts and by claiming that the *Victoria* deserves to be placed among the constellations, even more so than Jason's *Argo* (Transylvanus 1523, B viii v).

One might therefore expect von Sevenborgen to say very little about the breadth and emptiness of the ocean that Magellan discovered on the far side of the strait. Nevertheless, he does precisely the opposite, referring to that ocean as "vastum and ingens hoc pelagus" (that vast and extensive ocean) and "ingens maris spatium" (broad space of sea), and emphasizing the fact that no European ships had ever sailed its waters before (A viii r). He goes on to tell how Magellan and his men sailed for forty days from the mouth of the strait without seeing anything but ocean all around, before stumbling upon the Unfortunate Isles. After making this landfall, he continues, they sailed another three months and twenty days before sighting land again, covering a distance so vast that "nullum humanum ingenium caperet" (human skill cannot grasp it; A viii v). In this way, von Sevenborgen remains true to what he must have heard from

the expedition's survivors, sharing their sense of wonder at the dimensions of the ocean the expedition had traversed. In this, his account of the
Pacific crossing is consistent with that of Pigafetta.

Like *Le voyage et navigation*, moreover, *De Moluccis Insulis* celebrates
the fact that the fleet enjoyed fair winds and following seas all the way
across the Pacific, but there the similarity stops. Pigafetta, as we saw, figures the consistently favorable conditions as a boon from heaven that
delivered Magellan's men from what would otherwise have been certain
death on an ocean that was simply too wide to cross. Von Sevenborgen, however, says absolutely nothing about starvation, thirst, sickness,
desperation, loneliness, or mortality along the way (A vi v). The good
weather and favorable winds are elements in a larger story of smooth
sailing all the way across the South Sea that culminates in the successful
foundation of Christian empire on the island of Subuth (Cebu), which is
said to be rich in gold and ginger, and where Magellan celebrates Easter
mass, establishes friendly relations with the king, and converts over two
thousand islanders to Christianity, including the king himself (A viii v). In
the end, the vast ocean Magellan traverses is just like the native Cebuanos
he brings into the fold of church and crown. It is a defeated other, whose
function in the text is to magnify the achievement of the Spanish as navigators, conquerors, and propagators of the faith.

In order to render his account credible, von Sevenborgen avails himself of a strategy for conceptualizing maritime space and travel that had
developed when Europeans were first starting to imagine the possibility
of traversing the Ocean Sea westward to the Indies. As John Gillis has argued, oceanic islands were key to this project. One could imagine a long
ocean crossing only as long as one could count on the regular appearance
of islands along the way where one could stop to repair and provision
one's ships and refresh and restore one's crew. Hence the importance
of the imaginary islands that the late medieval imagination projected
into the western ocean. They spoke of the expectation that new islands
awaited those who dared to sail beyond the farthest-known islands to the
west of Europe, and they even suggested that it was possible to island-hop
as far as the easternmost islands off the coast of Asia (Gillis 2009, 45–64).
On the Behaim globe of 1492, which is believed to represent the world
much as Columbus imagined it, the imaginary islands of Brasilia and the
Isle of St. Brendan do precisely that. They are strategically placed in the
gap between the known to the west and the known to the east, making it
possible to imagine island-hopping one's way from Europe to the Indies
(see fig. 16). They were thus essential to the intellectual domestication of
the Ocean Sea during the years when Europeans were first learning to
sail its open waters.

Like Behaim's treatment of the Ocean Sea, von Sevenborgen's story of the first Pacific crossing hinges upon a crucial fiction. In his case, however, the fiction at issue is not an imaginary island, but an imaginary version of a real encounter with real islands, the Unfortunate Isles. Von Sevenborgen admits, as do other accounts, that these islands were the only land raised by the fleet on its long journey from the Strait of Magellan to the Marianas, and describes them as "steriles" and "inhabitatas" (barren and uninhabited). Yet unlike every other account of the expedition, *De Moluccis Insulis* tells us that Magellan and his men lingered there for two days, for "curationi et recreationi corporum" (to cure and refresh their bodies), since there was such good fishing to be had (A viii v). In this way, von Sevenborgen not only silences the examples of human suffering that stand out so powerfully in Pigafetta, but also provides the fleet with the way station that it never actually found. He does this despite having had access, as he himself claims, to written survivor reports that must have said something about the horrors of the crossing (A ii r, A ii v). The same ocean crossing that figures in Pigafetta as a nightmare survival story becomes in von Sevenborgen a tale of easy island-hopping, forty days to the Unfortunate Isles, three months and twenty days to Inaguagna (Guam), and then onward to "Acacan," "Massana," and "Subuth." The ocean itself, in turn, becomes a perfectly traversable basin, rather than a daunting expanse fundamentally hostile to domestication by European navigation.

This occurs, however, only as long as one does not look too closely at the three months and twenty days that it took for the fleet to sail from the Unfortunate Isles to Guam. This is the point where the whole strategy, at once narrative and cartographic, of figuring the success of the crossing and the overall navigability of the South Sea comes into crisis. Three months and twenty days was considerably longer than the time it typically took Spanish ships to cross the Atlantic, coming or going, although it was not unheard of on certain portions of the Portuguese Carreira da India. By any standard, however, this was a long time at sea, one that entailed considerable risk and cost a considerable price in pain and mortality. Von Sevenborgen simply does not address this fact. An attentive, informed reader could choose to interrogate the text on this point and thereby question its whole project. Alternatively, he or she could sail past it, either neglecting altogether its problematic nature or actively noticing the time involved and recognizing it as another trophy in Europe's ongoing conquest of the sea.

Enormous and almost empty, but perfectly traversable, this version of the South Sea fits well with the pamphlet's treatment of the New World and insular Southeast Asia as a continuous Indian space. Like Pigafetta, von Sevenborgen never uses terms like "America" or "the New World" in reference to what we call South America. Although he refers to the "incognitas terras" (unknown lands) recently discovered by the Castilians as a "continentem terram amplissimam" (very extensive continental land), we must remember that the early modern Latin equivalent of our own "continent" actually meant something quite different (A iii v).[16] To say a body of land was "continental" was to indicate that it was not an island, that it was instead part of the *orbis terrarum*, and not that the land in question was a "part of the world," one of the major divisions which together constituted the *orbis terrarum* (Lehmann 2013). It is therefore quite possible that von Sevenborgen, writing shortly after the return of the *Victoria* in full cognizance of the breadth of the Pacific, imagines the world in terms of Amerasian continuity, believing the *continentem terram amplissimam* to be an extension of Asia.

But whether or not von Sevenborgen believes America to be part of continental Asia, he most certainly believes it to be part of a singular Indies that spans Magellan's ocean. At one point, he mentions that the Spanish have discovered those *incognitas terras*, "per meridiem et occidentem" (to the south and west), thereby gesturing toward the overall southing of Castilian exploration, its movement not just west across the Ocean Sea, but also south into the tropics (A ii v). He is sure to mention that the fleet sails across the South Sea through the Torrid Zone and below the Tropic of Capricorn (A viii v). He also refers to the inhabitants of various locations in ways that speak to their shared "Indianness," or tropicality, despite the obvious differences among them. Unlike the savages of Brazil and Patagonia, the people of Borneo, in *De Moluccis Insulis*, are said to be civilized and to have imparted "mores et instituta paucis" (some customs and manners) to the people of other islands, but those people are all Indians, just like the inhabitants of the text's "American" locations (Pigafetta 1969, 31–32). Von Sevenborgen even treats the watercraft used by inhabitants of what we know as the Philippines as if they were interchangeable with other such craft seen elsewhere in the tropics. "Duas Canoas Indorum conspiciunt" (They saw two Indian canoes), he writes, "sic enim hoc genus navigii quod peregrinus est ab Indis nuncuparisolet" (for such is the Indian name for these strange boats; B i r). "Canoe," of course, is a name of Caribbean origin, one that Pigafetta knows to restrict to the cultures of the Atlantic, yet von Sevenborgen seems to believe that the word, not just the watercraft, is common to all "Indians" (compare Pigafetta 1969, 6).

Clearly, the architecture of the continents is not the principal metageography mediating the world-making work of *De Moluccis Insulis*. It is the theory of climates. In telling the story of the first European crossing of what we call the Pacific Ocean, von Sevenborgen is not writing about a voyage across a vast ocean that separates something called "America" from "Asia," but rather about a journey from one tropical location to another, across a large ocean aptly named the *South* Sea. That ocean, we are to believe, is so broad and empty that its traversal by Magellan's men represents a considerable accomplishment, but it is not so desolate as to truly present any obstacle to Europeans eager to cross it. It seems, then, that the South Sea that took shape in the European imagination after Balboa marched across Panama survived Magellan's supposed discovery of the Pacific. In fact, we can say with confidence that the first printed account to tell the story of the circumnavigation of the Earth by the *Victoria* does not register a "discovery of the Pacific" at all, at least not in the sense described by contemporary historians of exploration. There is no ocean separating two distinct geographical and cultural areas in *De Moluccis Insulis*. There is only an ocean that, in its navigability, has the potential to bring together under a single scepter mutually distant but fundamentally similar tropical locations.

Our next imperial narrative is that of Peter Martyr. As we saw above, the Italian historian was commissioned to write an official account of the Magellan expedition the same year the *Victoria* returned. That text was subsequently lost, but a version of it entitled "De orbe ambito" (On the circumnavigation of the world) appeared as part of Martyr's Fifth Decade, which was published in 1530, in Alcalá de Henares, as part of the first complete edition of the author's *De Orbe Novo Decades*.[17] We have no idea if the version of the text that appeared in print was identical to the one that was sent to Rome and destroyed, or if Martyr altered his story in the interim. I am of the opinion that Martyr very well may have revised it to respond to the challenge posed by *Le voyage et navigation*, particularly because it incorporates precisely those details that von Sevenborgen excludes regarding the suffering of Magellan's men on the Pacific crossing. If this is true, then Martyr's "De orbe ambito" represents an attempt to defend the imperial account of the South Sea and its crossing by co-opting the most problematic elements of the Pigafetta version.

As we might expect from a historian in his position, Martyr adamantly defends Spanish claims regarding the position of the Spice Islands relative to the line of demarcation. He claims that "Malacham ab ipsis usupratam, quod sit extra lineam illam dividentem, ab oriente in occidentem, ab utroque polo" (Malacca has been usurped [by the Portuguese], since it lies beyond the line drawn from pole to pole separating the east from the west), and he backs up this statement with some general assertions about global geography, made in the context of describing the route of the *Victoria* to a reader who might not have access to an appropriate map (Anghiera 2005, 2.266).[18] Following Ptolemy and others, Martyr insists that Eurasia extends across 180° of longitude (2.660–63). This claim was attractive in Spanish circles, because it placed a considerable amount of the Asian continent within the Castilian hemisphere, as we can readily appreciate by examining Juan Vespucci's map of the world on a polar projection (see fig. 9). Had Martyr chosen to give a longitude for the Spice Islands, he could have easily used the one provided by the Paris Pigafetta, of 171° west of Seville, or he could have derived one from Vespucci, whose positioning of Malacca just inside the Castilian demarcation corresponds with Martyr's own assertions, and whose work Martyr would certainly have known.

This claim implied nothing about a possible connection between Asia and the newly discovered landmass in the Ocean Sea. Vespucci, as we can see, equips Asia with a definitive eastern coastline, suggesting that he believes the New World is geographically separate. Martyr remains cagey on this point, which might come as a surprise to readers who are familiar with the pride of place he enjoys in the historiography about the discovery or invention of America. The Milanese humanist was the first to refer to the newly discovered landmass in the Ocean Sea as an *alterius terrarum orbis* (another world-island) when he wrote his First Decade in the 1490s, marking him as an early proponent of the insularity of the landmass and the fundamental otherness of its inhabitants (Anghiera 2005, 1.48). Yet as time went on, Martyr's views changed, along with those of so many others. By the 1520s, when he wrote the Fifth Decade, he was referring to the newly discovered landmass as the "putatum continentem" (supposed or putative continent), leaving the whole question of its status as either an island or an extension of the Old World in suspension (2.660).[19] The only thing of which he was certain was that those lands did not lie far from the easternmost reaches of Eurasia as they were mapped by Ptolemy, and perhaps Vespucci. This was because a broad Eurasia simply left no room for a broad ocean between the two.

Martyr's manner of describing the ocean Magellan sailed stands in marked contrast to those of both Pigafetta and von Sevenborgen. *Le voyage et navigation* figures the Pacific Sea as an expanse so vast that it is unnavigable. *De Moluccis Insulis* insists, paradoxically perhaps, that it is too large to measure, yet nevertheless traversable. "De orbe ambito," by contrast, treats it as a cartographic object that can be readily mapped and understood by way of existing frameworks of understanding. Martyr confidently asserts that the Pacific Sea discovered by Magellan is continuous with the South Sea discovered by Balboa (2.646). He explains that it runs into Ptolemy's "Sinu Magnu," or "Great Gulf," the watery gateway to the land of the Chinese (2.660). In this way, he maps the South Sea as an oceanic basin bounded on at least two sides, by the New World in the east and Asia in the west, and connected to other oceanic basins. He remains confident, furthermore, that the tropical latitudes of the South Sea contain numerous islands waiting to be discovered, and that, the tropics being what they are, these islands should be rich in spices and other forms of wealth (2.790).[20] Plainly put, Martyr maps Magellan's Pacific along the expectations of Ptolemaic cosmography and generates expectations about it that respond to the theory of climates.

Peter Martyr's account of the first Pacific crossing, in turn, does not lead one to imagine an ocean so vast that it is practically impossible to cross. Here is his version of the story:

Vastum eo tractu superato, capttarunt oceanum alius mare; id est nostril putati continentis a tergo marique illi iungitur, quot 'australe' in Decadibus apello, a Vasco Nuñez primum reperto, ex Dariene dirigentibus illum comogri regis filiis per id pelagus immensum. Aiunt tres se menses ac viginti dies egisse, coeli tantum et acquae salsae prospectu contentos; de summa rerum egestate deque infensis caloribus miseranda referent, oryzae quantum pugno capi queat, portionem quotidianam multis diebus fuisse solam, sine alterius ullius cibi mica, profitentur. Potabilis aquae penuria erat tanta ut cogerentur salsae maris aquae tertiam partem ad oryzam coquendam iniicere et, si forte puram quis bebere tentaret, claudere oculos prae viridi situ et nares prae fetore cogebatur. Per mare illud ingens tendentes ad occidentem et septentrionem, resumpsere denu lineam aequinoctialem, cui proximas reperere duas inertes insulas, quas appellavere *Infortunatas*, eo quod penitus inutiles ad desertas (2.646-48).

[After sailing through the strait [of Magellan], the Spaniards entered another vast ocean. It is the ocean on the opposite side of the supposed continent, and communicates with the sea, which I have called the Austral Sea in my Decades, and which was first discovered by Vasco Nuñez. . . . They [the expedition survivors] affirm that they sailed three months and twenty days on that immense ocean, seeing nothing but sky and salt water. They tell pitiful tales of their suffering, and of the tremendous heat they endured. For many days they had nothing to eat but a handful of rice, without a scrap of other food; potable water was so scarce that they were obliged to use one-third sea water for cooking their rice, and when a man drank this water he had to shut his eyes and stop his nose, so green was its color and so nasty its odor. Sailing through that great sea toward the Occident and Septentrion, they reached the equinoctial line and soon discovered two barren islands, which they named the Unfortunate Isles because they were deserted and useless.]

Martyr may have gotten his gory details from the expedition survivors he could have interviewed in his capacity as a member of the Council of Indies, or he may have borrowed them from Pigafetta after the latter's account was published in 1525. In any case, his treatment of the sufferings of Magellan's sailors looks abbreviated and toned down when compared to the Venetian's account. For one, Martyr reports the survival stories as second-hand information, raising the possibility that they might be nothing more than the exaggerated tales typical of returning travelers. He provides a time frame for the short rations, "for many days," suggesting that the lack of food was characteristic of some portion of the ocean crossing, but clearly not the entire three months and twenty days. He does not mention some of Pigafetta's most shocking anecdotes, like the desperate market for mice, and says nothing either in this passage or elsewhere about profound loneliness, sickness, or death. Martyr's account of the Unfortunate Islands does not mention the failed attempt to anchor there, and his account of the encounter with the inhabitants of the Ladrones, a little later on, does nothing to present their behavior as a mockery of the men's suffering. The pleas of the scurvy victims to be brought human entrails are nowhere to be found. There is no mention of the work of providence in preserving the fleet from certain ruin and the men from certain death. In sum, Martyr's version of the story leaves no doubt that the first Pacific crossing was a grueling experience, but it does not go so far as to suggest that the men barely survived, or that future attempts to follow in their wake would be doomed to failure. It tames Pigafetta's nightmare.

In this way, Martyr recruits the difficulties of the Pacific crossing to his overall rhetorical end, which happens to be the same as that of *De Moluccis Insulis*, the celebration of the first circumnavigation of the Earth as a triumph of epic scale. Martyr even appeals to the same classical example in making his point. "Quid super hac incredibili novitate confixisset Graecia, si Graeco alicui accidisit?" (What would the Greeks have invented about this incredible novelty, had a Greek done the same?), the historian asks rhetorically. They celebrated so proudly the voyage of the Argonauts, whose voyage from Greece to the Black Sea was nothing compared to the distance covered by the *Victoria* (2.658). In Martyr's account, the Pacific crossing is not a cautionary tale for those foolish enough to imagine repeating it, but a hard-won triumph in a heroic story of epic seafaring, an episode in the ongoing conquest of the Ocean Sea. As such, it does not suggest the Spice Islands are beyond the reach of Spanish navigation, but rather implies that a better-provisioned fleet could make the voyage without suffering the pains experienced by Magellan's men.

In all these ways, therefore, "De orbe ambito" goes even further than *De Moluccis Insulis* in its attempt to contain the Pacific and its dangerous implications for both Ptolemaic cosmography and Spain's imperial ambitions. Not only does it insist that the Moluccas are rightfully Spanish, but it even appeals to Ptolemaic ideas about the breadth of Eurasia in order to make this point. In this way, "De orbe ambito" maximizes the extent of Spain's territorial claims in East Asia and contains the waters of the Pacific within a basin of manageable size. It appeals to the theory of climates to populate that space with wealth ready for the taking, and it helps us imagine the ocean as perfectly navigable by co-opting the most disturbing details of the Pigafetta account in ways that neutralize their potential implications regarding the possibility of repeating the voyage. Martyr, in other words, denies the supposed "discovery of the Pacific" just as thoroughly as does Finé or any other of the European mapmakers whose cartographic products are so often presented as conservative, bookish, or even ignorant fantasies to be compared unfavorably to the charts produced in the Spanish circles to which Martyr was so closely tied.

None of the Spanish imperial historians, however, goes as far as Gonzalo Fernández de Oviedo in advancing the notion of a relatively small, perfectly navigable South Sea against Pigafetta's sublime expanse. As we saw in the previous chapter, Oviedo's 1526 *Sumario* argues that the Castilian Indies are not the same region described by Pliny and other classical authors, but a newly discovered region featuring previously unknown species of flora and fauna, as well as novel human societies. The text nevertheless groups all of the Indies together as tropical locations and sets them against places in the temperate zone, like Spain. So, while Oviedo's popular text was instrumental in disseminating the idea that the discoveries in the Ocean Sea were not part of the world as it had been known to the late Middle Ages, the historian himself did not conceptualize those discoveries as a "fourth part of the world." This was because he did not consider global geography primarily in terms of the architecture of the continents, but of the theory of climates. This meant that even though the distant Moluccas were part of "those Indies" rather than "these Indies," to use the language Oviedo himself sometimes favors, they were just as "Indian" for Oviedo as they were for von Sevenborgen and Martyr. That is, they were just as tropical and therefore just as available for colonization as the islands of the Caribbean. The only thing that mattered was their position relative to the antimeridian, and Oviedo had no doubt that they were on the Castilian side.

He specifically addresses Spain's South Sea ambitions in the final chapter of the *Sumario*, where he speaks to the prospects of the expedition captained by Jofre García de Loaysa, who had been dispatched from Spain in 1525 to consolidate the Castilian position in the Spicery. Oviedo is not optimistic. Having interviewed the survivors of the Magellan expedition, Oviedo knows just how difficult the Pacific crossing had been, and he expresses his doubts about the decision to send Loaysa to the Moluccas along the route Magellan had pioneered. The problem, as Oviedo explains it, was not the breadth of the South Sea, but the excessive distance introduced by the long southern detour required to pass through Magellan's strait. If you could cut the detour out of the way west to the Spicery, you were left with an ocean crossing of manageable length. Although the *Sumario* provides no numbers, Oviedo claims elsewhere that only 1,600 leagues separated the Moluccas from Panama City, about 70 percent of the distance that appears on the 1529 Ribeiro chart (1851, 3. 150).[21] According to the historian, it would therefore make sense to abandon the route via the strait and ship spices from the Moluccas to Panama City, then overland to the Caribbean and onward to Spain (1950, 270–73).

Time would prove him right. Eventually, Spain's attempt to establish a position on the far shores of the South Sea would shift from Old Spain to New Spain, and when that position was finally established in the Philippine Islands, it would function as a colony of the Mexican viceroyalty. In the meantime, however, what matters is the way Oviedo's argument contributes to controlling the potential damage that could arise from Pigafetta's account of the Pacific crossing. Like Martyr, Oviedo acknowledges that the crossing was a harrowing affair, but rather than translate the tale of the fleet's suffering into an epic register in the manner of the Milanese humanist, Oviedo remaps its relationship to the overall space of the South Sea. As we saw, the way Pigafetta tells the story, the hunger, thirst, and disease expand to fill the entire route. Oviedo blames it all on the demands of the southern detour, clearly implying that if ships do not use it, the men will not suffer. His more practical route from Panama to the Spicery promises to be smooth sailing all the way.

Yet the text in which Oviedo really takes aim at the challenge posed by Pigafetta is book 20 of his massive *Historia general y natural de las Indias* (General and natural history of the Indies). Written most likely during the 1530s, but not published until 1557, shortly after Oviedo's death, book 20 covers the history of Spain's South Sea projects from the genesis of the Magellan expedition through the immediate aftermath of the 1529 Treaty of Zaragoza. Its first few chapters take direct aim at Pigafetta. The first draws on survivor accounts to tell the story of the Magellan expedition, admitting, as do the other chroniclers, that the first European crossing of the South Sea was not what anyone expected it to be. He depicts Magellan, in the cold and hunger of a Patagonian winter, exhorting his tired and mutinous men to carry on by promising them that they will soon discover "un nuevo y nunca conocido mundo, rico de especería y de oro y de otros muchos provechos" (a new and previously unknown world, rich in spices and gold and many other advantages), once they penetrate the maritime passage he is certain they will find (Fernández de Oviedo y Valdés 1852, 12).[22] Yet once they make it through they do not discover a new world, but a new ocean, one that "otros cristianos nunca antes que estos navegaron" (no other Christians had ever sailed before), and that seems "cada día mayor é más amplíssimo le hallaban" (every day bigger and so very much broader), as the ships sail for three months without sighting any land except the Unfortunate Isles, "estériles e pequeñas e deshabitadas" (sterile, small, and uninhabited) (13).

Like the other imperial historians, Oviedo manages the story of the crossing so as to control the implications of this discovery. He does not fictionalize the encounter with the Unfortunate Isles, the way von Sevenborgen does, but he nevertheless imitates the secretary in remaining silent about the suffering of the men, saying nothing about hunger, thirst, sickness, or death along the way, yet making sure to mention the strong, favorable winds that bore the ships across the ocean, with no storms to trouble them along the way. He says nothing about distances or longitudes, leaving the reader with the impression that the South Sea is broad, but providing none of the information that he or she would need to map it.[23] One is left with the same impression one gets from reading von Sevenborgen, that the first European crossing of the Pacific Ocean presented no real difficulty, that it was smooth sailing all the way. Like the other narratives by authors closely connected to the court of Charles V, Oviedo's account of the crossing is a tale of the successful conquest of the Ocean Sea.

The next two chapters provide a second account of the same events explicitly drawn from Pigafetta. This often happens in the *Historia general*, creating the impression that we are dealing with a diligent historian who does not want to get in the way of his eyewitness sources, and eagerly provides different versions of events so that the reader can make up his or her own mind about whom or what to believe. In keeping with this practice, Oviedo dutifully notes that the Venetian was an eyewitness to the events he describes and should therefore be believed. Nevertheless, he then goes on to undermine Pigafetta's authority. The chapter becomes an act of discursive violence, a direct assault on the single source that posed the most difficulty for the imperial account of the Pacific crossing and its imperial cartography of the South Sea.

Oviedo notes supposed errors in Pigafetta's latitudes, and he calls into question the latter's reports of natural marvels, suggesting that the Venetian gentleman was a gullible amateur (23–30). Most importantly, however, Oviedo downgrades the Pigafetta text from a full-fledged chronicle to a mere mine of information. He does not reproduce the Venetian's text in its entirety, but only lists whatever information it provides about the lands and peoples visited by the expedition that does not appear in the first, presumably more authoritative, chapter. Oviedo gives no hint as to what he leaves out, nor does he explain his criteria for selection. Noticeably absent are the nightmarish details from Pigafetta's account of the Pacific crossing. While von Sevenborgen simply left them out, and Martyr tried to control their spin, Oviedo actually excises them from their most prominent source. Nowhere is the struggle over the story of the Pacific crossing more in evidence than in this high-handed treatment of Pigafetta's narrative. Nowhere is it more in evidence that Pigafetta's account of the agonies of the ocean crossing figured the lived experience of transpacific distance in ways that were potentially dangerous to Ptolemaic cosmography and Spain's imperial ambitions in the South Sea.

What we see, then, is a consistent, concerted effort to present Magellan's Pacific crossing in a particular light, one that maps the South Sea as an oceanic basin that was certainly on the large end of the spectrum, but was nevertheless clearly contained within continental boundaries, narrow enough to assure that the Spicery and its environs lay within the Castilian hemisphere, and, most importantly of all, was perfectly traversable by European vessels. This narrative cartography clearly served Spain's political purposes, but that does not mean that it was crassly deceptive. To say that would require us to identify an objective cartography of the world that these accounts falsified, but there was none. The cartography of the South Sea, like that of the world of which it formed a part, was irreducibly fictional, insofar as it relied upon faulty assumptions about the world's geography and upon deeply limited techniques with which to measure its daunting distances. In this context, a qualitative cartography of the Pacific, one that imagined its breadth as a lived experience rather than as a simple number, could hold considerable sway. The fact that the South Sea remained small on so many European maps testifies, perhaps, not only to the influence of the numbers in the Paris Pigafetta, but to the success of the imperial narratives in presenting that lived experience as an epic achievement rather than a nightmarish story of survival.

Cartographic Anxieties: Ribeiro's Ocean

I turn, finally, to the most famous of the Spanish planispheres that came out of the Magellan voyage and the subsequent squabble over the Moluccas. As maps rather than narratives, the Seville planispheres inevitably open up a two-dimensional space when they stretch the distance between the New World and the Spicery to reflect the information brought back by the *Victoria*. That space ends up occupying roughly one-quarter of the cartographic surface on these charts. Most of them (Anonymous, 1523; García de Toreno, 1525; Ribeiro, 1525; Vespucci, 1526) leave it almost entirely blank, save for some compass roses, the image of the lone *Victoria*, or a royal escutcheon.[24] The two charts Diogo Ribeiro made in 1529, by contrast, fill much of the space with images of nautical instruments and sailing ships in ways that directly engage with Magellan's crossing and the closely related question of the overall navigability of the South Sea. Like the imperial accounts of the expedition by von Sevenborgen, Martyr, and Oviedo, Ribeiro's iconography presents the first crossing of the Pacific by Europeans as a success, yet it also reveals anxieties about the period's ability to master the space involved on an intellectual if not a practical level.

Like the other Seville planispheres, the Ribeiro charts are centered on the line of demarcation and thus divide the world into a Castilian hemisphere on the left and a Portuguese hemisphere on the right. They place China and the Moluccas roughly 175° west of the line of demarcation, just inside Castilian territory. Unlike the other planispheres, however, the Ribeiro charts actually depict these territories twice, once on the left-hand side and once on the right. The antimeridian appears on both sides, marked by Portuguese and Castilian standards, in order to obviate any confusion this might cause. The Atlantic coast of what we call America stretches unbroken from Labrador to the Strait of Magellan, but only the southern landmass is identified as the "Mundus Novus." The upper landmass is left unnamed, and the map remains agonistic as to its possible status as an extension of Asia or a part of the New World. Near the right-hand edge of the map, the Asian continent trails off ambiguously, remaining open to the possibility that it continues eastward, into the empty space that occupies so much of the Castilian hemisphere.

That space is dominated by depictions of a quadrant and a large circular declination table, as well as cartouches explaining the use of these instruments. A third instrument, a nautical astrolabe, appears in the lower right. As Surekha Davies has convincingly argued, this particular repertoire of instruments seems to have been deliberately chosen to advertise Ribeiro's skill and value as an instrument maker, and more generally to endorse the novel art of astronomical navigation against the culture of ships' pilots who clung to traditional methods. This is why the compass, the technology central to the pilot's methods, is conspicuous for its absence among the instruments depicted (Davies 2003). The instruments, therefore, are not just there to speak of European mastery over the sea, but of the particular technologies and practices that in the opinion of cosmographers like Ribeiro were necessary to advance that mastery.

There is an irony, however, in making this statement on this chart, or any chart of this kind. As we saw at the beginning of the chapter, early modern mapmakers derived their longitudes by converting travelers' reports of distance traveled into coordinate degrees. The Ribeiro chart actually admits this to be the case, in a cartouche just above the quadrant, which identifies "la opinion y parecer de Juan Sebastián del Cano" (the opinion and judgment of Juan Sebastián del Cano), the man who had sailed the *Victoria* back to Spain, as its source for the longitude of the Spice Islands. In this way, Ribeiro admits that he is indebted to precisely the sort of person, using precisely the sort of techniques, that the visual rhetoric of the instrument illustrations rejects. Indeed, the whole chart, with its network of rhumb lines and compass roses inherited from medieval portolan charts and the practice of navigating the Mediterranean with compass in hand, speaks of Ribeiro's debt to the pilots and their navigational culture. The entire chart hangs suspended between the culture of the pilots, whose work made it possible, and that of the cosmographers, whose astronomical techniques promised to rid early modern cartography of its indeterminacy.

A similar irony attends the images of ships that populate the space between the New World and the Spicery. Six such images mark the route taken by Magellan's fleet across the ocean. Two of them appear just below and to the right of the declination table, sailing toward the northwest. Four sail from east to west near the equator, one just to the right of the declination table, a second within it, and two more to its left. None of them are identified as a particular vessel, but all of them bear the label "Voy al maluco" (I am going to the Moluccas). As we saw in the previous chapter, images like these perform a variety of functions. They help us distinguish between oceanic space and empty cartographic space, and they advertise European mastery of the seas, or of a particular sea, often by marking well-worn routes of maritime travel. Here, the ship images help us understand that there is a swath of ocean that cuts through the space between the New World and the Spicery along the route Magellan sailed, but also signal that we know nothing about the space beyond this track, out where the signifying power of the ship images gives out. That empty cartographic space to the north and south could contain land or water. The ships also commemorate Magellan's accomplishment, and in their anonymity, speak of a route of travel from Spain to the Spicery that has been opened up by his voyage and that could be exploited for future use. It is important to note that none of them seem to be tossed by storms, and none is on the verge of shipwreck. As reminders that the first Pacific crossing by Europeans had been smooth sailing all the way, the ships stand as graphic equivalents to the tale of Magellan's Pacific crossing as told by imperial historiography, monumentalizing Magellan's

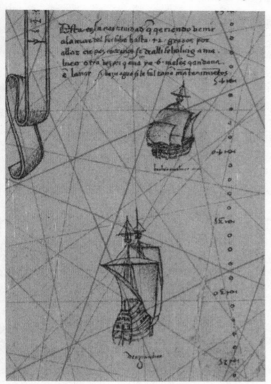

FIGURE 20. Magellan's *Trinidad* attempts and fails to discover the return route from the Spice Islands to the New World in this detail (see also fig. 7 above) from Diogo Ribeiro's *Carta universal en que se contiene todo lo que del mundo se ha descubierto fasta agora* (London, 1887). Photograph: Courtesy of the Geography and Map Division, Library of Congress, Washington, DC (call no. G3200 1529.R5 1887 MLC; control no. 85690293).

accomplishment, announcing that the South Sea had been conquered, and suggesting that the way was open for future commercial or colonial endeavors.

Even more interesting than these ships, however, are the two vessels we find northwest of the declination table, near the scale of latitude. One is marked "Vengo de maluco" (I am coming from the Moluccas) and the other, "Buelvo a maluco" (I am returning to the Moluccas). (Figure 20.) The inscription just above the images identifies the ship as the *Trinidad*, the vessel that was to sail back to the New World from the Moluccas at the same time that the *Victoria* made its way to Spain by way of the Cape of Good Hope. It failed in its attempt, only to return to the Spice Islands and fall into the hands of the Portuguese. The inscription tells how the ship reached 42° latitude north in its search of the westerlies that could carry it back across the ocean, but found only contrary winds. In this way, it provides a subtle hint as to the geographical relationship between the New World and Asia. Since the *Trinidad* encountered no land on its ill-fated voyage, the two landmasses could not possibly be connected, at least not south of the highest latitude the ship reached. Yet it also has something to say about the architecture of the oceans and the continuing project of conquering the sea. By admitting that Magellan's men had not entirely mastered the ocean they traversed, the inscription and the ships that go with it place a limit to the story of Spanish command of the sea told by the other vessels. The inscription adds, moreover, that the *Trinidad* sailed for six months in search of the return route, and turned back when the supplies began to run out. Shades of the Pacific as a nightmarish space of bare survival thereby make an unlikely appearance on Ribeiro's map. In the midst of Ribeiro's visual rhetoric of nautical mastery, we find an admission that the ocean between the New World and the Spicery remains defiant.

Conclusion

Magellan discovered that the voyage from the New World to the Moluccas was much longer than anyone had expected it to be, and that the way there showed no signs of the wealth he and Faleiro had expected to find. Where they had imagined rich islands to govern in perpetuity, the fleet found only open water and a few barren islets that mocked their expectations. Pigafetta gave potent literary form to this experience of the endless, deadly open ocean, effectively inventing the Pacific as a vast, empty watery expanse between the savage lands that had been discovered in the Atlantic and the more civilized places that lay on its farther shores. In this way, he provided fuel for the fire of the invention of America, and he suggested that the globe could not be mapped as a series of large continents that held the world's waters in tidy ocean basins. Neither Spain's imperial ambitions nor cosmography's Ptolemaic principles could stand for this, so an effort emerged to apply the cartography of containment and the rhetoric of smooth sailing, in different measures and at different moments, to corral Pigafetta's Pacific into the comfortable confines of Balboa's South Sea.

That the reasons for doing so could be primarily epistemic rather than political is evidenced by the tendency so in evidence among mapmakers outside Spain, like Finé, to map the distance from the New World to the Spicery in the shortest possible terms and to depict the South Sea as a whole as a basin contained on all sides by Africa, the *Terra Australis*, and Amerasia. In Spain, however, epistemic arguments joined hands with political ambition to limit the breadth of the South Sea and present the tale of Magellan's Pacific crossing as an epic of maritime conquest rather than as a harrowing survival story. Von Sevenborgen fictionalized the crossing to suggest that it had been smooth sailing all the way, while Martyr co-opted Pigafetta's gory details and combined them with the cartography of containment. Oviedo echoed von Sevenborgen and launched a head-on attack on the authority of Pigafetta as a credible source. Ribeiro, meanwhile, followed Reinel in deploying the iconography of smooth sailing to reiterate the official story of the Magellan crossing, but nevertheless allowed room on his map for a vignette that told of the Pacific's continued resistance to mastery by Spanish navigation.

By the time Ribeiro made this map, he would have been well aware that any claims that Magellan had effectively conquered the South Sea and opened it up to Spain's imperial ambitions were overly optimistic. In 1525, he had left Seville for A Coruña, in northwestern Spain, where Charles V had established his Casa de la especería, an institution modeled on the Casa de la contratación but designed to supervise Spain's new trade with the Spice Islands. The place also served as headquarters for the 1525 Loaya expedition, whose chances had so worried Oviedo. Ribeiro was to equip the expedition with the necessary charts and instruments. By 1529, however, news had reached Spain of the expedition's utter failure. Several ships had been lost to Pacific storms, and the survivors who reached the Spicery failed to discover the return route to the New World, just as the men of the *Trinidad* had before them. Stranded in the Moluccas with no hope of resupply or reinforcement, their fortunes waned as those of their Portuguese rivals waxed. So even as Ribeiro drew his map asserting his king's claim to the Moluccas and proclaiming Magellan's success in opening a route from Spain to the Spicery, he knew that the South Sea had not yet been conquered, that the pilots had failed in their task. The next phase in Spain's effort to map the Indies as a transpacific geography would have to confront this crescendo of failure. Oviedo would take it up in the balance of book 20, and so would Gómara, in his *Historia general de las Indias.*

4

Shipwrecked Ambitions

The profits from the Magellan expedition were meager, but as far as Charles V was concerned, they were enough to suggest it was worth pressing his claim to the Spice Islands. The emperor launched a diplomatic offensive in support of Spain's claim to the islands and authorized a series of expeditions meant to consolidate Spain's position there, as well as strike out for Cathay, Zipangu, and even the biblical islands of Ophir and Tarshish.[1] It must have seemed for a while that the world of Marco Polo was on the verge of falling under Spain's control, yet not a single one of these expeditions achieved its objectives. The only one that even got to its destination was the expedition of Jofre García de Loaysa (1525–27), who followed Magellan's route to the Spicery only to lose his life and most of his ships along the way. Although the survivors eventually made it to the Moluccas, they became embroiled in a lengthy low-level conflict with the Portuguese over control of the spice trade, receiving reinforcements only once, when a single ship arrived from Mexico under the command of Álvaro de Saavedra (1527), a kinsman of Hernán Cortés. The war did not end until news reached the islands that the emperor had pawned his claim to the Moluccas in exchange for 350,000 ducats from the crown of Portugal in the 1529 Treaty of Zaragoza. The treaty placed the antimeridian 297.5 maritime leagues east of the Moluccas, a position that left no room for Spain's claim to East and Southeast Asia, and required both kingdoms to construct maps reflecting this agreement. It would seem, therefore, that by signing the treaty, the emperor had abandoned East and Southeast Asia to Portugal and embraced Spain's destiny as a colonial power in the New World alone (Brotton 2004, 151).

Yet the crown of Castile does not seem to have constructed a single map that followed the dictates of the treaty. On the contrary, the evidence suggests that Spanish mapmakers continued to map the antimeridian precisely as they had during the decade after the return of the *Victoria*, placing it just east of the Portuguese city of Malacca (Martín Merás

1992, 89). Spanish cosmographers eventually came to speak of the region between their antimeridian and the one defined by the Treaty of Zaragoza as "el empeño" (the pawn), and recognized that Spanish ships could not enter it unless the crown returned the 350,000 ducats to Portugal, yet even this way of understanding the arrangement with Portugal was controversial, since no one could measure longitude with any accuracy, and the same geographical and geodesical variables that were at issue before 1529 continued to be contested afterward. As a result, Spanish dreams of discovery and conquest in the South Sea remained alive for almost a decade and a half after the Treaty of Zaragoza, fueling continued efforts to discover new lands in the blank spaces that framed the westward track of Magellan's fleet, particularly to the north.

Hernán Cortés himself began to build ships on the Pacific shores of Mexico shortly after the conquest of Tenochtitlán, with the intention of making for the Spice Islands. Interest in northwestern New Spain grew when Cabeza de Vaca and his companions appeared bearing rumors of sophisticated cities just to the north of their westward track across North America, and Fray Marcos de Niza apparently confirmed those rumors by claiming to have sighted fabulous Cíbola. From 1527 until 1544, Spaniards in New Spain devoted a great deal of energy toward the exploration of what eventually came to be known as "las Islas e Provincias que estoviesen en la mar del Sur hacia el Poniente" (the Islands and Provinces that might be in the South Sea toward the west), in other words, everything between the known parts of New Spain and the antimeridian, including whatever islands might be found and what was suspected to be the landmass of Amerasia (Navarrete 1971, 465). While the king's cosmographers continued to map the space between New Spain and China as empty *terra incognita*, Spaniards in New Spain shared the suspicion held by other Europeans that one could walk from Mexico City to Beijing, or follow the coast from Acapulco to Guangdong.

This second wave of South Sea ventures that originated in New Spain rather than in metropolitan Spain culminated in a three-pronged stab into *terra incognita* organized by Viceroy Antonio de Mendoza and Pedro de Alvarado, until the conquistador's untimely death in 1541, when the viceroy assumed full control. Francisco de Coronado was sent into the interior of "la India Mayor" (Greater India), Juan Rodríguez de Cabrillo was dispatched to chart the Amerasian coastline from Mexico to China, and Ruy López de Villalobos was charged with striking across the South Sea in order to conquer the Islands of the West. The effort fell apart when Coronado failed to find the Seven Cities of Cíbola (1542–44), Cabrillo found that the coast of California turned northward rather than westward

(1542–43), and Villalobos failed to find the return route from the Islands of the West, sealing the fate of his fledgling colony in the islands that he himself christened "las Filipinas" (the Philippines, 1542–44). After this series of costly disappointments, there would be no further attempt to extend Spain's empire from New Spain to the Islands of the West for another twenty years.

According to many historians, the failure of these expeditions clearly demonstrated that the South Sea was dauntingly wide and also suggested that the New World was indeed an enormous island separate from Asia (Nunn 1929, 30; Kelsey 1984, 328, 1998; León-Portilla 2005, 193; Flint and Flint 2008, 154–55). One of these scholars, Miguel León-Portilla, points to maps drawn during the 1540s by Alonso de Santa Cruz, Sebastian Cabot, and Battista Agnese as evidence that these expeditions had driven the last nails in the coffin of the theory of Amerasian continuity (figs. 21 and 22).[2] These maps depict Baja California as a peninsula and thus demonstrate that their makers were aware of the discoveries that had been made along the Mexican coast by Cortés and others, yet they also depict empty space between Baja California and continental Asia. In this way, argues León-Portilla, they attest to the dawning realization that the New World was geographically independent of Asia and that a broad empty ocean yawned between the two continents (193). Of course, we have already seen that empty space on charts of this kind did not necessarily express doubt regarding any particular geographical possibility, but I will leave discussion of this matter for the pages that follow.

For now, suffice it to say that the general histories of the Indies that were written during the same period seem to do the same thing as the maps, in León-Portilla's reading of them. The *Historia general y natural de las Indias* (General and natural history of the Indies), written by Gonzalo Fernández de Oviedo during the 1520s, 1530s, and early 1540s, and the *Historia general de las Indias* (General history of the Indies), written by Francisco López de Gómara during the late 1540s, lavish attention on the core regions of Spanish overseas activity, the Caribbean, Mesoamerica, and the Andes, and devote relatively little space to the Moluccas and adjoining areas. As we move from Oviedo to Gómara, moreover, the history of Spain's "Enterprise of the Spicery" incorporates more failed expeditions, thereby charting the shipwreck of Spain's South Sea ambitions. Like the maps León-Portilla cites, therefore, these histories seem to give up on the islands and provinces of the West and on Spanish dreams of transpacific empire. In so doing, they contribute to the ongoing invention of America as a continent separate from Asia, and the construction of a world image centered on the Atlantic, with (Spanish) America in the

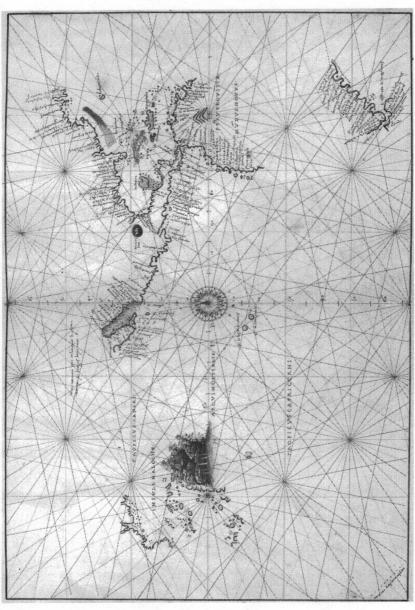

FIGURE 21. A highly fragmentary geography of the Pacific Basin. The Spice Islands are indicated by an image of a clove tree. Battista Agnese's Nautical Chart of the Pacific Ocean, from *Portolan Atlas of Nine Charts and a World Map, etc. Dedicated to Hieronymus Ruffault, Abbot of St. Vaast* (Venice, 1544). 21 x 29 cm. Photograph: Courtesy of the Geography and Map Division, Library of Congress, Washington, DC (call no. G1001.A4 1544; control no. 98687206).

FIGURE 22. Detail from map of the world probably derived from Spanish sources. The text spanning the space between Mexico and China describes the island of Zipangu, and the figure seated above depicts the Great Khan. Sebastian Cabot, Map of the World (1544). 220 x 125 cm. Photograph: Courtesy of the Bibliothèque nationale de France (registry c 06218; GE AA-582 [RES]).

absolute West, (Portuguese) Asia in the absolute East, and the South Sea along the margins, serving as the geographical and ontological boundary that separates America from Asia and defines their difference.

As we saw in the previous chapter, Oviedo's highly popular *Sumario* of 1526 insisted that the circum-Caribbean constituted a new discovery, a place that had remained unknown to the ancients, even while it identified the New World as part of the broader "Indies," that is, the lands that lay in the tropics. Over the course of the following decades as Oviedo continued to write his *magnum opus*, his ideas about the new discoveries changed, making it difficult to say anything definitive about the way he conceptualized the new discoveries. In general, however, it helps to distinguish his references to the "orbe nuevo," the "otro emisfério y mitad del mundo" (new orb, other hemisphere and half-world), from his occasional references to the *Nuevo mundo* (New World).[3] "Orbe nuevo" is an expression that Oviedo shares with Peter Martyr, who uses it to refer to the entire hemisphere that remained unknown to Ptolemy, whose map of the world spans only 180° of longitude. Hence Martyr's reference to South America as "nostri putati continentis in Orbe Novo" (our putative continent in the New Orb; 2.708). The "putative continent" of South America, in this sentence, can be understood as the geographical content of a geometrical container, the Orbe Novo. At times, both Martyr and Oviedo seem to be referring to the southern hemisphere, but if we follow the lead offered by the Italian mapmaker Giacomo Gastaldi, who constructed a map to accompany an Italian translation of Oviedo's writing, it could also mean what we call the western hemisphere (fig. 23). Indeed, it has been argued that Oviedo was one of the people responsible for popularizing this way of conceptualizing the world, in terms of its division into an eastern and a western hemisphere (Gerbi 1986, 258). The geographical content of Oviedo's *orbe nuevo* is *las Indias*, that collection of islands and a continental landmass that the historian calls "Tierra Firme," most of which lay between the Tropics of Cancer and Capricorn, and which he sometimes calls the *Nuevo Mundo* (New World), but never *America*. Following the planispheres of the Casa de la contratación, to which Oviedo had access as the crown's official historian of the Indies, the historian leaves Tierra Firme open-ended in the west, its precise geographical relationship with Asia unknown (Padrón 2004, 145–47).

The vast bulk of Oviedo's historical narrative unfolds here. East and Southeast Asia make a significant appearance only in book 20, which appears at the beginning of Part II of the three parts into which Oviedo divided the *Historia general y natural*. Part I had been published in 1535 and again in 1547, in a revised and expanded edition. Book 20 was probably written during the 1530s, but was not published until 1557, the only portion of the balance of the *Historia general y natural* to appear in print during the early modern period. It was the fruit of an abortive effort to publish Parts II and III that ended abruptly when the author passed away. The text covers Spain's attempt to gain control of the Spice Islands from the outset of the Magellan expedition through the immediate aftermath of the 1529 Treaty of Zaragoza, paying particular attention to the Loaysa

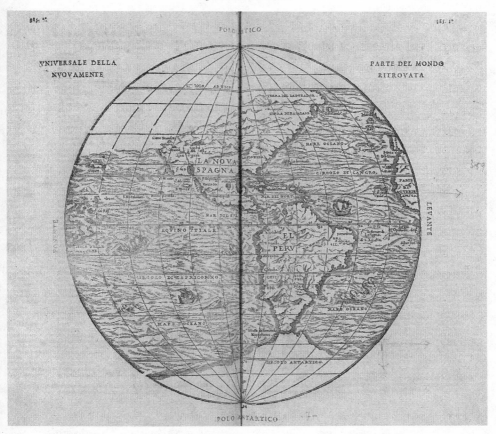

FIGURE 23. Map of the western hemisphere including New Spain, Peru, Japan, and the Moluccas. Giacomo Gastaldi, *Universale della Parte del Mondo Nuovamente Ritrovata* (Venice, 1556). 32 cm (fol.). This copy is from Giovanni Battista Ramusio's *Delle Navigationi et Viaggi*, vol. 3 (Venice, 1606). Photograph: Courtesy of the Albert and Shirley Small Special Collections Library, University of Virginia (call no. A 1563 .R35 v.3).

expedition, the war that the expedition survivors fought against the Portuguese for control of the spice trade, and Saavedra's failed attempt to reinforce their tenuous position in the islands. It also includes the most detailed description of insular Southeast Asia to appear in print during the first half of the sixteenth century, yet it has failed to draw much attention from scholars interested in Oviedo, who are overwhelmingly interested in the rest of the text and what it has to say about the New World.

Gómara's *Historia general de las Indias* had already been in print for five years when Oviedo's book 20 appeared. It was published in two volumes, the first covering the New World as a whole and the second focusing on the conquest of Mexico and the life of Hernán Cortés, whom Gómara served as secretary. The second volume soon began to appear separately, and became an early modern best seller, but it is in the first volume that we find Gómara's ideas about geography. There, the historian turns his back on his predecessor's open-ended approach to the physical geography of Tierra Firme. Unlike Oviedo, Gómara is not shy about using the term "Nuevo Mundo," nor is he ambiguous about its status as an island separate from Asia, as large if not larger than Europe, Africa, and Asia combined. In fact, he considers it to represent one of the three fundamental building blocks of global geography along with the Old World and the *Terra Australis*. Not only is it separate from the Old World, he insists, but it is also profoundly different. Its plants and animals are of kinds previously unknown and its people are bereft of the *policía* (polish, or police) typical of civilized societies, not to mention the light of the Faith. They are therefore in dire need of guidance from their Spanish betters. If any text analyzed in this book can be said to invent America along the lines imagined by Edmundo O'Gorman, it is Gómara's *Historia general*, where the Indies are America in all but name.

Gómara nevertheless dedicates sixteen of the 107 chapters that constitute the first volume of the *Historia general* to an account of Spain's effort to gain control of the Spice Islands, what he calls the "Empresa de la Especería" (Enterprise of the Spicery). Beginning with the Magellan expedition, Gómara covers the efforts by Loaysa, Saavedra, and Villalobos as well. He even provides a short history of the trade in spices from ancient times to the recent efforts of the Portuguese. Yet while Oviedo tries to end book 20 on an optimistic note by speaking of renewed efforts to establish a Spanish position on the far shores of the South Sea, Gómara's account ends in bitterness over the gag order that the emperor has imposed upon any discussion in the *cortes* of Castile about the possibility of returning the 350,000 ducats to Portugal and pressing Spain's claim to the Spicery anew. Having traced the history of the Enterprise of the Spicery from one failed expedition to the next, Gómara concludes with the shipwreck of Castilian ambitions in the South Sea.

This description of the South Sea episodes in Oviedo and Gómara captures, I believe, the way they have been treated by the existing historiography, as negligible appendices to texts that deal primarily with a place that we can call "America" with ever-increasing confidence, or as relics of Spain's ever-crumbling transpacific ambitions and the expansive vision of the Indies that went with them. Yet if these episodes were indeed as superfluous as they are in the eyes of modern scholarship, why did Oviedo and Gómara not see fit to leave them out altogether? However differently these two historians map the Indies, both of them seem to have believed that the story of Spain's attempt to assert its claim to the Spice Islands belonged in a general history of the Indies. Just as we can interpret their accounts of that attempt as relics of an obsolete geography or monuments to failed imperial ambitions, so we can also interpret them as efforts to hold on to a geopolitical imaginary that was creaking under the combined weight of failures on the level of practice and changes on the level of knowledge. When we look at Oviedo's book 20 and Gómara's "Enterprise of the Spicery" in this way, we find that neither of their general histories are so easily mapped into the historical teleology of the invention of America.

This chapter examines how Oviedo and Gómara keep Spain's South Sea ambitions on life support in the aftermath of the failures and changes of the 1530s and 1540s. Oviedo clings to the similarities and continuities among tropical locations imagined by the theory of climates in order to figure all of the Indies as conquerable territory, and he remains agnostic about the geographical relationship between Tierra Firme and continental Asia, yet he also allows the South Sea to open up as a broad, problematic ocean basin that has frustrated Spain's South Sea ambitions, and he reveals anxieties about the future of those ambitions that lurk just beneath the surface of his putative optimism. Gómara, meanwhile, abandons the theory of climates for the architecture of the continents as his preferred metageographical scheme, embraces the theory of American insularity, and thoroughly recapitulates the failures Spain has experienced in the South Sea, yet he nevertheless works even harder than Oviedo to contain the Pacific and the material challenges it poses to Spain's future ambitions. He may invent America in all but name, but he does not give up on the Enterprise of the Spicery. Instead, he finds a new way to dream of transpacific empire.

Damage Control in Oviedo's Book 20

As we shall soon see, the Oviedo of book 20 continues to think in terms inspired by the theory of climates, by mapping the New World and the Moluccas into a single tropical zone, particularly on the level of human society and culture. Nevertheless, whatever confidence Oviedo might have had about the natural coherence of the region, he has no choice but to relate the difficulties that the Spanish have experienced in attempting to extend their power across the South Sea to the Moluccas. The Loaysa and Saavedra expeditions were failures, no matter how one spun their stories, and they failed in large measure because of the challenges posed by Pacific storms and the continued puzzles of Pacific navigation. Oviedo thus finds it necessary, at last, to reckon with the enormity of that ocean and the difficulties it presented. However successful he might have been in his effort to contain the Pacific on the pages of the *Sumario* or in his account of the Magellan expedition, as we saw in the previous chapter, when it comes to the stories of Loaysa and Saavedra, he has no choice but to allow Pigafetta's indomitable ocean to come roaring forth. Oviedo's narrative thus becomes an exercise in damage control, turning a troubled Pacific crossing into a sign of heroic endurance and a military defeat into a moral victory, but ultimately failing to control for the unsolved mysteries of Pacific navigation, which threaten to keep Spain from realizing its transpacific ambitions.

Like the rest of the *Historia general y natural de las Indias*, book 20 rests on the privileged access to period sources that Oviedo enjoyed in his capacity as the crown's official chronicler of the Indies, a position he held from 1532 until his death in 1557, yet it also relies on the oral account provided by two men who had sailed with Loaysa and participated in the war with the Portuguese and their Moluccan allies, Andrés de Urdaneta and Martín de Islares. Along with the other Castilian survivors, they had surrendered when news of the Treaty of Zaragoza reached the Moluccas and returned to Iberia on Portuguese ships, but these two chose to continue on to the Indies in pursuit of new opportunities, in the company of Pedro de Alvarado. On their way to New Spain, they stopped in Santo Domingo and spoke with Oviedo about their experiences in the South Sea. Although we have no record of the interview, we do have a copy of Urdaneta's official report about the Loaysa debacle that can serve as a touchstone for understanding Oviedo's choices and the rhetorical strategy they are meant to serve (Urdaneta 1954).[4]

Book 20 invites us to picture these men swapping stories about exotic customs in distant parts of the Indies, particularly toward the end where it describes insular Southeast Asia in some detail. There we read that the locals sacrifice the servants of a dead king, so that they might serve him in the afterlife. Oviedo claims that Urdaneta witnessed this practice on an island in the Moluccas and then adds that he too had seen it carried out in "nuestras Indias y Tierra firme" (our Indies and Tierra Firme), specifically in Castilla del Oro and Cueva (in modern Panama) (103). The truth behind these ethnographic claims matters less to us than does the ideological effect of this moment of recognition or misrecognition. Oviedo insists that the same custom has been observed by reliable eyewitnesses on opposite sides of the South Sea, but does not feel compelled to explain why groups of people who live so far apart from each other should share the same custom. This is because the explanation is already there, in the shared tropicality of the Indies and the Moluccas, which is too obvious to require mention. Barbarity of this kind is precisely what one would expect from the sort of subpar human that is born of a tropical climate. It is the condition of possibility for the remark, which in turn serves as tacit confirmation of the validity of the theory.

The theory of climates may have appealed to Oviedo because of its epistemic power but there can be no doubt that it was also politically convenient. When one maps human difference according to latitude rather than landmasses, one need not worry about the possible implications that the breadth of the South Sea held for the architecture of the continents and any human geography that might go with it. If the Moluccas were just as tropical as Castilla del Oro, if their inhabitants were just as deficient, then the Spice Islands were just as available for conquest and colonization as Panama, no matter how much water lay between them. Nevertheless, as an official historian of the crown, Oviedo was well aware of the fact that it was not enough to "tropicalize" the Moluccas in order to defend Spanish claims to the islands. He also had to separate them from the equally tropical, transasiatic Indies of Portugal, and suggest that they could indeed be reached and governed from the New World, in order to convert them into a plausible western outpost of Spain's transpacific Indies.

Oviedo knew the cards were stacked against him. In his account of Magellan's efforts to win the support of Charles V, the historian notes that the Portuguese *fidalgo* brought the emperor "noticia de la India oriental, y de las islas del Maluco y Especería" (news of Oriental India, and the islands of Maluco and the Spicery), only to immediately qualify this statement by explaining that "aunque dije oriental, entiéndese que a España es oriental, pero aquí en estas nuestras Indias tenemos la Especería y el Maluco e sus islas al Occidente" (although I said oriental, it should be understood that it [India] is oriental from Spain's point of view, but that here in our Indies we hold that the Spicery and Maluco and its islands lie toward the west) (8). With this remark, we learn something about how Oviedo perceives the geographical imagination of his reader. He clearly assumes that the reader imagines the world along the lines that were rapidly becoming hegemonic in early modern Europe, placing America in an absolute West and Asia in an absolute East, treating the North Sea as an integrative body of water, and the South Sea as a geographical obstacle and an ontological boundary. He then strikes at this assumption by reminding the reader that the cardinal directions are relative, not absolute, and that in a global world, notions of what is near or far, here or there, east or west, change with one's own position. In this way, he unties the Spicery from its moorings in conventional notions of the world's geography and makes the islands available for remapping into a transpacific imaginary.

He does this in other ways as well, including the choices he makes when deciding how to refer to the inhabitants of the Moluccas. Like Urdaneta, Oviedo attends to their cultural particularity in myriad ways, but while Urdaneta indiscriminately refers to the islanders as either "indios" (Indians) or "moros" (Moors), Oviedo uses "indios" almost exclusively and even insists the Muslims of the Moluccas were few in number (Urdaneta 1954, 232, 235, 236, 237; Fernández de Oviedo y Valdés 1852, 101). By Oviedo's account, all of the inhabitants of the Moluccas are Indians, but few are Moors. In this way, Oviedo minimizes their cultural and religious ties to the absolute East and by implication, to Portugal.

Yet gestures like this are not enough, thanks to the very real challenges posed to Spain's transpacific ambitions by the active resistance of the Portuguese and their indigenous allies to Castilian hegemony, and by the continued difficulties that the South Sea posed to European navigation. Drawing on his interview with Urdaneta and Islares as well as written sources, Oviedo tells how Loaysa's fleet of seven vessels was scattered by a storm shortly after debouching from the Strait of Magellan, how the tiny patache, a small vessel not designed to cross open water by itself, made for New Spain, its fifty-person crew surviving on the daily egg laid by their solitary hen, how the flagship crossed the South Sea alone, its men constantly working the pumps while disease decimated their numbers, taking first Loaysa himself, then his pilot, who just happened to be none other than Sebastián del Cano, the man who had sailed the *Victoria* into Seville, and then at least thirty-five other men. Just as Magellan sighted the Unfortunate Islands but found no respite there, so the Loaysa survivors raise the Island of St. Bartholomew, but cannot drop anchor (50–52, 58–60). By the time the vessel reached the Moluccas, it bore a complement of only twenty-four survivors. The fleet had left Spain with 450 men. If the Pacific was placid but deadly in Pigafetta's account, here it is a roaring monster of an ocean that devours ships and men.

Like Pigafetta, however, Oviedo turns all this suffering on the South Sea to literary advantage by attaching it to one of the stated purposes of the *Historia general y natural de las Indias*. As he explains in his prologue to Part II, the text was meant to celebrate the achievements of the emperor's Spanish vassals on land and sea, "como valerosos y experimentos varones, sin excusarse del cansançio, sin temor de los peligros, con inumerables y exçesivos trabaxos é no pocas hambres, nesçesidades y enfermedades incontables" (as valorous and experienced men, without avoiding fatigue, with no fear of danger, suffering innumerable and excessive trials and not a few bouts of hunger, want, and too many illnesses to count) (1–2). Storms, hard work, sickness, and death are part of the job description of conquistadors. The greater the suffering, the greater the glory, and the greater the example provided to others. The problem with this rhetorical strategy, however, is that shipwreck stories tend to break apart "the monolithic practice of state writing"; they disrupt the forward momentum of "expansionist historiography," thereby exposing its operation as ideology, as Josiah Blackmore has argued (2002, 28). Thanks to the survival of Loaysa's flagship and its skeleton crew, Oviedo does not have a total shipwreck on his hands, but he has something dangerously close, a debacle of an ocean crossing with ominous implications regarding the navigability of the South Sea and Spain's potential to effectively control the Moluccas. Oviedo's response to the challenge is to press on just as the survivors did and to tell the tale of the military conflict that began when the Loaysa survivors reached the Moluccas, only to discover that the Portuguese had built a fort on the island of Ternate and had established an alliance with the island's indigenous rulers. Oviedo tells how the Spanish allied themselves with the Rajah of Tidore, who had traded with Magellan's men, sworn loyalty to Charles V, and was currently suffering retribution for his actions at the hands of the Portuguese and the Ternatans, and how they joined their allies in doing everything they could to disrupt the activities of the Portuguese in the islands.

The conflict that extended from the arrival of the Loaysa survivors until news of the Treaty of Zaragoza reached the Moluccas superimposed the rivalry between the two Christian kingdoms onto what Barbara Andaya has called the long-standing history of "complementary dualism" or "friendly hostility" between the two islands and their respective networks of allies built by marriage ties, ritual relationships, and economic exchanges. The two networks interacted sometimes as friends and sometimes as enemies (B. Andaya 2015, 164–65; see also L. Y. Andaya 1993). Conflict between Tidore and Ternate, therefore, was nothing new, and the war probably served indigenous ends just as much as it furthered European ones. Nevertheless, as we might expect, Oviedo treats the Spanish and Portuguese as its real protagonists, saying little about the agency and agendas of the Moluccans on either side, except when it serves his narrative agenda. Yet while his war story has clear protagonists, it does not have a clear plot. Oviedo presents the conflict as a series of dramatic but discontinuous moments of extraordinary violence, without making any causal links among them, much less investing the whole with a beginning, middle, and end. Anyone can tell that the situation of the Spaniards is becoming more desperate as time goes on, but there are very few prominent turning points, and no clear narrative arc tracing the overall progress of the war. The final surrender does not stem directly from the dire military situation, but from the intervention of a *deus ex machina*, the arrival of the news of the treaty.[5]

The annalistic structure may be the result of the origins of Oviedo's text in a conversation among military men, which could very well have impressed Oviedo with tales of individual adventures but given him little understanding of the larger strategic context. It is certainly not an inevitable consequence of the nature of the war itself. Urdaneta had no problem plotting his account of the conflict in his written report, by choosing to emphasize his own supposedly heroic involvement. Yet whatever the origins of Oviedo's annalistic approach, we can nevertheless detect its ideological effects. Even annals, after all, are meaningful in their lack of conventional plot structures.[6] Here, the lack of narrative structure can be understood as a strategy of avoidance. By resisting the effort to tell a clear story, Oviedo avoids placing an embarrassing emphasis on the reality of Portuguese victory and Castilian defeat. In other words, he avoids constructing a narrative that could have looked like an epic about the enemy, and that would have decisively mapped the Moluccas into the Portuguese East.

He also manages to snatch a certain kind of victory from the jaws of an obvious defeat. Although his discontinuous episodes do not develop a clear story, they nevertheless communicate a clear message: the Spaniards fought heroically and dealt with the locals honestly, while the Portuguese schemed perfidiously and acted cruelly toward the indigenous inhabitants. Time and again Oviedo figures Spanish conduct as examples of *osadía*, or "daring," that quality that was thought central to Spanish military prowess. The Spanish fight battles against unfavorable odds, refusing to retreat for fear of dishonor. They miraculously evade capture and take decisive action that turns the course of battles. Their daring impresses the locals and even the Portuguese (73–75, 85, 90–102). Brave almost to a fault, they are also fair in their dealings with their indigenous allies. When the hostilities end and the Spanish find out that their allies are now plotting to kill them, the Castilians forgive them for it, because, in the end, the allies did not get to carry out their plan (99). Although the Castilians eventually lose the military struggle, one is left with the impression that it is not because of their own lack of ability as warriors or colonizers.

The Portuguese, meanwhile, are portrayed as neither cowardly nor inept, but nevertheless emerge as inferior to the Spanish in both respects. While the text never accuses the Portuguese of cowardice, it also never praises Lusitanian bravery or daring. Instead, it points to the reputation the Portuguese have for cruelty toward the locals. The Castilians encounter that reputation when they first arrive in the islands, when a local ruler believes them to be "faranguis," or Portuguese, and thus knows them to be bad men, for "donde quiera que allegasen los faranguis, hacían mucho mal" (wherever the faranguis went, they did bad things) (63). Later, we read how the inhabitant of an island under Portuguese attack decides to kill his wife and children with his own hands rather than allow them to become slaves of the invaders (84). The text also emphasizes the treachery of the Portuguese in their dealings with the Spanish. The Lusitanian governor goes so far as to infiltrate the Castilian fort on Tidore and have its captain poisoned. The act earns one of Oviedo's few direct editorial interventions, which marshals the Bible and classical history to preach that there is no greater treachery than to kill by taking advantage of another's trust (80). Oviedo even returns to the incident at the end of his account, equipping book 20 with a moralizing conclusion about how only God can truly protect one from one's enemies, in this case the perfidious Portuguese (109).

Oviedo's Spanish heroes may have lost the war, but they have won the battles, and most importantly, they have demonstrated that they are the superior men, not primarily by comparison with the Moluccans, but rather by comparison with the Portuguese. This moral victory suggests that the outcome of the conflict, Portuguese control of the Spice Islands, is unjust, undesirable, and in need of correction. It also has the effect of assuring the reader that even though the project of transpacific empire poses serious challenges, Spain, or rather that military masculinity at the heart of Castilian collective identity, can make it across and can accomplish heroic deeds on its far shores, just as it can everywhere else. Whether or not the South Sea basin actually holds together as a naturally coherent space, Spanish men, Oviedo implies, are up to the task of uniting it into a single political space. The narrative thus ends on an optimistic note, with Oviedo's report that Alvarado, Urdaneta, and Islares are hatching plans for a new South Sea venture, aimed at other islands that promise to be just as rich as the Moluccas themselves (109–10).

Yet beneath Oviedo's optimism, I argue, lies a barely concealed anxiety. Blame for the Castilian defeat ultimately falls on the failure of Álvaro de Saavedra to find the return route to New Spain. Saavedra arrived in the Moluccas after the conflict had been raging for fifteen months. He had sailed from Mexico after news of Loaysa's troubles arrived in New Spain with the patache, under orders from Hernán Cortés and ultimately, Charles V himself. Oviedo says nothing about his troubled Pacific crossing, in which he lost two of his three ships. We only learn that a single galleon arrived in the Moluccas and was soon refitted for the return journey (84). We also learn that Saavedra, like the men of the *Trinidad* before him, failed to find the return route both on that occasion and on a subsequent attempt made shortly afterward. Oviedo states explicitly that his failure rendered the Spanish position in the Moluccas untenable. Cut off

from New Spain, the Castilians could not send for supplies and reinforcements, while every year the Portuguese received help from Malacca (89).
Saavedra's failure, therefore, is the real disaster of the entire narrative,
not Loaysa's troubled ocean crossing and not the failure of the military
struggle against the Portuguese. The failure of these voyages in search of
the return route decides the outcome of the war, and it calls into question
the very possibility of realizing Spain's transpacific ambitions. Oviedo,
however, says almost nothing about what happened on either the first
journey of roughly six months duration or the second of seven months,
which cost Saavedra his life.[7] His reserve may reflect the poverty of his
sources, but it nevertheless creates a poignant silence around what were
arguably the most important Spanish failures of the entire enterprise, as
well as a navigational challenge that remains to be solved if the new South
Sea venture is to have any hope of success.[8]

The problem of the return voyage, with all the questions it raises about
how Spain's valiant heroes are to survive on the far side of the South Sea
when they have no way of sending for help from home, makes a veiled re-
appearance at the end of the text. After finishing his tale, Oviedo provides
a lengthy description of insular Southeast Asia based upon Urdaneta's detailed knowledge of the region (100–107). It banishes the monsters and
marvels of past descriptions, and covers the region's geography, politics,
ethnography, and commerce with a level of detail and accuracy that had
no precedent in European print sources (Lach 1965, 1.2: 495). The concluding chapter, however, turns to a natural marvel, a small island where
one can gather inexhaustible amounts of a nut resembling an almond,
despite the fact that the island has none of the trees that produce them.
The text explains that the nuts are deposited by birds that consume them
on another island nearby that is covered with such trees and then fly to
the almond island to roost. The final sentence of book 20 relates how the
bedraggled Spaniards often turned to the island for sustenance during
the war with the Portuguese: "E aun entre las fatigas y nesçessidades que
los castellanos, à causa de la guerra con los portugueses, padesçieron en
el Maluco . . . muchas veçes les fué buen socorro, y parte de bastimento,
para su sustentaçion, estas almendras que tengo dicho" (And even among
the trials and wants that the Castilians suffered in Maluco on account of
the war . . . many times these nuts came to their aid, and provided the
supplies they needed to nourish them) (50).

In this way the description of the islands wraps back to the story of the
war and the difficulties caused by the inability to send for help to New
Spain. On the face of things, Oviedo provides an image of the Spanish
sustaining themselves with an endless supply of nutritious food, provided
by the islands without any real effort on their part. The marvel is natural,

but the tone is Edenic. Mention of it here, at the end of the text, provides some measure of hope that new expeditions will succeed, that the islands themselves will somehow provide. But the promise is a weak one, and the account of this marvel turns out to be a double-edged sword. While the almond island promises that the archipelago will provide, it also evokes an image of emaciated Castilians picking through bird droppings for the food they need to survive. In this way it turns into an emblem of what the South Sea had so often meant for those eager to carry empire to its far shores: hunger, desperation, and death. Beneath the surface of the marvel, as so often happens, we can glimpse the anxiety that it tries but fails to conceal, that the South Sea cannot be conquered, and that future expeditions will meet with disaster, just like their predecessors. We remember that the Moluccas are no Eden, at least in the Spanish experience so far. They are islands of death, islands of shit. Beneath the historian's wonder at the plenty of the almond island lies the tacit acknowledgment that the Pacific Ocean yawns between Oviedo's Indies and the Spicery, threatening to reduce daring Castilians to abject desperation and to tear the Moluccas away from Spain.

Gómara's World of Frustration

In Oviedo, the South Sea yawns between the New World and the Spicery, swallowing ships and men, forcing the historian to scramble for ways to keep them together. He makes implicit appeals to the shared tropicality of the Indies and the Spicery in order to keep the two places together as part of a single theater of empire, but knows that the gesture is not enough, so he attempts to convince his reader that Castilian masculinity is up to the challenge of forging an empire across the storm-tossed waters of the world's largest ocean. Gómara, by contrast, keeps the ship of Spain's flagging South Sea dreams afloat by jettisoning useless cargo. He abandons the theory of climates altogether, and with it any attempt to suggest that a transpacific empire can be naturalized by appealing to the shared tropicality of the people and places on either shore of the South Sea. As we shall see, he even proposes that the architecture of the continents as an alternative framework for mapping human difference. In this way, Gómara becomes a full participant in the ongoing invention of America, even while eschewing the name coined by Waldseemüller and Ringmann.

Much less invested than any Spanish imperial historian before him in constructing a sense of "Indian" continuity across the South Sea, Gómara can tell the truth about the sufferings endured by the Magellan expedition during its Pacific crossing. He can also avail himself of the general contrast that we saw in Pigafetta between the savage Brazilians and Patagonians on one side of the South Sea and the relatively civilized Asians on the other. He nevertheless writes as a Castilian patriot who is loath to give up on the kingdom's claim to the Spice Islands, so while he participates in the ongoing invention of America as a continent separate from Asia, he does not participate in the invention of the Pacific, or at least not fully. In Gómara's general history, the South Sea becomes what Oviedo cannot allow it to be, an ontological boundary between the New World and Asia, but it never becomes what Oviedo admits it to be, an almost insurmountable physical obstacle between the New World and the Spicery. While Oviedo's account of Loaysa's fraught Pacific crossing brings Pigafetta's Pacific roaring out of the domesticated confines into which von Sevenborgen, Martyr, and Oviedo's *Sumario* had stuffed it, Gómara tucks it back into the straight jacket of imperial cartography, insisting on its relatively small dimensions, and aggressively reasserting Spain's claim to the Moluccas and much else besides.

Climatic theory gets short shrift at both the beginning and end of the first volume of the *Historia general*. In one of a series of chapters dedicated to general principles of cosmography and geography, Gómara expounds upon and rejects the notion that there exist regions of the globe that are uninhabitable because of extreme climates, citing both Scripture and the evidence of experience. Spanish voyages through the Torrid Zone and into the Frigid Zone have made "espantajos de los antiguos" (scarecrows of the ancients) and of their baseless ideas about the geographical distribution of life on Earth (chap. 3).[9] Near the end of the volume, Gómara returns to the theory of climates in a chapter devoted to the skin color of the Indians. It begins by marveling that human beings are not just black or white, but every shade in between, thereby setting us up for an argument that maps gradations of skin color onto gradations of latitude, only to take a different tack altogether by noting that Seville, the Cape of Good Hope, and the Platte River are all equidistant from the equator, yet their inhabitants are white, black, and chestnut in color, respectively. It also adds that the inhabitants of the Torrid Zone in Africa and Asia are black, while those in parts of the Indies that he lists by name are not. Although Gómara never says so explicitly, he is clearly calling into question the explanatory power of the theory of climates, by pointing out that observed realities about human diversity and its geographical distribution do not correspond to the expectations that the theory sets up. In a bit of pious irony, he suggests that the variety and distribution of human skin colors is a mystery of God's inscrutable design (chap. 216).

Gómara is therefore free to follow Pigafetta in emphasizing the overall contrast between the people who live on one side of the South Sea and those who live on the other. In his account of the Magellan expedition, the encounters with the "indios" of Brazil and Patagonia bear all the hallmarks of encounters with New World savages: cannibalism, nudity or near nudity, the use of hammocks, the skillful wielding of bows and arrows, body painting, unequal economic exchange, Indian wonder at European ships and fear of European firearms, communication by signs, and, in the case of the Patagonian giants, an appearance so fierce and brutal "que no semejan hombres" (they do not resemble men) (chap. 92). The encounters on the other side of the South Sea, however, read very differently. None of the islanders are "indios," and no one paddles a "canoe." Gómara compares the thievery of the inhabitants of the Ladrones to that of the gypsies and suggests that the inhabitants might have actually come from Egypt. Although the inhabitants of "Zebut" (Cebu, Philippines) wear little or no clothing and paint their bodies, they use porcelain dishes, and, more importantly, they are organized into recognizable polities ruled by a king with a name, Hamabar, with whom Magellan can communicate via his Malay interpreter and slave, Enrique. The sultan of Tidore is a Muslim, who swears allegiance to Charles V on his copy of the Koran and finds the smell of pork repugnant. The people of Brunei, in turn, write on pages made from tree bark, like those used by the Tartars, and in fact get their writing material from the Tartars. There are marvels (large pearls) and novelties (coconuts) to report, but there is no sense that everything is new and unprecedented, as there is in the New World (chap. 96). This is because we are no longer in the New World, but back in the Old. Later chapters on the natural history of the spices and the history of the spice trade in ancient times reinforce this impression that the places Magellan reached have always been part of the reader's universe, but even without this help any reader can recognize that the cannibals and savages of the New World are quite different from the relatively civilized people of the Archipelago, as Gómara calls the islands of Southeast Asia (chaps. 97, 107).

This emphasis upon the difference between politic islanders and savage Indians reaches a head when the expedition visits the Sultanate of Brunei. The episode figures in von Sevenborgen, Pigafetta, and Oviedo as well, but Gómara adds key details that lend the episode unique significance. Gómara devotes an entire chapter to it, dwelling on the size and sophistication of the capital city, and the obvious wealth of the sultan, Siripada. We read how the eight men who seek an audience with the sultan spend the night in his palace, sleeping on cotton mattresses. They ride through the city on elephants and gawk at the sultan's retinue of gentlemen dressed in silk, wearing golden rings studded with precious stones, and sporting daggers finished in gold, gems, and pearls. Their reaction to all this pomp and finery is the most telling detail of Gómara's whole account:

> Viendo los españoles tanta majestad, tanta riqueza y aparato, no alzaban los ojos del suelo, y hallábanse muy corridos por su vil presente. Hablaban entre sí muy bajo, de cuán diferente gente era aquélla que la de Indias, y rogaban a Dios que los sacase con bien de allí. (Chap. 95.)

> [Seeing such majesty, so much pomp and wealth, the Spaniards could not raise their eyes from the floor, and they felt deeply ashamed of their filthy appearance. They whispered among themselves about how very different these people were from those of the Indies, and prayed to God to get them out of there safely.]

When their turn comes to present their gifts to the sultan, "abrieron su presente con harta vergüenza" (they opened their gift with deep shame). I know of no other moment anywhere in Spanish writing about the Indies that reports embarrassment on the part of Spaniards appearing before a non-European, not even before Moctezuma or Atahualpa. This reaction to the opulence of the sultan of Brunei tells the reader that Magellan's men are not in Kansas anymore, or anywhere else in America for that matter.

As we saw in the previous chapter, both Pigafetta and Martyr paint a similar contrast between the peoples of the New World and those of the Archipelago, but neither of them maps that contrast onto a continental architecture that insists upon American insularity. Gómara, by contrast, takes precisely this step. After rejecting the theory of climates as the foolishness of the uninformed ancients, Gómara's introductory chapters remap the world by first casting the uniform space of the Ptolemaic grid over the entire globe and then mapping the three world-islands into its empty space. The text mentions legendary voyages from antiquity that proved one could sail from Cádiz to India by going around either the northern or southern periphery of the *orbis terrarum*, and then traces an imaginary voyage, point by point, around the shores of the continental New World. The text adamantly insists that the coast of California turns toward the northeast and eventually connects to the coast of Labrador, "para llegar a cerrar la tierra en isla" (in order to enclose the land and make an island). It then adds that there are other such islands, including the one that is believed to occupy much of the southern hemisphere (chap. 12). The Old World, the New World, and the *Terra Australis*, this is the basic architecture of Gómara's globe.[10]

This does not mean, however, that Gómara abandons the more conventional division of the world into Europe, Africa, Asia, and America. On the contrary, this is the architecture that matters when he returns to the issue of the geographical distribution of human difference at the end of the volume. Inconsistency, apparently, was not a problem for Gómara, particularly when he wanted to score points for the moderns against the ancients. Just before dealing with the puzzle of human skin color and its unpredictable arrangement in global space, Gómara offers a brief reflection, "On the Bread of the Indians," noting that the Indians, too, have their bread, even if it is made from corn rather than wheat. This leads him to remark that bread is universal, but that different branches of the human family make it from different grains. While Indians make it from corn and Europeans from wheat, Africans make it from rice or barley and Asians from rice (chap. 215). Gómara thereby maps human cultural variation onto the four parts of the world, rather than onto the climatic zones. Although he does not suggest that the architecture of the continents can account for the causes of that variety, as the theory of climates claims to do, he suggests that it provides a better framework for mapping human difference, thereby pointing toward the eventual consolidation of the architecture of the continents as the dominant framework of European geography in the work of Abraham Ortelius (fig. 18).

Willing as he is to sunder America from Asia both geographically and anthropologically, Gómara has no trouble relating the failure of all those efforts to follow the Mexican coastline to Asia or to discover golden cities in what is now the American southwest.[11] More to the point, he also has no trouble sharing details about the difficulties that have marked Pacific crossings. His brief account of Magellan's crossing is not as lurid as that of Pigafetta, but it is nonetheless honest about the deaths from scurvy:

> Navegó cuarenta días o más sin ver tierra. Tuvo gran falta de pan y de agua; comían por onzas; bebían el agua tapadas las narices por el hedor, y guisaban arroz con agua del mar. No podían comer, de hinchadas las encías; y así murieron veinte y adolecieron otros tantos. Estaban por esto muy tristes, y tan descontentos como antes de hallar el estrecho. Llegaron con esta cuita al otro trópico, que es imposible, y a unas isletas que los desmayaron, y que las llamaron Desventuradas por no tener gente ni comida. (Chap. 93.)

> [He sailed for forty days without sighting land. He suffered great shortages of bread and water. They ate by the ounce. They pinched their noses when they drank water, because of the stench, and they cooked their rice with sea water. They could not eat, for their gums were swollen, and twenty men died of this illness, while another twenty suffered from

it. They were thus quite sad, and as despondent as they had been before discovering the strait. Impossibly, with all these cares, they reached the other tropic, and some islets that disappointed them greatly, and they called them Unfortunate because they lacked people or food.]

Subsequent chapters tell briefly of the Loaysa, Saavedra, and Villalobos debacles, and they mention that Hernán Cortés searched for but failed to find the rich islands in the South Sea (chaps. 102-3). A simple pattern emerges involving the death of men, the loss of ships, and the successful interference of the perfidious Portuguese. The various expeditions launched during the 1520s that never left the Atlantic Basin, like those of Estevão Gomes and Sebastian Cabot, receive no mention at all until the end of the narrative arc, when the sudden accumulation of fruitless enterprises serves to communicate quite emphatically that the Enterprise of the Spicery has cost Spain a great deal but has won it nothing of value.

Given his embrace of American insularity, his willingness to depict the difficulties suffered by Magellan, Loaysa, and the rest, and his clear acknowledgment that Spain has failed to extend its empire across the South Sea, one might expect Gómara to say something about the Pacific's indomitable breadth and emptiness, yet, surprisingly, he does nothing of the sort. Instead, he echoes Oviedo's belief, as expressed in the *Sumario*, that the distance from Central America to the Moluccas is quite manageable, and he suggests that Spanish ships should head to the Spice Islands from there, avoiding the route through the Strait of Magellan. The long southern detour through the strait, after all, was the reason that Magellan and Loaysa had suffered so much on their way to the Spicery. Gómara, therefore, embraces the theory of American insularity but also contains the Pacific. Although he separates America from Asia, he keeps the two continents quite close to each other, arranging them on either side of a narrow, navigable South Sea. Historians who find a causal connection between Magellan's discovery of the Pacific and the triumph of the theory of American insularity would therefore do well to reread Gómara.

There was precedent for this in European cartography. During the 1540s, the influential Swiss cosmographer Sebastian Münster mapped the South Sea as a very narrow expanse and filled it with islands that bridged the waters between the New World and Asia in the latitudes north of Magellan's route (fig. 24). Yet even if we set transpyrenean cartography aside and focus exclusively on the Spanish context, we find that, contrary to León-Portilla's argument, belief in a narrow South Sea and in the theory of Amerasian continuity survived the failures of the 1530s and early 1540s, just as they had survived Magellan's putative discovery of the

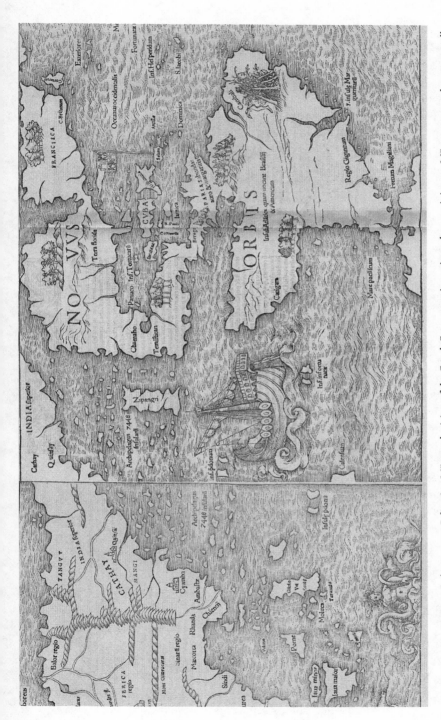

FIGURE 24. As Thomas Suárez notes, Sebastian Münster's vision of the South Sea comes into view when we join two Münster maps that are usually considered separately, the *Tabula orientalis regionis* and the *Tabula novarum insularum*. The copies joined here are taken from Claudius Ptolemy, *Geographia universalis, uetus et noua* (Basil, 1545). Photograph: Courtesy of the Renaissance Exporation Map Collection, Stanford University Libraries (57069282).

Pacific. The most important chronicle of the Coronado expedition was not written until the 1560s, yet it still refers to North America as "la India Mayor" (Greater India), and it likens the indigenous groups encountered by the expedition to Bedouins, Turks, and Moors, in order to suggest that Cathay and Zipangu had been close at hand all the while (Flint 2005, 381–82). The principal chronicle of the Villalobos expedition, of uncertain date, makes a passing reference to the geography of the South Sea that suggests its author imagines it as a large embayment in an Amerasian coastline stretching from New Spain to China (Anonymous 1983, 48).

The Villalobos expedition had ended in failure, but not because of any trouble on the Pacific crossing. The fleet made it from Mexico to Mindanao without incident, pioneering the transpacific route that would later be used by Miguel López de Legazpi. It then fell victim to the same trap that had snared its predecessors, the failure to find a return route to New Spain. The surviving pilots reported that the distance across the South Sea at the latitude Villalobos had sailed was shorter than had been previously thought.[12] No surviving nautical chart of Spanish origin registers that abbreviated distance, but several maps derived from Spanish sources do. Among them are two of the maps that León-Portilla interprets as evidence of increasing awareness of the insularity of America. The Cabot world map of 1544 puts about 90° of longitude between Peru and the Moluccas, and it clearly remains open to the theory of Amerasian continuity (see fig. 22). Battista Agnese's chart of the Pacific from around the same time puts the distance between Peru and the Moluccas at about 100° of longitude (see fig. 21). Its treatment of the Gulf of California makes the progress that Cortés made along the Pacific shores of Mexico look like a victory in the ongoing discovery of the coastline of Amerasia. By abbreviating the distance separating New Spain from China, Gómara was marching to the same beat as other Spanish mapmakers. It was only in his insistence on the insularity of the Indies that he was listening to a different drummer.

As the South Sea shrank in the Spanish imagination, Spaniards sought new opportunities to expand westward to the Moluccas or adjoining areas. In New Spain, a suit was filed against the king of Portugal, claiming he had violated the terms of the Treaty of Zaragoza by building new fortifications in the Moluccas (Varela 1983, 95). In metropolitan Spain, the Cortes of Castile petitioned the emperor to redeem the Moluccas from Portugal by returning the 350,000 ducats that had been presented to him in 1529. Gómara himself is our only source about the incident (chap. 105). According to him, Charles V sent word from Flanders to Valladolid ordering all discussion of the matter to cease immediately. The *Historia general* criticizes the decision by reporting the chilling effect of the gag order, "de lo cual unos se maravillaron, otros se sintieron, y todos callaron" (at which some marveled, others became resentful, and all shut their mouths) (chap. 105). In making this extraordinary public criticism of the emperor, Gómara shows his hand as a Castilian patriot who sees continued adherence to the arrangements made at Zaragoza as an imperial policy at odds with the interests of the kingdom of Castile. According to the new understanding of the South Sea that emerged out of the Villalobos catastrophe, the Moluccas were indeed within the Castilian hemisphere. According to Castilian patriots like Gómara, Spain would do well to renew its claim to the islands, despite all the failures to colonize them.

Gómara even provides a solution for the problem of how to get from Spain to the Spicery. In a chapter that has earned a well-deserved place in the prehistory of the Panama Canal, Gómara presses Charles V to build a maritime passage through one of four spots in Central America, so that ships can sail from Spain to the Spicery unimpeded (chap. 104). This was not the solitary fantasy of a Spanish priest who had only seen Panama on a map and who knew nothing first-hand about its rugged terrain. To begin with, Gómara is not talking about building infrastructure that could accommodate the large, deep draft vessels of the industrial age, but rather the small, relatively shallow draft vessels of his own time, ships that could make their way far up the rivers of Central America, which would form an integral part of the passage he was proposing. The project was undoubtedly ambitious, but not beyond the pale of capabilities of the day. "Sierras son, pero manos hay" (There are mountains, but there are also hands), Gómara insists. He was not alone, furthermore, in believing the project was feasible and in urging the crown to carry it out. By 1550, several individuals had made similar proposals, going so far as to survey the terrain and work out many of the technical details pertinent to different sites in Mexico and Central America.[13] Gómara, therefore, was actually endorsing in the very public venue of print a project that had been bandied about in official circles for several decades and that would continue to attract the attention of the crown for several more.[14]

Nevertheless, there is something shrill about Gómara's writing in this part of the *Historia general*. His narrative of Castilian failure and Portuguese success in the contest over the Spice Islands seems out of place in a volume originally titled *Hispania Victrix* and dedicated to celebrating the triumph of Spanish imperialism in the Indies. Within the larger context of the volume, the Enterprise of the Indies looks like an unfinished project. Gómara's eagerness to cut through the brute physical reality of the New World by building the maritime passage that nature failed to provide speaks of a deep sense of frustration with the hand Castile had been dealt by both nature and emperor. Most importantly, however, his account of the distance from Mexico to the Moluccas abandons for desperate hyperbole sober assessment of the available sources. Gómara provides no numbers and cites no maps to support his contention that the distance is short enough to be practicable, despite the access he enjoyed to the planispheres of the cosmographers in Seville. Instead, he simply insists that if the original line of demarcation had been drawn far to the east of its actual position, the Moluccas would still be in Spanish territory (chap. 101). For this to be true, the islands could not possibly be located where the Seville planispheres place them, just inside the Castilian hemisphere, but would have to lie deep in Castilian territory, deeper perhaps than they appear on any extant map. The assertion, therefore, is no serious assertion about the world's geography, but rather an exaggeration deployed for rhetorical effect, meant to convince the reader that everything, including global geography, is on the side of Spain's transpacific ambitions. Perhaps it is also a sign of Gómara's own frustration with that very geography and with the disappointments of the Enterprise of the Spicery to date.

Let us return, then, to the remarks I made at the outset of this chapter about the ways that Oviedo and Gómara map the Indies. As we saw, Oviedo leaves the geography of Tierra Firme open in the west, thereby revealing that he shares the agnosticism of the Seville planispheres regarding the question of its insularity or its continuity with Asia. He also clings to the theory of climates as a way of constructing similarities and continuities among territories separated by an indomitable Pacific that finally makes an appearance in his writing, disrupting his earlier attempts to contain it. Gómara, meanwhile, insists on the insularity of the New World, as well as upon its alterity, thereby participating in the invention of America with a degree of clarity that we never find in Oviedo, yet he insists even more adamantly than any imperial historian we have seen so far that the South Sea is relatively narrow and well-contained by the continents that define its opposing shores. While he gives up on the idea that the Moluccas are a western extension of the same tropical world that the Spanish have encountered in the Caribbean and Tierra Firme, he insists even more adamantly than anyone before him that global geography is on Spain's side, and that the Moluccas not only lie on the Castilian side of the antimeridian, but deep inside the Spanish hemisphere. If my initial remarks seemed to plot Oviedo and Gómara into a Whiggish narrative about the ongoing invention of America, it should now be clear that the relationship between the two historians is better understood as a peculiar discursive exchange. Oviedo allows the Pacific to break out of the confines that imperial historians, himself included, had created for it, but cannot therefore admit to any fundamental difference between the people and places on its opposing shores, or to the possibility that those shores cannot be brought together under the same crown. Gómara, by contrast, does admit to existence of fundamental differences between the New World and Asia, but must therefore contain the Pacific with even greater insistence than anyone before him, holding on to Castile's legal claims even when action has been rendered impossible by imperial decree. It is discursive exchange rather than progress that marks the movement from Oviedo to Gómara, a discursive exchange that speaks of their shared interest in holding on to the Indies of the West and thereby keeping Spain's transpacific ambitions alive.

The Cartography of Shipwreck

The contrast between these two authors can be captured by comparing two maps of the New World published outside of Spain, but constructed out of source materials provided by the Casa de la contratación. The first is the "La carta universale della terra firme e isole delle Indie occidentale, cio è del mondo nuovo" (Universal map of the mainland and islands of the West Indies, that is, the New World), designed by the Venetian printer Giovanni Battista Ramusio to accompany an anthology of Italian translations of Martyr, Oviedo, and an anonymous chronicler of the conquest of Peru (fig. 25).[15] Based on the work of Diogo Ribeiro, it was the first widely available print map to depict a continuous coastline stretching from Labrador to the Strait of Magellan. The second is the "Americae sive qvartae orbis partis nova et exactissima descriptio" (New and exact description of America, also known as the Fourth Part of the World), printed in Antwerp by Hieronymous Cock on the basis of a prototype provided by the Seville cosmographer Diego Gutiérrez (see fig. 8). It is the only sheet map of the New World printed during the sixteenth century with authorization from the Spanish crown. Like the general histories I have been discussing, both maps emphasize the key areas of Spanish exploration, conquest, and colonialism, that is, the Caribbean, Mesoamerica, and the Andes, although they inevitably devote a lot of space to other areas as well, most notably Brazil. Yet while the general histories hold on to the Indies of the West by including narratives of Spain's efforts to claim and control the Moluccas, these maps simply crop those Indies out of the picture, depicting only that part of Spain's empire that was coming to be known as America. Like those histories, however, the maps do not entirely let go of the Indies of the West, even as they seem to participate in the ongoing invention of the fourth part of the world.

Cropping the Moluccas out of the picture must have made economic sense to Ramusio. To reproduce the whole of Spain's Indies, he would have needed either a larger sheet, which would have raised costs, or a

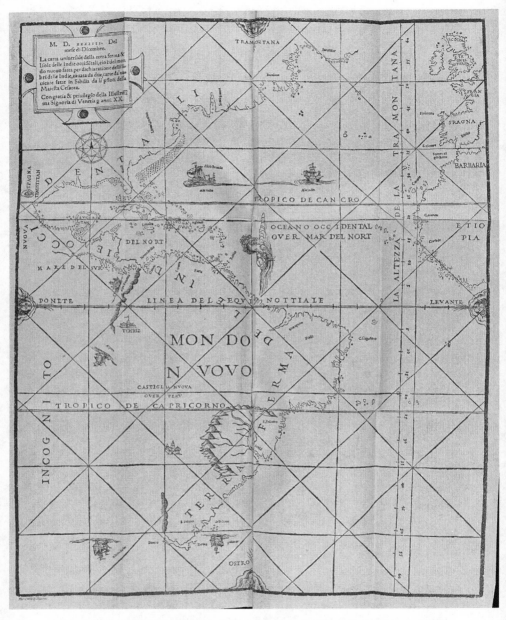

FIGURE 25. The first widely available print map depicting a continuous coastline from Newfoundland to the Strait of Magellan. Giovanni Battista Ramusio, *La carta universale della terra firme e isole delle Indie occidentale, cio è del mondo nuovo,* from *Libro primo della historia del' Indie Occidentali* (Venice, 1534). Photograph: Courtesy of the Tracy W. McGregor Library of American History, Albert and Shirley Small Special Collections Library, University of Virginia (call no. A 1534 .A54).

smaller drawing, which would have sacrificed detail or legibility for the sake of including a lot of empty space, all to inevitably take a side in a territorial dispute in which Ramusio had no stake, and which could potentially hurt sales. This does mean, however, that he rejects the Spanish concept of the Indies as a transpacific expanse and produces instead a map of "America." The typography clearly identifies only South America as the "New World" and identifies this landmass as part of a larger whole, the "Terra Firma delle Indie Occidentali" (The Mainland of the West Indies), suggesting that what we call North America is not part of the New World. It nevertheless takes no position on the question of whether or not this landmass is an extension of Asia. Not only does Ramusio crop the image to eliminate the space at issue, but he also includes the word "Incognito" along the map's left-hand edge to eliminate any doubts as to his agnosticism regarding the matter.

More importantly, Ramusio also includes a small but significant assemblage of words and images that serves to tie the Moluccas to the "Indie Occidentali." The Strait of Magellan is named as such for the first time on any printed map, while two ships appear on either side of the strait, one labeled "alle Moluche" (to the Moluccas), recalling the ships that traverse Ribeiro's Pacific. The assemblage invokes the transpacific voyages of Magellan and Loaysa, reminding the reader of the existence of the Moluccas somewhere offstage, in that largely unknown expanse to the west, and perhaps of Spain's ambition to control the islands and to trade in spices along the route Magellan pioneered. The map, therefore, may not depict the entire hemisphere claimed by Spain, as its sources must have, but it nevertheless invokes the most important place in that hemisphere outside Tierra Firme, the Spice Islands. More importantly, it depicts those islands as a destination that one reaches by traveling west, through the Strait of Magellan, and makes no reference to the possibility that the Moluccas might be considered oriental. In the end, the Moluccas are indeed a part of the geography this map depicts, as a western extension of the Indie Occidentali. So too are whatever other places might lie to the west, in the *terra incognita* between Mexico City and the Spicery.

The Gutiérrez/Cock map is much more reticent about the Indies of the West, and much less ambivalent about the status of the New World. Like Ramusio, Cock and Gutiérrez leave the unknown reaches of the northern landmass out of the picture, by cutting the image off at the longitude of the Gulf of California, but unlike the Ramusio map, the Gutiérrez/Cock clearly refers to the resulting geography as "America" and calls it the fourth part of the world. The map does nothing to draw attention to the Moluccas or to any other transpacific objective of Spanish imperialism. Yet even more telling than this cartographic silence is the

loud, clear statement made by the iconography. The fearsome giants of
Patagonia watch over the Strait of Magellan, guarding the entrance to the
waters west of America. There we find ships foundering in storms, others facing menacing sea creatures, and one vessel sitting becalmed in the
heat of the tropics. The *Insule unfortunate* appear just above the Tropic
of Cancer, reminding the reader that the way west is bereft of wealth, but
rich in trouble. The waters east of America, meanwhile, feature allegories
that speak of the European conquest of the seas, as well as numerous
vessels sailing the routes pioneered by Spain and Portugal. Although a
naval battle rages in the southern ocean, no ship faces peril from the dangers of the ocean itself. The map, therefore, does not just figure the New
World as America, but also maps the fourth part of the world into the
dominant world image that presents the Atlantic as a domesticated ocean
basin that successfully integrates the Old World with the New, while simultaneously figuring the Pacific in terms that recall the old Ocean Sea,
an impassable maritime boundary between the world-islands.

We know very little about this map and why it was made, but we
can nevertheless imagine both epistemic and political reasons for its approach to New World cartography. If the source materials provided by
Diego Gutiérrez looked anything like the surviving maps produced by
cosmographers in the Casa de la contratación, they are unlikely to have
called the New World "America" or have referred to it as the fourth part
of the world. The name does not appear on any of the surviving Seville
maps from the first half of the sixteenth century, not even on the 1550
chart of the Atlantic by Gutiérrez himself, or the heavily inscribed 1551
planisphere made by his brother and fellow mapmaker Sancho.[16] It is
therefore reasonable to assume that the title of the Gutiérrez/Cock map,
which calls the New World "America, or the Fourth Part of the World,"
was imposed by the Flemish engraver, following the prevailing trends
of northern European cartography, where the America idea had always
enjoyed more popularity. The map thus exemplifies what could happen
when knowledge traveled from Seville to Antwerp, from southern Europe to northern. Geographical knowledge makes the journey unaltered,
we presume, but it gets interpreted differently by the community that
imports it. A perfect example of cartographic translation.

Politics, however, do not get lost in the translation. Gutiérrez's cartography may have moved away from Spain, and its unique interest in a
New World empire, to the Low Countries, where cartography was becoming a more commercial enterprise and therefore a more politically
neutral one, but it has not left the domain of Philip II and is still very
much invested in the politics of Hapsburg imperialism. The map seems
to have been designed to make a clear and very public statement about

Hapsburg territorial prerogatives in the New World, particularly with regard to the competing claims of France and Portugal, in the wake of the
1555 Treaty of Cateau-Cambrésis, which ended a protracted war between
France and the Hapsburg Monarchy. Its use of "America," and its silence
regarding so much as the possibility of Spanish transpacific empire, could
thus be understood as a diplomatic gesture, meant to avoid ruffling the
feathers of the Hapsburg's Portuguese relations. This may be the reason
why the map also remains silent about the single feature that figures so
prominently in so many other Spanish maps, the line of demarcation in
the Atlantic.

Nevertheless, the Gutiérrez/Cock map cannot shed its debts to the
cartography of the Casa de la contratación, debts that speak in subtle
ways of the South Sea dreams that the map attempts to erase. Specifically, the map's scale of longitude suggests that it was built atop some very
Spanish assumptions about global geography. The Gutiérrez/Cock map
follows Ptolemy and the general trend in Renaissance cartography by
counting longitude from the Canary Islands eastward all the way around
the world, placing the mouth of the Amazon at roughly 317° of longitude.
This is the approximate location of the line of demarcation on most maps
that depict it. If we extrapolate from this number to a notional global
geography, we find that the antimeridian, in the implied world of the
Gutiérrez/Cock map, falls somewhere around 137° east of the Canaries.
We cannot say with any certainty what Gutiérrez or Cock would have
placed at this longitude had they constructed a map of the world rather
than one of America, but we can make an educated guess. On a modern
map of the world, when we move 137° east from the island of Hierro, we
find ourselves very near the longitude of Manila, but since early modern
cartography tended to overestimate the breadth of Eurasia, the same displacement on just about any European world map made between 1540
and 1580 puts us somewhere in Asia, but west of the Philippines. Depending on the map, it might put us near the Moluccas, or near the mouth of
the Ganges River, or anywhere in between. On Spanish maps like those of
Sebastian Cabot (1544), Alonso de Santa Cruz (1540s), and Sancho Gutiérrez (1551), we find ourselves somewhere in the Bay of Bengal. A privileged reader familiar with Spanish cartography of the globe, therefore,
would recognize in the longitudes of the Gutiérrez/Cock map continued
adherence to a Spanish vision of the world, one that mapped the entire
South Sea, from Peru to Malacca and possibly beyond, as an enormous
lake within the Castilian half of the world.

What would that reader have made of the map's agnosticism regarding
the geography of the *terra incognita* in the northwest? As we saw above,
Spaniards in New Spain continued to speculate, well into the 1560s, that

the country where they found themselves might very well be continuous with continental Asia, so it is not unreasonable to imagine a reader
in metropolitan Spain or the Low Countries wondering the same thing
when faced with the Gutiérrez/Cock map. For that reader, the map's embrace of the America idea, if not the theory of American insularity, did
not foreclose the possibility of new, more properly Asiatic conquests at
some time in the near future, in the unknown landmass that stretched
westward from the area depicted by the map. Yet even if that reader, like
Gómara, embraced the idea of American insularity and had no hope for
the prospects of a new march into the west, he or she would have realized
that East Asia was not too far from America, that it was, in fact, no more
than about 100° west of California, within Castilian territory, closer to
Mexico City than that metropolis was to the Cape Verde Islands. Like
Gómara, he or she may have found that distance seductive and wondered
whether the dire warnings about South Sea navigation posed by the iconography should be heeded or ignored. The crown, in fact, was already
ignoring them. As the ink was drying on the Gutiérrez/Cock map in Antwerp, a new South Sea venture was being prepared in New Spain, by
order of Philip II. It would succeed where others had failed and would
transform Spain's relationship with the Indies of the West.

5

Pacific Conquests

The arrival of the galleon *San Pedro* in Acapulco on an October day in 1565 marked a turning point in the history of the Spanish Pacific. It was piloted by Andrés de Urdaneta, the same man who had sailed with Loaysa and traded war stories with Oviedo years before, only to then join the Augustinian order and live as a friar for almost three decades before answering the call of Philip II to assist with the colonization of the Philippine Islands. Urdaneta's task was to discover the return route from insular Southeast Asia that had eluded previous expeditions, while the fleet's commander, Miguel López de Legazpi, established a foothold in the archipelago. Urdaneta succeeded, providing Legazpi's fledgling colony with the lifeline it needed to survive. Within a few years, Legazpi conquered the sultanate of Maynila, securing its spectacular harbor, its rich hinterland, and its established commercial ties with Ming China for exploitation by Spaniards. Urdaneta's lifeline became the pipeline that funneled Asian luxury goods to New Spain, where they were exchanged for American silver at tremendous profit to those involved. The friar's accomplishment made it possible to transform Spanish Manila from a modest port of call on a local trade route into a major entrepôt of global commerce. It made it possible, in other words, to build the world of the Manila Galleons, with its rich new patterns of exchange between Asia and the Atlantic.

The Spanish had finally accomplished what Columbus had originally set out to do, yet the event met with little fanfare. Unlike Columbus's letter to Luis de Santángel or Cortés's letters to Charles V, Legazpi's narrative of the expedition was never published.[1] No poet rose to become the Homer of Urdaneta's odyssey. Only a brief anonymous pamphlet, published in Barcelona in 1566, announced to the Spanish reading public that an expedition had sailed from New Spain "para descubrir las islas de la Especería que las llaman Philippinas por nuestro Rey" (to discover the Spice Islands that are called the Philippines after our King; Hidalgo

Nuchera 1995, 210).[2] It tells a dramatic story of successful skirmishes with the region's Muslims, of the everlasting friendship forged between the Spanish captain and the king of the island of Cebu, and of the wondrous discovery of a Flemish Christ figure that had been left in these very islands by Magellan and his men, a certain sign of the providential nature of the whole enterprise. Although the pamphlet is clearly based on Legazpi's narrative of the expedition, it names neither him nor Urdaneta, emphasizing instead the promise of the new colony itself. The Philippines, claims the text, sit at the heart of a vast archipelago of roughly 75,800 islands populated by civilized people and brimming with gold and spices. Ginger, cinnamon, gold dust, jewels, and other riches, it adds, have already reached Seville from the islands, in the ship that brought news of the conquest.[3] The pamphlet thus assures its readers that all the old dreams surrounding conquest in the South Sea have come true, but we can only wonder how many those readers might have been, since only a single copy of the pamphlet has come down to us. We have no way of knowing if this is because few were ever printed or because the many others were lost to the ravages of time. In any case, the decades that followed failed to follow up on the Barcelona pamphlet. Few publications specifically devoted to the Philippines would appear in either Spain or Spanish America until the early years of the seventeenth century, and these would not circulate widely.

There were various reasons for this silence. One was the tight control that Philip II was known to have exercised over knowledge of his overseas possessions. Even if Legazpi had wanted to publish his narrative of the expedition, the king would not have allowed it. Another reason, however, was the sense of disappointment that soon surrounded the new colony. Legazpi and his men approached the conquest of the Philippines as previous conquistadors had approached the colonization of the New World, establishing *encomiendas* and harnessing indigenous labor to extractive ends, but they failed to locate the archipelago's rich sources of gold, and they found no spices on which to grow wealthy. The conquest of Manila and the associated shift from resource extraction to long-distance trade as the basis of the colonial economy allowed some to succeed, but theirs was not the kind of success that lent itself to the military ethos at the heart of conquistadorial narrative. Camões may have found a way to write epic poetry about a commercial venture, but no Spanish poet ever did. The crown, meanwhile, soon found itself spending more to defend and maintain the colony than it earned from tribute. Rather than celebrate the new acquisition, the king's counselors debated whether or not it should be abandoned or else traded to Portugal for Brazil.

Knowledge about the Philippines nevertheless made it into writing and onto maps, sometimes in the context of print works dedicated to other parts of East and Southeast Asia, namely China, but more often in manuscript materials of numerous kinds that circulated in both secular and ecclesiastical circles. The conquest of the islands, moreover, had a clear impact on the way that Spanish officialdom mapped Spain's overseas possessions and imagined Spain's imperial future. This chapter examines the impact of the conquest of the islands on the geography and cartography prepared for use by the king and the Council of Indies by Juan López de Velasco, who served as the council's first "Cosmographer and Chronicler Major of the States and Kingdoms of the Indies, Islands and Mainland of the Ocean Sea." It examines how Velasco remapped the Spanish Indies as a transpacific space in the wake of Legazpi and Urdaneta.

Velasco's post was created during the 1570s as part of a flurry of institutional reforms that were meant to better harness early modern science to the needs of empire. Philip II may have been wary about publishing what Spain knew about its overseas possessions, but he was keenly aware of the importance of gathering such knowledge and putting it to good use in official decision making.[4] Velasco was charged, among other things, with composing an official chronicle and an official cosmography of the Spanish Indies.[5] Four years after accepting the job, he completed his *Geografía y descripción universal de las Indias* (Universal geography and description of the Indies, 1574). Drawn from depositions of Church officials, the documentation accumulated over the years by the Council of Indies, the papers of the royal cosmographer Alonso de Santa Cruz, documents sent to Velasco from the Indies in response to a 1572 request, and a number of privately authored works, as well as numerous print materials, the *Geografía y descripción* was the first synthesis of Spanish knowledge about the Indies written under official aegis. It surveyed the physical and political geography of the Indies, emphasizing the institutional apparatus of empire, secular and ecclesiastical, but also dabbled in natural history and ethnography.[6]

The *Geografía y descripción* has often been mined for what it has to tell us about Spain's overseas empire during the 1570s, and more recently, it has been studied as a crucial piece of Spanish "secret science," but it has never been subjected to analysis as a piece of cartographic literature shaped by imperial ideology. In this chapter, I argue that the *Geografía y descripción* represents an official effort to reconceptualize the Indies in the wake of Legazpi and Urdaneta. With the colonization of the Philippines and the conquest of the Pacific, the faded dream of transpacific empire shone brightly once again. Some Spaniards even saw in these events the harbinger of a new wave of imperial expansion, perhaps even the conquest of China itself. Times like these demanded a new geographical vision that acknowledged the existence of this incipient transpacific empire and that spoke of future opportunities in the South Sea. Velasco's writing and the maps that he constructed to accompany it provided the Spanish crown with that vision.

Built from disparate materials, the *Geografía y descripción* shows all sorts of cracks, and therefore it allows us to imagine how Velasco cobbled this new vision out of the archive he inherited and the opportunities that were emerging in his day. The text combines a cosmography of the Indies that identifies *las Indias* almost exclusively with the New World, as Gómara did, a hydrography that contains the Pacific even more radically than do earlier Spanish cartographic projects and a brief description of the lands and peoples across the South Sea figured as a new imperial frontier. The result is not a seamless body politic, but a Frankenstein monster of barely compatible parts, each of them marked with the scars of internal contradictions. This chapter examines the three parts in turn, unpacking those contradictions and Velasco's attempts to contain them, in order to create a new map of Spain's expanded and expanding empire. It then turns to a précis of the *Geografía y descripción* that Velasco prepared for the use of the Council of Indies around the year 1580 and examines how this later work attempts to erase the scars of the earlier monster, by sacrificing any reference to Spain's expansionist ambitions in the transpacific west. In the end, Velasco has little trouble mapping the newly acquired islands into his expanded vision of the Spanish Indies, but he struggles with China, that rich and sophisticated kingdom that had always looked like a sideshow to Spain's interest in the Spicery, but which was rapidly becoming the main attraction.

Velasco's America

The invention of America had come a long way by the time Velasco put pen to paper. As we saw in the previous chapter, the failure of Coronado to find golden cities in what his chronicler called "Greater India," and the failure of Cabrillo to follow the California coast to China, suggested to some Europeans that what we call North America was indeed separate from Asia. Gómara, as we saw, was quick to embrace this idea, and over the course of the next twenty years, other intellectuals both in Iberia and beyond the Pyrenees followed suit. Some found encouragement in rumors that the French had discovered the Northwest Passage, and they began to separate North America from Asia on their maps with a body of water known as the Strait of Anian (Nunn 1929, 33). Others noted that no one had ever sighted typically Asian fauna, like elephants or camels, anywhere in the New World (Nunn 1929, 19). Ethnographic evidence added fuel to the fire, particularly as Portuguese and Jesuit accounts of the civility of Japan and China provided a stark contrast to the perceived savagery of the native inhabitants of the Americas. By the time Abraham Ortelius published his *Theatrum orbis terrarum* in 1570, an atlas that was just as influential in the court of Philip II as it was anywhere else in Europe, the invention of America was more or less complete. Ortelius's maps unambiguously depicted America as an insular continent, separate from Asia, and his notorious title-page allegory adopted the architecture of the continents as the basic framework for the global geography of humanity, depicting the native inhabitants of America a particularly savage, distinct branch of the human family. (See fig. 18.) His friend Gerhard Mercator followed suit, finalizing the process of invention for many a European intellectual, despite continued ignorance as to the actual geography of the northern Pacific.

In this context, it might seem strange that López de Velasco never comes entirely out of the closet on the issue of American insularity, but only peeks out by cagily admitting that America and Asia were probably connected in the past but may or may not be connected any more. The ancient connection, he explains, is a necessary supposition of the ongoing effort to account for the origins of the native inhabitants of the New World. They were believed to be human, which meant that they had to be descendants of Adam and Eve by way of Noah and his family. This meant that the ancestors of the Amerindians must have participated in the repopulation of the world after the Flood, making their way from Mt. Sinai across Asia and into the New World. For that to be possible, the two parts of the world had to have been physically connected, at least in biblical times (López de Velasco 1894, 2).[7] Although Velasco does not cite a source for this argument, we can speculate that he knew of it through the work of Benito Arias Montano, a Spanish priest and humanist heavily involved with the production of the Antwerp Polyglot Bible who had close ties to the court of Philip II. Building on work by earlier intellectuals, mostly French Huguenots, Arias Montano equipped the Polyglot Bible with a supplementary volume that advanced this very argument, and he even included a map that depicted the hypothetical land bridge linking America to Asia in the North Pacific (fig. 26).[8]

Superficially, these arguments look like a latter-day version of the theory of Amerasian continuity, but they can also be understood as symptomatic of a significant shift in the whole debate over the geography of the new discoveries. The theory of Amerasian continuity, as we have seen it in previous chapters, did not construct North America as a landmass physically connected to Asia, but rather denied its "American" identity altogether, identifying it instead as a previously unknown extension of

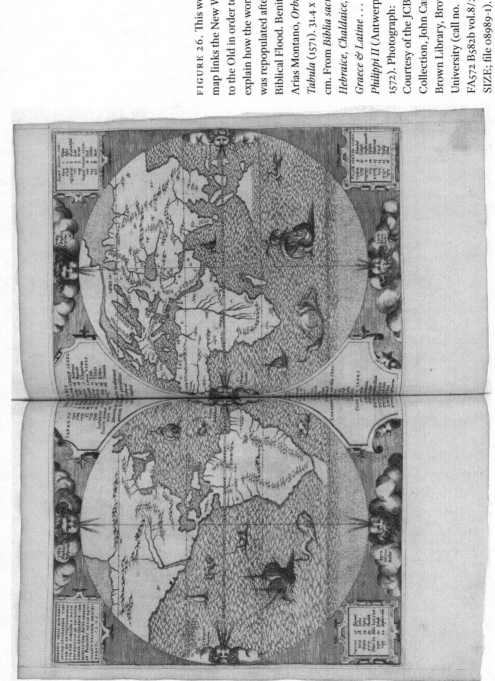

FIGURE 26. This world map links the New World to the Old in order to explain how the world was repopulated after the Biblical Flood. Benito Arias Montano, *Orbis Tabula* (1571). 31.4 x 52 cm. From *Biblia sacra, Hebraice, Chaldaice, Graece & Latine . . . Philippi II* (Antwerp, 1572). Photograph: Courtesy of the JCB Map Collection, John Carter Brown Library, Brown University (call no. FA572 B582b vol.8/2-SIZE; file 08089-1).

India extra gangem, a part of "Greater India," or quite simply, as East Asia. The civility of the Mexica was thought to be of a piece with the civility of Cathay, which was thought to lie just over the horizon from Mexico City, and the body of water we know as the Gulf of Mexico was called the Gulf of Cathay (see fig. 6). There was nothing truly controversial or unusual about the people of this region, who were really nothing more than Asians who had failed to appear in previous accounts of that part of the world. Only what we call the Caribbean and South America counted as the New World, and it was there that the European imagination located the sort of novelty, alterity, and even savagery associated with the term.

By contrast, as Giuliano Gliozzi argues, the attempt to explain the origins of the native inhabitants of America by writing them into the narrative of postdiluvian human migration that began among French Huguenot intellectuals during the middle of the sixteenth century and was taken over by Spanish Catholic intellectuals shortly afterward, presupposed the novelty of what we call North America (Gliozzi 2000, 317–18). It did not treat North America as an extension of Asia, and its inhabitants as previously unknown Asians, but rather as the northern half of a New World that was thought to extend up and down the entire western hemisphere. It therefore identified the inhabitants of the entire landmass, north and south together, as a distinct branch of a universal humanity, as defined by biblical anthropology and its monogenetic imperative, that needed to be written into the history of the repopulation of the world after the Flood. When its proponents insisted that there must be a land bridge that connected North America to Asia in the vicinity of what we know as the Bering Strait, therefore, they were not taking over the old theory of Amerasian continuity, which altogether denied that "North America" was separate and its inhabitants different, but rather reattaching an insular New World to the Old World, but without thereby erasing its uniqueness.

Velasco, however, only summarizes these arguments briefly, without engaging in them, as a way of avoiding the issue of whether or not the New World was separate from Asia, and its people a distinct branch of the human family. This suggests that we should interpret his caginess about the possible insularity of the New World as a symptom of a larger ambivalence that cuts through his various attempts to describe the Indies cosmographically and cartographically. On the one hand, he maps the *Indias occidentales* exactly as we would expect the crown's official cosmographer to map them, as the whole Castilian hemisphere, including most of the New World as well as East and Southeast Asia. On the other hand, he equips the *Geografía y descripción* with a general cosmographical introduction that mentions only New World examples, never Asian ones, not even when there are obvious ways to fit them in. The result is a text shot through with contradictions, suspended between a vision of the Indies that seems to participate in and even contribute to the ongoing invention of America and one that breathes new life into established ways of imagining Spain's potential transpacific empire.

When Velasco actually defines the Indies in explicit terms, he calls them the insular and continental lands of the Ocean Sea between the original line of demarcation and the antimeridian (1). It is clear from the sketch map of the Castilian hemisphere that Velasco drew for the *Sumario*, a précis of the *Geografía y descripción*, that the definition includes most of the New World, as well as East and Southeast Asia and a variety of islands and possible headlands in the ocean in between (see fig. 11).[9] The chart, which I examine in greater detail below, converts the South Sea into a Spanish lake, and it maps the Far East as the Far West. The text of the *Geografía y descripción*, meanwhile, grabs at even more territory. In likely response to pressure from an official censor, Velasco insists that the Castilian hemisphere actually extends far to the west of the position indicated on the map, all the way to the Bay of Bengal, recalling Fernández de Enciso's assertion that the antimeridian cut through the mouth of the Ganges River (9; Fernández de Enciso 1948, 25). So, while Gómara had struggled to hold on to a Spicery that was receding over the horizon of Spain's effective reach twenty years earlier, Velasco gobbles up as much of Asia as any Spanish cosmographer had before him, clearly registering renewed confidence in the possibility of realizing Spain's transpacific dreams in the wake of Legazpi and Urdaneta.

Yet the general cosmographical introduction that opens the *Geografía y descripción*, a lengthy section entitled "Descripción universal de las Indias" (Universal description of the Indies), makes no mention of East and Southeast Asia (23–39).[10] New World place names like "Peru," "Honduras," "Guatemala," "Quito," etcetera, abound in these pages, but names like "Filipinas" or "Nueva Guinea" are nowhere to be found (10–34). The chapters on botany, zoology, and the like include no Asian examples. The discussion of mines and metals does not mention the silver produced by Japan (23–25). The chapter "The Government and Republic of the Indians" says nothing about the complex and sophisticated government of China (28–29). We have to wait until we get to the section specifically devoted to the transpacific territories of the Castilian demarcation, near the end of Velasco's manuscript, to learn anything about the people and places on the far side of the South Sea. That section, therefore, is not just the only part of the *Geografía y descripción* where knowledge about Asian people and places appears; it is the only place where such knowledge matters. In the general cosmographical introduction, as in the bulk of

Velasco's text, "the Indies" are the Spanish New World, that is, the Caribbean, Mesoamerica, and the Andes.

The silence of Velasco's cosmographical introduction regarding all things Asian speaks volumes about the extent to which the New World had come to be constituted as a place apart in the Spanish imagination, as represented by the sources Velasco had at his disposal. Even as Spanish officialdom, and Spaniards actively involved in Spain's transpacific enterprises in one way or another, continued to map the Far East as the Far West of the Spanish Indies, Spain's extensive and intensive colonial involvement with New Spain, the Caribbean, and the Andes constituted *las Indias* as "America" in everything but name, abandoning East and Southeast Asia to the Portuguese and the Jesuits, and to the continent-spanning Orientalism that was beginning to hatch in their own attempts to map places like China, Japan, and the Spicery. This is the geography that Velasco inherits, and that comes through in his own "Descripción universal de las Indias." Whatever his maps and official definitions may say, his Indian universe clearly does not stretch across the Pacific.

Despite this apparent participation in the ongoing invention of America, however, the "Descripción universal" nevertheless makes sense of the natural world by appealing to the theory of climates rather than to the architecture of continents that is so salient in Gómara's *Historia general*. The cosmographer explains that the portion of the Indies effectively governed by Castile lies "entre los dos trópicos" (between the two tropics), and he allows tropical thinking to seep into what he has to say about any number of topics, including the climate, agricultural fertility, botany, and zoology of the Indies, as well as the temperament, government, and religion of the native inhabitants. Within the tropics, Velasco explains, the nights are always equal in length to the days, making for lands of a certain temperament (temperamento) (10). The heat in these lands is ameliorated by the abundance of rains and fresh air, making for a climate that is neither too hot nor too cold, and generally salubrious. There, rich vegetation abounds, and the land is fertile and productive almost everywhere, although local factors can make for significant exceptions to the rule, as they do in the deserts of Peru (11, 16). Metals are found where the sun shines brightest, on the western slopes of tropical mountain ranges, and no one bothers to look for them in the higher latitudes, where the temperatures are lower (23, 92). In the parts of the New World that lie outside the tropics, meanwhile, "todo sucede por la orden que en Europa" (everything occurs according to the same principles at work in Europe) (16).

Climate matters when it comes to human beings as well, although the only explicit link Velasco makes between latitude and human complexion comes as a brief observation that the inhabitants of higher latitudes are whiter and taller than those of the tropics (27).[11] Nevertheless, Velasco's generalizations about the native inhabitants of the Indies rehearse stereotypes that had grown out of European encounters with the tropical populations of Brazil and the Caribbean. The skin of the natives, Velasco writes, is darker than that of Europeans, the color of "membrillo cocho que tira a leonado y bazo" (cooked quince, tending toward tawny and yellowish brown), but it is not black (27). Their material culture is very modest, they have few possessions, and their diet is simple. They sleep in hammocks and live off maize and root vegetables. They are naturally servile and thus incapable of governing themselves. This means that, left to their own devices, they live for the most part in *behetrías*, lawless groups responding only to strongmen, or under tyrannical governments. Either way, they end up at war with each other for the most trivial reasons, fighting with primitive bows and arrows. They wallow in all sorts of monstrous vices, particularly human sacrifice, cannibalism, polygamy, and incest. And just as they are easy prey to human tyrants, so they are also prey to the tyranny of the devil, whom they worship in ignorance (25–36).

Velasco does not allow the cultural sophistication of Mesoamerica and the Andes to challenge the general applicability of these stereotypes. He reduces the clothing of the Mexica and the Inca to simple cotton coverings that represent only a minor improvement over the otherwise universal nakedness of the Native American (27–28). He describes their systems of government as tyrannies, based upon the brutality of the rulers and allowing no freedoms to the ruled (338). He reduces their cities to nothing but a smattering of "pueblos desordenadamente poblados" (poorly organized towns) (28). He dismisses their writing systems as mere *aides de mémoire*, minor exceptions to the general rule that the Amerindians lacked letters and sciences (30). He refuses to recognize their use of cacao and coca as currency as real "money," and he offers no admiration for their achievements in gold and silverwork, featherwork, and weaving (31).

This tendency to downplay the sophistication of the Mexica and the Inca is borne out by the portions of the *Geografía* devoted specifically to New Spain and Peru. The description of Mexico City, for example, admits to the city's great size at the time of its conquest, putting its population at sixty thousand households. It describes the city's location on an island in a system of lakes, and it alludes to the markets and fine houses of the Mexica nobility mentioned in the famous descriptions of the city, among them that of Hernán Cortés himself. Nevertheless, it minimizes the significance of all this by deflecting the reader's attention and admiration from the old Mexica capital to the modern Spanish one. We read, for example, that the houses of the defeated Aztec lords were quite fine, but everyone else's house was quite mean, and that by contrast the houses of today's Spaniards are all solidly built of stone, wood, and mortar. We learn nothing about the markets of Moctezuma's day, except that they existed, but learn that the Spanish have rebuilt the market squares and that the commercial life of the city is abuzz with all sorts of merchandise. We read that fresh water arrives in the city by way of an aqueduct, but do not find out that this was just as true in pre-Hispanic Tenochtitlán as it is in sixteenth-century Mexico City (191–92).

Velasco is more generous with Cuzco, the old Inca capital. We learn that pre-Hispanic Cuzco had "grandes edificios, guarnecidas las paredes de mucha chapería de oro y de plata, y una fortaleza fabricada de piedra" (great buildings, their walls covered with gold and silver, and a stone fortress), as well as a great temple to the Sun, home to more than four hundred virgins and "rico de grandes tesoros, ropa, plata, piedras de esmeraldas y oro" (rich in great treasures of clothing, silver, emerald stones and gold) (478–79). Nevertheless, while these and other details speak of past glory, they do not contradict the earlier assertion that everything that smacked of well-ordered government among the Indians failed to achieve European levels of *policía*. Velasco is explicit on this point:

> Hubo en estas partes del Mediodía un imperio, que fue el de los Ingas, como en las Indias del Norte el Mexicano, aunque este parece haber procedido de tiranía; y así, en todo lo que alcanzó, que fue desde Quito hasta chile, los señores del reinaban muy absoluta y tiránicamente y no sin alguna grandeza y magestad. En todas las otras partes, los indios carecían de república y gobierno porque todos eran beheterías. (338–39)

> [There was once in these southern parts (the Indies of the South) an empire, which was that of the Incas, just as in the Indies of the North there was once an empire of the Mexicans, but this empire seems to have been rooted in tyranny, and so, throughout its reach, from Quito to Chile, its lords ruled absolutely and tyrannically, although not without some grandeur and majesty. Everywhere else, the Indians lacked any kind of republic or government.]

Beneath the veneer of Inca civility lies the fundamental servility of the Amerindian, that same inability for true self-governance that produces savagery and violence in some places and tyrannical empires in others. The reader is to conclude that everywhere throughout the Indies, the native inhabitants exhibit the complexion of people born and raised in the tropics. What looks like European-style civilization, at least in some descriptions of Mexico and Peru, is really nothing of the kind.[12]

Velasco's handling of the advanced civilizations of the Americas represents a typical response to the challenge that they posed to the theory of climates and the ideology of empire. In the New World, Europeans encountered advanced civilizations precisely where they were not supposed to exist, in the tropics, and failed to find them where they were supposed to be, in the temperate zones of North and South America. Velasco's response was similar to that of many European mapmakers, who eagerly populated South America with images of Brazilian cannibals, but rarely featured the cities of the Mexica (Davies 2016, 254–55). He forces the Mexicans and Peruvians to fit into the metageographical mold of the theory of climates. In so doing, he remains consistent with the view of such notorious apologists of empire as Juan Ginés de Sepúlveda and Pedro Sarmiento de Gamboa, who did not appeal specifically to theories of tropicality, but who nevertheless denied the civility and even the full rationality of the Mexicans and Peruvians by making arguments quite similar to those we find in Velasco, that their cities were not really cities at all, or that their systems of government were tyrannical.[13]

So, while Velasco parroted conventional claims that the Spanish Indies by right spanned the South Sea and included the Far East as their Far West, his implicit concept of the Indies was consistent with that of other Europeans of the last quarter of the sixteenth century. The New World was a place apart, home to a branch of the human family that was inherently savage. Despite the apparent accomplishments of the Mexica and the Inca, they were really no better than the naked cannibal that had come to stand for America in Ortelius's frontispiece allegory, and in other similar illustrations that were beginning to grace the visual and material culture of early modern Europe. Yet unlike Gómara, Velasco does not ground the difference of the Amerindian in the theory of American insularity. He never insists, as Gómara does, that the New World is an enormous island, a place apart. As we have seen, he remains ambivalent about the whole issue and appeals to the theory of climates rather than to the architecture of the continents as the implicit metageography of his decidedly imperialist cosmography.

Whether by accident or design, Velasco's decision to ground his description of the Indies in the theory of climates rather than in the architecture of the continents proves convenient for his attempt to suture East and Southeast Asia onto his geography of the Indies in the final forty of the 609 pages that constitute the first edition of the *Geografía y descripción*. When one mapped America as an island, as Eviatar Zerubavel insists, one drew a line that split Americans from Asians, and which lumped each group together as a distinct human population (1992, 23–30; see also 1996, and 1993). When one mapped the Indies by way of climate, by contrast, one drew the lines differently, distinguishing between civilized northerners and savage southerners, but leaving the overall figure open to the east and west. Oviedo and von Sevenborgen had availed themselves of this characteristic of the theory of climates in order to map the Spicery and the New World together as different parts of a single tropical expanse, both equally available for conquest by civilized people from temperate climes, like the Spanish. In the 1570s, after Legazpi and Urdaneta had breathed new life into dreams of transpacific empire that were at their nadir when Gómara wrote his *Historia general*, it would be Velasco's turn to utilize the theory of climates in order to remap East and Southeast Asia into a cartography of empire that had all but given up on those regions as potential theaters of conquest. Climate would prove to be a crucial concept in Velasco's attempt to put the brakes on the ongoing invention of America, to reopen the western frontier that had cut the New World off from Asia in the European imagination, and to map East and Southeast Asia as a part of the West Indies that he called "las Islas del poniente" (The Islands of the West).

Mapping a Novel New World

By the time Velasco was writing, the name "Islas del poniente" was well established in Spanish usage. It had circulated among Spaniards in New Spain during the decade following the cession of the Moluccas to Portugal in the 1529 Treaty of Zaragoza, replacing alternatives like "las Malucas" (the Moluccas) or "la Especería" (the Spicery) as the preferred name for the purportedly rich islands that lay west of New Spain but east of the antimeridian, by Spanish reckoning. According to Miguel Rodrigues Lourenço, it served to assert Spanish prerogatives in insular Southeast Asia at the expense of Portuguese claims (2010, 14). Velasco's use of "Islas de poniente" comes across as deeply ambivalent, just like his use of "las Indias." In the *Geografía y descripción*, he uses the term to refer to the islands of Southeast Asia, but also to the entire transpacific region, islands and continental territories together. His ambivalence suggests that in 1574, he was grasping for a way to map the broader, more inclusive Indies that with the conquest of the Pacific and the colonization of the Philippines had passed from pipe dream to administrative reality, and that he never thought the problem all the way through before declaring his geography finished. Nevertheless, despite his ambivalence, his description of the people and places of East and Southeast Asia serves a clear political agenda. In Velasco's major work, the paper empire of Spain's transpacific territories becomes the new frontier of Spain's imperial ambitions.

At one point, the *Geografía y descripción* identifies the "Islas del poniente" as an Asian archipelago that happens to lie within the Castilian demarcation. It explains that the name refers to the same territory that was once commonly known as "la Especería," a group of islands that neighbored "las provincias de Asia" (the provinces of Asia) (569, 572). Over the course of two chapters, one devoted to the Philippines exclusively and the other to the rest of insular Southeast Asia, the text describes the islands one by one, locating them in space, saying something about their size, outlining their general topography, enumerating their natural resources, and so forth (573–91). It mentions their discovery by Magellan and Loaysa, and it describes the novel route that connects the islands to New Spain by way of the South Sea. Together, these chapters are heir to previous Spanish descriptions of the region by Urdaneta, Oviedo, and Alonso de Santa Cruz, although they do not seem to be based on any of the three (Urdaneta 1954, 247–49; Fernández de Oviedo y Valdés 1852, 100–109; Santa Cruz 2003, 343).

All of this material, however, sits within a larger series of chapters that covers the transpacific region as a whole, including mainland China, the islands of Japan, the Ryukus, the Solomons, and New Guinea, whose status as a large island or a continental headland remained undetermined. An introduction to the region, which is meant to function as a local analogue to the "Universal description of the Indies" that opens the *Geografía y descripción* as a whole, clearly states that the "Islas del poniente" are not just the archipelago once known as the Spicery, but rather all the insular and continental territories that lie within the Castilian demarcation west of "las Indias de la Nueva España y el Perú" (the Indies of New Spain and Peru), from 13° or 14° latitude south to 35° or 40° latitude north.[14] That introduction, moreover, bears the title "Descripción de las Islas del Poniente" (Description of the Islands of the West). The toponym thus applies both to the archipelago in particular and to the region as a whole.

Velasco does nothing to explain this ambivalent double reference. In fact, he does not even mention it, suggesting that there we may be dealing with an unfinished portion of the *Geografía y descripción*. Yet this very ambivalence helps us understand the ideological impetus of his text. Velasco had no lack of an alternative terminology, one that could distinguish between the archipelago and the larger region of which it formed a part. In 1570, for example, the crown was already referring to Spain's transpacific territories as the "Indias del Poniente" (Indies of the West), using the term that Velasco would later adopt in the *Sumario*. It was not, therefore, that Velasco lacked the language he needed, but that he was holding on to the language that was already in play. This may have been because he found something attractive in referring to Spain's transpacific frontier as "islands" rather than as "Indies."

The attractiveness of the term comes through in the balance of his general introduction to the region. Velasco admits that its "islas y provincias" (islands and provinces) are not yet known in their entirety, that the Spanish presence there is slight, and that the region is so large and internally diverse as to make generalization about it quite difficult (569, 571). He generalizes anyway. After defining the region in the manner described, and saying something about the territorial dispute with the Portuguese, Velasco notes that "estas islas" (these islands) "por caer las mas dellas debajo de la Equinocial y paralelos cercanos a ella, son todas calientes y húmedas conocidamente, aunque no por ello dejan de ser sanas y de buen temperamento" (because most of them lie on the equator or on nearby parallels, are all famously hot and humid, although they do not for that reason cease to be healthy and of good climate) (571). With this remark, it is clear that he has abandoned the expansive sense of "Islands of the West" for the restrictive one, that he is no longer talking about the region as a whole, but about the archipelago. After all, the region extends far into the northern temperate zone, as anyone could see on the accompanying maps, but this passage refers to the location of the islands on the equator or near it, in other words, in the tropics. The attraction of the language of "islands," therefore, has nothing to do with notions of insularity, but with tropicality. By treating the archipelago as the heart of the region, the key to its general character, Velasco signals his desire to tropicalize it, just as he had tropicalized the whole of the New World.

The general description of the native inhabitants of the islands that follows this remark attempts to do precisely that. That description responds to the reader's likely expectations about people who live in the tropics, and it echoes what Velasco has already said about the inhabitants of the New World. The islanders have no kings or leaders, we read, but rather live in "behetrías" (leaderless bands) that wage war against each other constantly, using bows and arrows, and sometimes spears or poisoned darts. They live in humble huts grouped into small, poor communities, and they are liars with a tendency toward theft, given over to all sorts of debauchery, including polygamy, but not sodomy (572). Like the inhabitants of the New World, the people of the islands are available for conquest and are even in need of the civilizing and evangelizing influence of Hispanic colonization. By emphasizing the tropical islands rather than the temperate mainland of the territories across the Pacific, Velasco has

converted the region into a novel New World, and he has implicitly characterized the conquest of the Philippines as a continuation, or a reprise,
of the conquest of the Caribbean.

Like the New World, however, the transpacific west features people
who are not so readily described as tropical savages. Velasco admits that
the islands exhibit a greater diversity of physical appearance than the New
World, including individuals with black skin and others with white skin
and blonde hair. Temperate types, it seems, have somehow made their
way into the tropical islands. The culture of some of the islands, moreover, exhibits signs of civility quite unlike anything that has been seen
in the New World. Some islanders use harquebuses and bronze cannon,
while others sail watercraft that are as large, well built, and seaworthy as
any caravel. Although the cosmographer is clearly describing a Southeast
Asian *proa*, he calls these vessels by the name usually used to refer to the
sophisticated watercraft of the Chinese, "juncos" (junks) (572–73). In
this way, a description that began by painting an image of the islanders
akin to Vespucci's account of the Brazilians in the *Mundus Novus* letter
finishes by recognizing a considerable degree of civility among them and
even associating the islanders with the highly advanced culture of China.

When Velasco later turns to China and Japan, his attempt to figure the
Islands of the West as a duplicate or extension of his thoroughly tropicalized New World runs off the rails. Clearly following sources of Portuguese and Jesuit origin, Velasco affirms that the Chinese do not live in
towns, but in large cities worthy of the name, packed with fine houses of
brick and mortar. They are fully literate, have "escuelas y ciencias" (formal academic institutions and organized bodies of knowledge), and have
been printing books for many years. They wear elegant clothes of cotton and silk, and they produce fine luxury goods that they export to the
Philippines. They have a sophisticated system of government, and most
incredibly of all, their kingdom is protected along its northern frontier
by a very strong wall, one thousand leagues long (592). The Japanese,
meanwhile, have white skin and possess fine features. They are monog-
amous, and the women are beautiful, resembling those of Castile. Like
the Chinese, the Japanese dress well. They eat simple meals from fine
porcelain plates. They are literate, "like the Chinese," and their language
sounds like German. The towns visible from the sea are small, but others farther inland are large and feature well-built houses with tile roofs
like those of Spain. Although they are "gente soberbia y escandalosa" (a
proud people, prone to conflict), they are also fine horsemen who wear
armor skillfully wrought from iron and steel, and they handle the bow as
well as the English. They respect their king, and thanks to the efforts of
the Jesuits, many of them have become Christians (596–99). Unlike the

islanders or the Indians, these people exhibit all the trappings of civilized life as Europeans know it, except, of course, the true faith.

One would think that after admitting such things, the cosmographer would give up on his attempt to figure the inhabitants of the region as a whole as doubles of the tropical Indians, but Velasco remains undaunted. Having reproduced ample evidence of China's high degree of civility, he modulates that evidence in ways that keep the Chinese from looking like true peers of Europeans. Although the Chinese army is large, he asserts, the Chinese are no good at war, and their artillery and cavalry are inadequate by European standards. Although they have white skin, as one would expect of inhabitants of the temperate zone, the Chinese are greedy, suspicious, and gluttonous. Most importantly, they are "gente soberbia y muy cobardes, y vil y afeminada y tan avasallados, que en pasando cualquier gobernador por la calle se arriman todos a las paredes mucho antes que llegue, y ninguno le habla sino de rodillas" (an arrogant but cowardly people, vile, and effeminate, and so submissive that whenever a governor passes through the street they all press themselves against the wall long before he appears, and no one dares speak to him without bending the knee) (592–93). While the Japanese, in all their martial virility, look like the Castilians of Asia, the Chinese look like anything but. In their cowardice, effeminacy, and despicable subservience to authoritarian rulers, they look more like the inhabitants of a hot climate than those of a temperate one.

Velasco's treatment of China recalls his treatment of Mexico and Peru. Just as he could not allow the Mexica or the Inca to stand out as civil exceptions to the general savagery of tropical America, so he cannot allow China to stand as an unqualified exception to the general savagery of the inhabitants of the Islands of the West. Note that China's rulers govern through fear, like tyrants, just like the ancient rulers of Mexico and Peru. Yet while the American context allowed Velasco and others to appeal to the theory of climates in order to downplay Mesoamerican and Andean civility, the Asian context does not allow him to do the same. As one could see on any map, including Velasco's own, most of China lay in the temperate zone, where one would expect to find fully rational people perfectly capable of governing themselves. There could be no theoretical rationale for Velasco's strained attempt to deny that the Chinese were just as civilized as Europeans, at least not in appeals to latitude. Luckily for Velasco, there were plenty of dystopian accounts of China and the Chinese circulating in manuscript, providing a stark contrast to the utopian vision of the Portuguese and Jesuit materials that were available in print. We shall learn more about them in the next chapter. For now, I simply suggest that these accounts were what allowed Velasco to do with China what he had done with Mexico and Peru, to downgrade its civility so it would fit, however imperfectly or incompletely, into his account of the Islands of the West as a region available for conquest.

Unlike the authors of those dystopian accounts of China, Velasco never proposes that Spain should attempt to subdue the Middle Kingdom by military means, but he nevertheless hints at an invasion in the making. According to Velasco:

> Sábese cierto y averiguado, por las cartas de Cosmografía antiguas y modernas, que toda aquella tierra viene a caer y está dentro de la demarcación de los Reyes de Castilla, y así como es perteneciente a ella, aunque no está descubierta ni tomado posesión della en nombre de los Reyes de Castilla. (592)

> [It is well known and has been demonstrated, by Cosmographical maps both ancient and modern, that all that territory falls within the demarcation of the Kings of Castile, and thus belongs to that kingdom, even though it has not yet been fully discovered, and no one has taken possession of it in the name of the Kings of Castile.]

At first, this looks like yet another assertion of Castile's *de jure* authority in East and Southeast Asia at the expense of Portugal, but when we get to the final phrase, with its assertions that China has "not yet been discovered" and that no one has yet "taken possession of it," everything changes. Velasco is not just affirming his king's *de jure* authority over China in the present, but imagining the assertion of *de facto* authority in the future, through a ceremony of possession just like the ones that so often marked the Spanish experience in the New World. Such a claim would presumably require military action to back it up. Evidently, a new way of connecting America to Asia had emerged in the Spanish imagination. It was no longer a matter of mapping the New World as an extension of Asia, but one of mapping East and Southeast Asia as a new, transpacific America, a novel New World, a new frontier for Spanish imperialism.

This was only possible thanks to a sea change in Spanish attitudes toward the Pacific, as manifest in Velasco's *Geografía y descripción*. The text does not just deal with the geography of the Indies, but also with its hydrography, which describes the oceans and seas of the Spanish hemisphere in general terms, covers the basic patterns and practices of oceanic navigation in the North and South Seas, and traces the maritime routes used by Spanish vessels in both oceans in considerable detail.[15] This "Hidrografía general de las Indias" (General Hydrography of the Indies) figures the South Sea as a double of the North, containing it as never before, and thereby effecting a discursive conquest of the Pacific that reflects its real-life conquest by Urdaneta and the regular sailings of the Manila Galleons.[16] It is to this Pacific conquest that I now turn.

The Hydrography of Containment

Velasco's efforts to contain the Pacific responded to political and intellectual trends that had developed in the court of Philip II over the ten years prior to the completion of the *Geografía y descripción*. Controversy over the dimensions of the South Sea marked Spain's new Pacific venture from the initial conceptualization of the Legazpi expedition through the years following Urdaneta's return. When Philip II began to entertain the idea of dispatching an expedition to the islands that Villalobos had christened the Philippines, he sought the advice of people with experience on the South Sea, in an effort to determine whether or not he held a legal claim to the islands. Andrés de Urdaneta argued that he most certainly did not. O. H. K. Spate insists that Urdaneta was the only person at the time who appreciated the real dimensions of the Pacific Ocean, and their implications for Spain's ambitions in East and Southeast Asia (1979, 105). According to the friar, the islands lay on the Spanish side of the antimeridian, but within the *empeño*, the part of the demarcation that Charles V had pawned to Portugal in the Treaty of Zaragoza, so Spain could not sail there unless the monarch returned the 350,000 ducats that had been paid to his father. Urdaneta urged the king to send his expedition to New Guinea instead or else to look for new opportunities in the northern latitudes of the South Sea, on the way from the New World to Japan (Urdaneta 1560, and 1561). A veteran of the Villalobos expedition, the pilot Juan Pablo Carrión, argued the opposite case. He insisted that the Philippines were indeed close enough to Mexico to put them safely in Castilian territory, and that Spain could occupy them without reservations and certainly without returning any money to Portugal (Audiencia de Nueva España 1564, 17). Faced as his father was before him with competing expert accounts of the breadth of the South Sea, Philip II chose to believe the one that best served his interests and sent Legazpi to conquer the Philippines. Urdaneta was recruited under false pretenses and did not learn what the fleet's destination was until the expedition's secret orders were opened at sea.[16]

Upon his return from the Philippines, the Augustinian immediately departed for Spain, where he participated in an official *junta* of cosmographers that was supposed to settle the matter of whether or not Spain was entitled to colonize the Philippines. Each of the participants, including Urdaneta, reached the same conclusion, that the Philippines did indeed lie within the Castilian hemisphere, but that they also lay within the *empeño*, and so Spain had no right to conquer, settle, or trade in the islands until Philip II returned the 350,000 ducats to Portugal. Clearly, the king must not have approved of what they had to say, since several of the cosmographers later submitted wary retractions of their original opinions (Goodman 1988, 59–71). Philip II decided to keep his new colony, while Urdaneta returned to New Spain, eager to press on to the Philippines to do missionary work. He was forbidden from doing so by his Augustinian superior, and he remained in Mexico until his death.

The king, meanwhile, faced considerable opposition from Portugal to Spain's presence in the Philippine Islands, but he found a solution to his difficulties in the arguments of his ambassador Juan de Borja and a collaborator of his, the Neapolitan astrologer and mathematician Giovanni Battista Gesio. Borja and Gesio had been trying to penetrate the veil of Portuguese cartographic secrecy, in order to determine what the Portuguese really thought about the position of the antimeridian. The ambassador sent Gesio to Madrid to convince the court that the Portuguese had been lying all along, that they knew the true position of the antimeridian to be far to the west of where they had publically insisted it lay.[17] The Hapsburg monarch, Gesio insisted, not only could keep the Philippines, but could even seize the Moluccas without returning a single *real* to Portugal and press on to mainland Asia with complete impunity (cited in Goodman 1988, 61). Gesio made his case, María Portuondo explains, by arguing that the Atlantic was wider than it was generally thought to be, and the Pacific, narrower. This had the effect of dragging the antimeridian westward across Asia so that China and Japan lay on the Spanish side, even when one cut the region of the 1529 *empeño* off the western edge (Portuondo 2009, 189).

The impetus to contain the Pacific, therefore, did not originate with Velasco, but with the interests of the crown and the convenient arguments of a freelance Neapolitan cosmographer whose ideas about global geography coincided so neatly with the king's imperial ambitions. Either too convenient or too well acquainted with sensitive information to be allowed to wander, Portuondo argues, Gesio was kept in Madrid and granted a modest stipend (187). He put himself to work reviewing Velasco's work, including the *Geografía y descripción* and its maps, which are now lost. The Neapolitan excised numerous passages from the original text, including several lines of praise for the cosmographical skills of Martín de Rada, an Augustinian who had accompanied Urdaneta to the Philippines and whose estimate of the longitude of the islands informed Urdaneta's opinion in the 1566 *junta*, as well as three assertions that the Philippines were part of the *empeño*, one of which mentions the unanimous decision of the 1566 *junta*.[18] Gesio's influence, however, was probably not limited to redacting passages. It is easy to imagine, for example, that the Neapolitan was behind the text's assertion that the Portuguese knowingly falsified the longitudes on their maritime charts in order to justify their presence in East and Southeast Asia (1). But even if we believe Gesio's influence lay in cutting material rather than motivating its production, one thing is clear, that he wanted Velasco to adopt the smallest estimates available for the breadth of the South Sea so as to assure that East and Southeast Asia appeared as part of the Spanish hemisphere in Spain's official description of its overseas empire.

The "General Hydrography of the Indies" suggests that Velasco embraced this point of view wholeheartedly. The first section of the hydrography describes the general layout of the Ocean Sea in the Spanish demarcation, explaining how Tierra Firme divides the ocean into two bodies of water, the North Sea that stretches eastward from Tierra Firme toward Spain and the South Sea that stretches westward to "la India Oriental" (Oriental India) (54). It then divides these two oceans into their constitutive seas and basins, each with its unique name, like the "golfo de Nueva España y la Florida" (The Gulf of New Spain and Florida), and the "golfo de la China" (Gulf of China). Velasco makes it a point to mention that the greater number of subdivisions provided for the North Sea stems from the fact that it is better known than the South Sea, and in this way he prevents this particular difference from disrupting the overall implication of the description, that the geography of the Indies (India Oriental | South Sea | Tierra Firme | North Sea | Spain) is symmetrical. The passage says nothing about the breadth of the two oceans, not even in relative terms. In these ways, it begins the work of shrinking the Pacific down to the size of the Atlantic, of figuring the South Sea as a double of the North.

A second section of the hydrography figures the two oceans as functionally identical maritime spaces. A chapter on the prevailing winds in the North and South Seas describes how the principal sailing routes on the two oceans follow the same counter-clockwise pattern. On both oceans, the text notes, Spanish vessels avail themselves of the easterlies that blow surest near the equator to sail from east to west and to climb to roughly 40° north longitude to find the westerlies that carry them back east (79–80, 84). On both oceans, we read, pilots rely on the same constellations to determine their position, because what matters in celestial navigation is not the ocean one sails, but the hemisphere, north or south, in which one finds oneself (63). Even though he has alluded to the fact that the North Sea is better known than the South, Velasco says nothing about why. He does not mention the very different histories behind these two patterns of navigation, how the North Sea was conquered quickly and easily, and how the South Sea consumed so many lives and so much treasure before yielding its secrets. We learn nothing about how Spain's past experience of the two oceans has been so different, only about how the current experience of both basins is alike.

The third section of the "General Hydrography" provides a glimpse into the differences that Velasco has so far elided (63–88).[19] It provides a detailed, leg-by-leg description of the routes that Spanish vessels follow to cross the North and South Seas. Derived from existing rutters, verbal accounts of sailing routes that most pilots preferred to charts as working references, the text describes each leg in terms of length (leagues) and time (days or months), and it provides latitudes for key points of inflection along the route (64–85). If we dig into this material, we find plenty of details that speak of the marked difference between sailing the North Sea and sailing the South. We learn, for example, that the transatlantic voyage from Spain to New Spain took about sixty-five days overall, but that it was broken up into distinct legs of travel by customary stops at the Canary Islands, Dominica, and various ports in the Greater Antilles before reaching San Juan de Ulúa on the Gulf coast of Mexico (68–73). The transpacific voyage from New Spain to the Philippines took only slightly longer, two-and-a-half months, but provided no stops along the way.[20] The difference between the return trips is even more striking. While the return from San Juan de Ulúa to Sanlúcar de Barrameda took between fifty-five and seventy days and featured stops in Havana and the Azores, the return from Manila to Navidad took over four months, all of them on open water (77–85). It is clear from this material that the North and South Seas were nothing alike, at least from the perspective of anyone trying to cross them. Transpacific navigation required much more time on the open ocean, making for a more miserable trip and a far riskier one. The return trip from Manila, unmatched in length by any of Spain's routes of travel in the North Sea, must have looked like a particularly harrowing prospect.

Yet Velasco was not writing for mariners or anyone else making the trip to the Indies. He was writing for imperial administrators at the highest level of power, and he wanted them to see the big picture, in which the South Sea was just as thoroughly domesticated as the North Sea. So he introduces his detailed accounts of the two transoceanic routes with generalizations that state how long it takes to get from one endpoint to the other, and the total distance traversed. The reader learns that it takes two-and-a-half months to cover the 1,600 leagues that stretch between Sanlúcar and San Juan de Ulúa, in New Spain, and two good months to traverse the 1,400 leagues between Sanlucar and Nombre de Dios in Tierrafirme (modern Panama) (64). Later, he or she learns that it takes two-and-a-half months to sail the 1,600 or 1,700 leagues from Navidad, in New Spain, to Manila (84). At neither point does Velasco mention that the westbound Atlantic routes include a wealth of way stations to break up the voyage, while the westbound Pacific route features no such stopping places; nor does he mention the dramatic difference in the time it takes to make the two eastward voyages. By burying the differences in the details, and emphasizing the similarities in the more general remarks, Velasco contains the Pacific, converting it into a rough double of the thoroughly domesticated Atlantic.

Velasco rendered this in visual form on the general map that accompanies the *Sumario*, and that may very well have been copied from the earlier map drawn for the *Geografía y descripción*. (See fig. 11.) I turn to this map now to see how it supports Velasco's effort to render the South Sea as a double of the North. Gesio hated this map, just as he hated all of Velasco's cartographic efforts. According to the Neapolitan, the cosmographer-chronicler of the Council of Indies was an incompetent mapmaker who relied on inaccurate sources, mishandled all sorts of technical issues, and remained ignorant of the principles of mathematical geography (Portuondo 2009, 189). The nautical chart Velasco had made for the *Geografía* was plagued with insuperable limitations, Gesio insisted, and the regional charts were so badly constructed that they did not even merit sustained criticism (188). The maps in the *Sumario*, in Gesio's opinion, were not even real maps. They were more like the work of a painter who captures the "memory" of an event, as opposed to the work of a scientist (*científico práctico*) who captures the event as it actually was (cited by Portuondo 2009, 202).

Gesio was right. As we saw above, Velasco paid no attention to the data in his own description of the maritime routes of the Spanish empire that might be used to contradict his ideologically charged hydrography. He was a humanist and an imperial administrator, not a mathematician or even a professional cosmographer. His maps, such as the one in question, should thus be understood as acts of translation from the mathematically precise cartographic mode of the fully trained cosmographer to the more notional mapping of the educated layman.[21] They do not model geography in geometrically precise ways, although neither do they acknowledge their lack of precision and thus admit to any lack of authority. Instead, they map the empire in broad strokes, shaping space along the lines suggested by actual historical developments, such as the conquest of the Pacific, and by vague imperial ambitions. Unlike other maps of Spain's empire *de jure* that effectively marshal the rhetoric of mathematical accuracy, this one abandons such pretense in favor of a clear ideological fantasy that is taken for a reality.

For our purposes, the key fantasy here has to do with Velasco's portrayal of what we call the North Pacific. Here it becomes the "Golfo de poniente," a lake-like, almost triangular embayment of the South Sea clearly bounded on two sides by the converging coastlines of mainland Asia and America, and on the third by the row of coastlines that stretch from east to west just south of the equator. At the eastern edge of the assemblage, the track of Álvaro de Mendaña's 1567 voyage from Peru to the Solomon Islands links the insular coastlines to the shores of the New World, completing the impression that the *Golfo de poniente* is a distinct, nearly enclosed body of water. North of the islands, the route followed by the transpacific galleons crisscrosses the waters between New Spain and the Philippines, speaking of their domestication by Spanish navigation. No numbers appear anywhere on the map. There is no scale of latitude, longitude, or distance, yet we can nevertheless tell from visual inspection that the *Golfo de poniente* is only slightly wider at the Tropic of Cancer than the North Sea. The Pacific appears as a double of the Atlantic, attesting to the ease of access that Spain has to all parts of its empire.

We should not be too eager, however, to identify this as the world of the Manila Galleons, rendered in cartographic form for the very first time, particularly not if we are treating this map as a stand-in for the chart that is missing from the 1574 *Geografía y descripcion*. In the world of the Manila Galleons, the relationship between America and Asia is primarily a commercial one, mediated by Spain's sole Asian colony, the Philippines. In 1574, however, it was not yet apparent that commerce would predominate over conquest in Spain's relationship with the Asian mainland. Many Spaniards thought China was the next Mexico or Peru, an exotic empire ready to fall to Spanish arms. A series of Spanish incursions into Cambodia still lay in the future, as did Spanish entanglements in Formosa, Borneo, and the Moluccas. This map tells its original reader that the way west across the South Sea is now open, that the tropical *Islas del poniente* are now just as accessible from New Spain as New Spain is from metropolitan Spain. The symmetry of the arrangement suggests something about the nature of the expectations that underpin it, that the *Islas del poniente* are now open, not for business, but for conquest.

Inventing the Indies of the West

Velasco's new cartography of empire was far from stable or permanent. Four years after finishing the *Geografía y descripción*, the cosmographer produced his *Sumario*, by jettisoning most of the material that serves to raise the *Geografía y descripción* to the level of a full-blown Renaissance cosmography, including the "Universal description of the Indies" and the first two sections of the "General Hydrography."[22] According to María Portuondo, the treatise was meant to serve as a "geographical primer" for incoming members of the Council of Indies, providing them with the minimum essential knowledge they needed to run the empire (194). This much leaner description of Spain's overseas empire moves through the Indies one administrative unit at a time, listing the various centers of population with the number of Spanish and indigenous inhabitants in each, distinguishing fertile from infertile landscapes, noting the location of mines, harbors, navigable rivers, and impassable mountains, cataloging the institutions of the state and the church, including tribunals, cathedrals, and monasteries. Attention to physical geography is minimal, just enough to make sense of the political and economic geography at the heart of the project. Nowhere does the text mention the prehispanic past or provide details about Amerindian cultures.

By eliminating what he had to say in the *Geografía y descripción* about the history and temperament of the Amerindians, Velasco also eliminates any concern with the history of the conquest, its justification, and its potential future. The Indians of the *Sumario* are not tropical savages, whether the naked sort characteristic of the Caribbean and Brazil or the slightly dressed-up versions of Mesoamerica and the Andes. Velasco does nothing to comment on their temperament, not even at the most obvious junctures. When he calls the Windward Islands the "Caníbales" (Cannibals), for example, he says nothing about their notoriously anthropophagous inhabitants (432).[23] When he describes Mexico City and Cuzco, he has no need to downplay the civility of the Mexica or Inca, because he says nothing about the prehispanic past (446–49, 501–9). The Indians of the *Sumario* have no history. They are just "indios," the subject population of Spain's overseas empire, cultivating the fields here, working the mines there, and paying tribute everywhere.

The transformations reach across the Pacific. We read that China is one of the largest, most opulent, and most fertile kingdoms in the world. The Chinese, we read, live in "grandes ciudades y muchos pueblos" (large cities and many towns), and that they are "gente política prevenida para su defensa, pero no guerrera" (a politic people prepared to defend themselves, but not warlike), but we learn nothing about the supposed feebleness of its military and we come across no negative insinuations about its system of government (533). The deprecatory remarks about China and the Chinese that we saw in the *Geografía y descripción* are quietly set aside by a remark explaining that China is now much better known than it was before, thanks to the commercial ties that have been established between the Philippines and the Middle Kingdom (532). The empire of the Ming is no longer a double of the empires of Moctezuma and Atahualpa, but something entirely different from what the Spanish have encountered in the New World. It is a fully civilized, powerful country with whom the Spanish can profitably trade, but can never conquer. The regional map of the Indies of the West reinforces this impression, by using color to clearly distinguish temperate China and Japan from the tropical islands (fig. 27).

Ironically, it is only in the *Sumario*, where Velasco clearly acknowledges the difference between the tropical and temperate portions of the transpacific west, and gives up any dreams of conquest on the Asian mainland, that he manages to fully integrate the region into his cartography of the Indies. While the *Geografía y descripción* held on to the problematic but well-established language of "islands" of the west, the *Sumario* adopts the more recently coined language of "Indies of the West," rendering the transpacific territories commensurate with the New World. In fact, the *Sumario* goes even farther than that, abandoning the binary distinction between the New World and the Islands of the West that marks the *Geografía* for a tripartite distinction among the *Indias del mediodía* (Indies of the South, i.e. South America), the *Indias del septentrión* (Indies of the North, i.e. North America), and the *Indias del poniente* (Indies of the West, i.e., East and Southeast Asia). The Islands of the West are no longer opposed to the Indies, the New World, or whatever one wants to call the place that Velasco describes in the cosmographical introduction to the *Geografía y descripción*. They enjoy full membership in the *Indias occidentales* as one of its three major regions. The general map adopts this nomenclature, and the use of color, whether original or not, emphasizes the equal status of the three parts. There, we do not see America and Asia divided from each other across the Pacific, but the three commensurate parts of the Indies gathered around the cozy *Golfo del poniente*.

The map, moreover, does nothing to acknowledge the very real differences that exist among the three parts of the Indies. It does not, for example, divide the Indies of the North and South into their constitutive *audiencias*, nor does it identify them as the geographical homes of the two viceroyalties New Spain and Peru, since doing so would inevitably single out the Indies of the West as the only one of the three parts that was not home to a viceroyalty. Neither does the map depict any centers of population, since their names would only draw a contrast between the widespread distribution of Spanish toponyms in the Indies of the North and South and their localized concentration in the Indies of the West. Through these silences, the general map in the *Sumario* attempts to erase the suture that so clearly cuts through the *Geografía y descripción*, inviting the reader to imagine the tripartite Indies as a single imperial expanse.

The speed with which Velasco's geography of conquest, past and future, was transformed into a vision of peaceable empire speaks to the

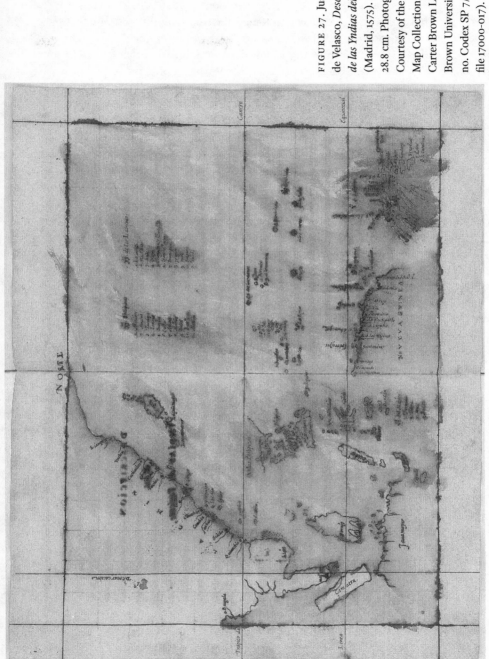

degree of controversy that swirled around Spain's future in East and Southeast Asia during the years after the Legazpi-Urdaneta voyage. As I have mentioned, the two cartographies Velasco provides, one that figures China as the new frontier of a still-expanding empire, and entertains dreams of further transpacific conquest, and the other that figures China as a wealthy trading partner, and portrays the empire as a settled, peaceable affair, were not unique to Velasco. The chronicler-cosmographer was simply synthesizing the possibilities that were already out there, in the debates among Spanish colonists, missionaries, and merchants, who were invested, in one way or another, in Spain's Chinese future. The next chapter delves into their various cartographies, further exploring the various ways that Spanish cartographic literature located China in the transpacific west of Spain's geopolitical imaginary.

6

The Location of China

By the time the Spanish colonized the Philippines, the Portuguese had been active in the South China Sea for roughly four decades, engaging in the illicit trade that flourished along the shores of southern China despite the efforts of the Ming government to close its borders to outsiders. In 1555, as the Chinese began ever so slightly to open the doors to foreign commerce, the Portuguese gained permission to reside in the trading post of Macau year-round and to participate in the annual commercial fairs in Guangzhou, thereby consolidating their position in the region. Macau soon became the most successful center of private maritime trade anywhere in Portuguese East Asia, and by the time the Spanish conquered the sultanate of Maynila, it was on its way to becoming the crucial link between the *Estado da India* and Nagasaki (Disney 2009, 2.183). Asian luxury goods made their way back to Europe on Portuguese carracks by way of Goa, and the trickle of knowledge about China and the Chinese that had begun to flow early in the century turned into a steady current.

That knowledge began to appear in print around midcentury, when the crown of Portugal lifted the veil of secrecy from its privileged knowledge of the East Indies. Some of the earliest accounts of China based on sixteenth-century contacts appeared in the general histories of Portuguese expansion published by men who had themselves spent time in the East Indies, João de Barros and Fernão Lopes de Castanheda, and who drew on the authority of the eyewitness to underwrite what they had to say about Portugal's empire in the East. Other accounts, like the description of China attributed to the Portuguese captive Galiote Pereira, appeared in collections of Jesuit letters printed in Portugal, Italy, or Spain.[1] This tradition reached an important milestone in the first Jesuit volume on China, the *Historiarum Indicarum libri XVI* (History of the Indies, book 16), by Giovanni Maffei, published in Venice in 1588, which lay the groundwork for the influential Jesuit Sinography of the seventeenth century that began in earnest with the publication of the journals of Matteo Ricci.

In the meantime, Iberian writers built upon the early interventions of Barros and Castanheda to develop their own Sinography, which responded to a variety of intellectual opportunities and served a diversity of purposes. The Portuguese Dominican Gaspar da Cruz borrowed heavily from both Barros and Pereira to produce the first European book devoted entirely to China, the *Tratado em que se contam muito por extenso as cousas da China* (Treatise in which the things of China are related at great length), published in Évora in 1569. The Spaniard Bernardino de Escalante then cribbed Cruz's book to produce the first Spanish book about China, the *Discurso de la navegacion que los portugueses hacen a los reinos y provincias del Oriente, y de la noticia que se tiene de las grandezas del reino de la China* (Portuguese navigation to the kingdoms and provinces of the Orient, and news of the greatness of the kingdom of China), published in Seville in 1577. The Augustinian friar Juan González de Mendoza then borrowed heavily from Escalante to produce the book I mentioned in my Introduction, the *Historia del gran reino de la China* (History of the great kingdom of China), published first in Rome in 1585 and then in its definitive edition in Madrid a year later. Of all these publications, Iberian and Jesuit alike, the Mendoza book was the most successful, going through some forty editions and translations before the century was out. It was the most influential European book about China of the entire sixteenth century, and it continued to be read even after the appearance of Ricci's work, well into the seventeenth.

In this chapter, I analyze the Escalante and Mendoza texts as pointed interventions in an ongoing debate over the location of China, in two senses of the term. On the one hand, both Spanish authors echo their Portuguese sources and march in step with their Jesuit contemporaries, by locating China near the top of the hierarchy of human cultures that was emerging from the European encounter with the extra-European world, below Europeans themselves on account of their false religion, but above almost all other non-Europeans, including the native inhabitants of the Americas, on account of their high degree of *policía*. On the other hand, the two Spanish authors part company with the Portuguese and Jesuits by locating China in the transpacific West rather than the Portuguese East. While Escalante and Mendoza certainly recognize that China forms part of the continent of Asia, they minimize the significance of this fact, and they emphasize instead its role as the western terminus of a Castilian route of travel across the New World and the Pacific. In this, they join hands with a series of figures whom they considered to be ideological rivals, Iberian Sinophobes who thought China was just another

Mexico or Peru, ripe for conquest by daring conquistadors. By mapping the China of the Sinophiles into the transpacific space of the Castilian Sinophobes, Escalante and Mendoza banish military conquest from the map of Spain's transpacific empire in the making, converting the South Sea into a theater of spiritual conquest and commercial opportunity.[2]

The New World of Iberian Sinophobia

Most discussions of early modern European Sinology create the impression that European knowledge of China developed in a more or less linear fashion from the early and highly influential reports of medieval travelers, Marco Polo especially, through the scant rumors of early sixteenth-century travelers like Ludovico de Varthema and the early syntheses of Portuguese historians like Barros and Cruz, to finally flourish in the massive output of the Society of Jesus. While this tradition grew in scale and sophistication, it also elaborated on a single overarching theme, that China was an extremely old, prosperous, and sophisticated society worthy of admiration and even emulation by Europeans in all matters save religion. The key elements of this argument were already in place in the Iberian Sinography of the sixteenth century, which generally portrays China as a highly urbanized country, its cities large, handsome, well-fortified, and connected to each other by an elaborate fluvial system that made it possible to move people and goods cheaply and easily. This discourse tells of urban markets brimming with the products of a large and industrious artisanal class and of a peaceable, fully cultivated countryside, free of brigandage, thanks to the country's highly rational and effective system of government. In early modern accounts of China, we read how the monarch in Beijing ruled through a governing class of mandarins, chosen through a meritocratic system of examinations. These mandarins exercised justice by wielding pitiless juridical punishments and operating a harsh penal system, but with admirable fairness and transparency. As the elegant and sophisticated products of China's elaborate educational system, they were fully versed in the country's literature and learning, which not only featured a system of writing but even used printed books of domestic manufacture. Although the Chinese practiced a false, idolatrous religion, they were so clearly reasonable that they could be converted to Christianity entirely by persuasion. All that was needed were competent missionaries to bring the Middle Kingdom into the Christian fold.[3]

While this image of China enjoyed broad circulation, it was not the only one that was available to sixteenth-century Europeans. A very different version of the Middle Kingdom and its people emerged from the fraught milieu of the South China Sea and circulated entirely in manuscript sources. It responded to the very real threat to Iberian life and limb posed by Chinese pirates, the terrifying possibility of ending up on the business end of Chinese justice, and the ample experience of corruption on the part of China's governing officials, who allowed smuggling to take place under their very noses as long as they got their cut. It was also nourished by first-hand acquaintance with the Chinese expatriate communities that dotted the trade routes of the South China Sea in defiance of Ming bans upon emigration and contact with foreigners, putting the lie to notions of Ming governmental effectiveness. None of this encouraged Iberians on the ground in Southeast Asia to view China with rapt admiration, much less to idealize its system of government. On the contrary, experience tended to nourish fear and contempt of China and the Chinese.

Both emotions were certainly characteristic of Spanish attitudes toward the burgeoning Chinese population of the Philippine Islands. During the last quarter of the sixteenth century, Manila became home to the largest Chinese expatriate community in Southeast Asia, as the opportunities provided by the new transpacific trade triggered an explosion in the size of the resident Chinese population from the few dozen present before the arrival of the Spanish to many thousands at the turn of the century. Known as "sangleys," these Chinese residents of the Spanish Philippines were indispensable to the life of the colony, yet they were generally thought to be crafty, dishonest, materialistic schemers, probably disloyal to their Spanish rulers and certainly apt to rebel in support of an invasion from the mainland. If they converted to Catholicism, the sincerity or depth of their devotion was viewed with as much suspicion as that of other New Christians elsewhere in Spain's empire (Horsley 1950; García-Abásolo 2012; Gil 2011). The Spanish authorities thus tried to confine them to a walled ghetto known as the Parián, which was eventually situated outside the walls of Manila, but within sight of its guns. While some sought to convert and assimilate the *sangleys*, others spoke of expelling them, just as the mother country had expelled the Jews and would eventually expel the *moriscos*. On various occasions, the relationship between Spaniards and *sangleys* exploded into violence, leaving countless Chinese dead. Under such circumstances, it is difficult to imagine the Spanish colonists in Manila giving much credence to the flattering account of China and the Chinese that was beginning to circulate in print half a world away.

In fact, numerous Spanish colonists advanced a markedly Sinophobic account of the Chinese, usually in the wake of frustrated efforts to approach China diplomatically, and in the context of proposing a military approach to Chinese affairs.[4] Generally speaking, the Sinophobes admit that China is indeed large, populous, productive, and highly urbanized, but they also insist that it is not at all the highly civilized or well-governed place that the Sinophiles make it out to be. Everything that the Sinophiles hold up to be a sign of Chinese civility becomes in the discourse of the Sinophobes mere appearance, a pleasant mask for a cruel and sordid reality. Behind the good order exemplified by the ruler-straight boulevards typical of Chinese cities, for example, one finds the tumbledown warrens of the teeming masses. Chinese books, upon close examination, are not as well made as their European counterparts, and Chinese learning is superficial at best. The Chinese army is large, but the soldiers are cowardly and the officers effeminate, the gunpowder weapons poorly manufactured and little match for their European counterparts. Most importantly, the mandarins are not at all the measured men of letters that the Sinophiles make them out to be, but hedonistic tyrants, even atheists, who terrorize the populace into submission. That populace, in turn, is a servile, cowardly, superstitious lot that lives in fear of its tyrannical masters. The mandarins, argue the Sinophobes, would never allow missionaries to enter the Middle Kingdom because the conversion of the people to Christianity would undermine their grip on power, so China had to be taken by force of arms, preferably soon, before the Chinese realized the danger they were in and began to apply their considerable resources to improving their defenses.

Eventually the Sinophiles won the day, and the discourse of the Sinophobes was largely forgotten until late in the seventeenth century, when it became a useful weapon in the Chinese Rites controversy, and again in the eighteenth, when it was pressed into service to construct modern Orientalism, and particularly the notion of the Oriental despot.[5] As a result, if historians of the early modern European encounter with China attend to the Sinophobes at all, it is only to tell the story of their quick and utter defeat by the Sinophiles (e.g., Romano 2013, 103–12). The Catalan historian Manel Ollé is among the few who have taken the Sinophobes seriously, plotting the development and demise of their vision of China, as well as the plans to conquer the Middle Kingdom that went along with it. He reminds us that the peaceful, commercial world of the Manila Galleons did not spring fully formed from the head of Miguel López de Legazpi, but emerged instead from a period of uncertainty, when it was unclear who would step through the door that Legazpi had opened, the merchant, the missionary, or the military man. Ollé's work also reminds us that the development of European knowledge about China was not a linear process, nor was it univocal. It was instead both multivoiced and dialectical, as the competition between hawkish Sinophobes and dovish Sinophiles lead to the expansion and development of both discourses.[6] In order to understand Escalante and Mendoza, therefore, we would do well to begin with the writing of the particular Spanish Sinophobe that triggered Escalante's Sinophilic response.

That man was Francisco de Sande, a native of Cáceres who studied law at the University of Salamanca, served for five years as a member of the *audiencia* of New Spain, and as governor and captain-general of the Philippines for six years, from 1575 to 1580.[7] A few years before Sande arrived, Manila had managed to repel a devastating attack from a powerful Chinese pirate, variously known as Lin Feng (林 風) or Limahon, and holed him up in his improvised stronghold on the coast of Luzón, northwest of the colonial capital. The Spanish authorities attempted to leverage their provisional success against Lin Feng, dispatching an embassy to China, the first of its kind to leave Manila, to extract permission from the Chinese authorities to establish a Spanish trading post and a Christian mission on the mainland. When the pirate unexpectedly escaped, the diplomatic effort fell apart, and Sande, who had come to power while the embassy was away, concluded that Spain had no choice but to deal with China militarily. He was not the first to suggest invasion as an option, but he was the first to develop a more or less serious military proposal, which he sent to Philip II in 1576 (Sande 1576, 1580).

Sande's letter constitutes one of the centerpieces of the hawkish Sinophobic discourse that emanated from Manila during the 1570s and 1580s. Needless to say, it locates China on the Castilian side of the antimeridian, and it even raises the alarm of Portuguese encroachment. It also admits that China is large and populous and that it produces foodstuffs and artisanal goods in abundance, but everything else aims to disparage the Middle Kingdom and its people.[8] According to the governor, the Chinese are cowardly, superstitious, lazy sodomites who only work when compelled and who are given to drunkenness and gluttony. The men dress and wear their hair just like the women, suggesting that that the Chinese do not observe proper gender distinctions and that the men themselves are effeminate. Their only form of learning is reading and writing, none other, and they recognize European writing to be superior to their own. Although their cities are large, their buildings are small and ruinous. They do not have properly minted currency. Remarkably, they are capable sailors, given "cuan bárbaros son" (how barbarous they are), but they do not dare ride horses, do not generally carry weapons, and fight with crudely made cannon and harquebuses. They fear soldiers armed with pikes.

Most importantly, however, Sande avails himself of the stereotype about the government of China that was already established in some Portuguese descriptions of the country, that it is tyrannical. At the center of this stereotype lie accounts of a specifically Chinese form of torture and of the brutality of the Chinese penal system. Both were first described in print by Galiote Pereira, a Portuguese soldier-merchant who found himself in a Chinese prison, having violated the Ming ban on foreign trade. The Jesuits published his *Tratado da China* (Treatise on China), which was written during the 1550s, in a 1562 collection of letters about the Indies (Pereira and Cruz 1989, 13). Pereira's account of what would later come to be known as the *bastinado* is worth citing in full:

> São os açoutes destes homens huns pedaços de bambus partidos polo meio, afeiçoados pera aquillo; não ficão agudos mas rombos e damnos nas coixas diguo nas curvas. Deitão hum deste açoutados no chão e alevantão a quana com ãbas as mãos, dão tam gramdes açoutes que espantão quem os ve da crueza delles. Dez açoutes tirão muito samgue, e se são vimte ou trimta fiquão as curvas todas espadaçadas, e cinquoemta ou sesenta ade estar hum homen muito tempo em cura, e se he cemto não tem nenhuna, mas morre disto (D'Intino 1989, 121).[9]

> [The whips used by these people are bamboo rods split down the middle, specifically fashioned for this purpose. They do not sharpen them, but leave them rounded, and they strike us on the thighs. They lower the person to be whipped to the ground, and they raise the rod with both hands, and they deal such great blows that anyone watching will be frightened by their cruelty. Ten blows draw a great deal of blood, and if there are twenty or thirty, they leave the thighs torn to pieces, and if fifty or sixty, a man will be a long time convalescing, and if a hundred, then there is no cure known to man, and one dies from it.]

According to Jonathan Spence, this gruesome passage would be used in later centuries to argue that the Chinese had a unique capacity for cruelty (1999, 21).[10] At the time, however, it did not point to a racialized tendency toward brutality, but rather helped identify the nature of China's government. In the discourse of the Sinophobes, the cruelty of the *bastinado* and the prisons served to characterize the mandarins as tyrants.

The issue was not just that the mandarins used torture or that the prisons were gruesome. Early modern Europeans, after all, generally had no objections to the use of torture *per se*, and they were accustomed to prisons being unsavory. The argument was that the mandarins applied torture indiscriminately and imprisoned people on the slightest whim, in order to cow the populace into submission. The argument went back to the earliest surviving eyewitness account of China written by a sixteenth-century European, a letter by one Christóvão Vieira, a member of the ill-fated embassy to China led by Tomé Pires in 1517. Describing the Middle Kingdom from the perspective offered by the Chinese prison where he was held captive, Vieira claims that the ruling mandarins have no love for the people they govern, and so,

> Não fazem senão roubar matar açoutar por tromentos ao povo e ho povo mais mal tratado destes mandarins do que he o diabo no inferno, daqui vem o povo não teer amor ao Rei e aos manderins e cada dia se andão allevantando e fazem se ladrões. (D'Intino 1989, 25)

> [They do nothing but rob, kill, whip and put to torture the people. The people are worse treated by these mandarins than is the devil in hell: hence it comes that the people have no love for the king and for the mandarins, and every day they go on rising and becoming robbers.] (Ferguson 1902, 124)

The mandarins, in other words, exhibited the signature characteristics of the tyrant, as that figure was defined by classical and medieval political theory, serving their own private interest rather than the public good and using fear to maintain power.[11] Another Portuguese captive, Amaro Pereira, must have recognized the connection. Writing in 1562, he refers to the abuse of the population by the ruling mandarins as "tiranias e roubos" (tyrannies and robberies), thereby becoming the first person on record to use the term "tyranny" explicitly to characterize mandarin rule (D'Intino 1989, 94). Roughly ten years later, Sande repeats the case made earlier by Vieira and Amaro Pereira, claiming that "es gente [los chinos] muy sujeta a los mandadores y sufren con gran paciencia los castigos, que *por nada* cortan una oreja y dan cien azotes" (The people [of China] are very submissive to the rule of their masters and suffer their punishments with patience, for they cut off an ear or give a hundred lashes *at the slightest provocation*) (Sande 1576. Emphasis added). According to the governor, the mandarins exercised cruelty indiscriminately in order to inspire fear and thereby safeguard their power. This is what made them tyrants.

Although later generations would appeal to early modern descriptions of mandarin cruelty to construct the notion of oriental despotism, there was nothing specifically "oriental" about any of this, as far as the sixteenth century was concerned. Mandarin cruelty may, in fact, have represented a form a "despotism," if by that term we mean a particular kind of tyranny. "Tyranny," since the Greeks, was thought to be an aberrant form of monarchy, whereby a ruler treated his subject as a master would a slave. It could occur anywhere, in any society. "Despotism," by contrast, was a systemic version of more or less the same type of government. It was not the exception but the rule in those societies where it was identified. According to Lucette Valensi (1993), the term made one of its first appearances in Venice during the 1570s, when ambassadors to the Sublime Porte began to apply it to the government of the Ottoman Empire. Yet despotism was not the sole property of the Ottomans, or even of Asians. According to Joan-Pau Rubiés, the sixteenth century could and did identify this kind of systemic tyranny among all sorts of people who were supposedly incapable of governing themselves, including the native inhabitants of Africa and America, even though they did not necessarily call it "despotism" (Rubiés 2005, 116). Hernán Cortés does it in the "Second Letter from Mexico," where he brands Moctezuma a tyrant precisely because of the fear that he purposefully inspires in the Cempoalans (1993, 163). Systemic tyranny is a central theme of Pedro Sarmiento de Gamboa's history of the Incas, and the whole Toledan school of Andean historiography it helped spawn (Parra 2015). As we saw in the previous chapter, López de Velasco picked up on this, finding tyranny to be typical of the so-called governments of the New World. Despotism, therefore, was to be found everywhere during the sixteenth century, not just in the Orient.

One of its key consequences, according to the Iberian Sinophobes, was military vulnerability. Both Vieira and Amaro Pereira argue that the people of China so resented the arbitrary cruelty of the governing class that they were certain to greet potential invaders as liberators and to rise up against their tyrannical masters in support of an invading army (D'Intino 1989, 35, 94). Vieira, like his fellow captive Vasco Calvo, thus urges Portugal to launch an attack on Guangzhou that would reprise earlier conquests of other strategic port cities like Malacca, Goa, and Hormuz. Sande, by contrast, counts on a more generalized Chinese rebellion that will make it possible to take all of China. Clearly, he does not expect the few thousand men that he requests to be sent to Manila from Mexico and Peru to be equal to the task. He expects the Chinese to rebel, "porque son muy mal tratados" (because they are very ill-treated), and thereby swell the numbers of his invading army. If only he could take one of the maritime provinces, he assures the monarch, the rest of the kingdom will fall like a house of cards (Sande 1576). That is how fragile the Ming government's hold on power actually is, at least in the imagination of the governor of the Philippines.

In writing to Philip II, therefore, Sande was drawing upon an established tradition of hawkish Sinophobia that had been circulating among Iberians in the South China Sea for decades, albeit along channels that we do not fully understand, since the publication history of the Vieira and Amaro Pereira letters does not allow us to establish a pattern of influence from one Sinophobe to the next. In the shift of scale from taking Guangzhou to conquering all of China, however, we can see that Sande does not just borrow from the discourse of Portuguese Sinophobia. He also adapts it to the general pattern of Castilian colonialism, placing China at the end of Spain's long trajectory of territorial conquest. Never does he invoke the Turks, the Mughals, or any other Asian people in his attempts to describe the Chinese to Philip II. His touchstones of comparison are always American. The governor claims that the smallest of China's provinces is home to more people than Mexico and Peru combined, and that its arable land is richer than that of New Spain. He insists that China is rich in gold, silver, mercury, copper, and lead mines, and thereby he implies that it will yield precious metals on a scale comparable, at least, to Spanish America. He also suggests that the Chinese will be quick to come around to Spanish ways. According to Sande, the Chinese recognize the clear superiority of European letters to their own system of writing and hold no real devotion to their religion, making them easy to convert to Christianity. The implication, never stated outright, is that China will not only give up at least as much wealth as America, but promises to more readily become the sort of thoroughly Catholic and Hispanicized colony that the Spanish have been trying to build in the New World. China, in the eyes of the governor, is not only a novel New World, but a more promising one (Sande 1576).

Sande even adapts the stereotype of mandarin tyranny to an argument taken straight out of the Spanish experience with the conquest of America. Having studied at the University of Salamanca, Sande would have been familiar with the arguments of Francisco de Vitoria, the Neo-Scholastic theologian responsible for rethinking the theory of just war in light of the Spanish experience in the Indies. According to Vitoria, it was justifiable to take military action against a foreign state with the purpose of liberating its people from tyrannical rule (Vitoria 1991, 225–26; also Pagden 1990, 21). Sande explicitly appeals to this argument when he insists that "la guerra con esta nación de chinos es justísima, por librar personas miserables, que matan y toman hijos ajenos para estupros, y las justicias y mandadores y el Rey hacen tiranías nunca oídas [he war against the nation of the Chinese is most just because it liberates a miserable people, for they (the mandarins) kill and take the children of others as concubines, and the judges and leaders and the King commit unheard of tyrannies]" (Sande 1576). In doing so, the governor of the Philippines does not just provide an abstract legal basis for his proposed invasion: he also maps mandarin tyranny into the historical experience of Spain in America. China, for Francisco de Sande, is clearly a transpacific Mexico or Peru, and he hopes to be its Cortés or Pizarro.

In his letters to Philip II, the governor asks for six thousand men to be sent across the South Sea from New Spain and Peru in order to conquer China. The number may seem laughably small to us, but it was enormous by the standards of the day. Never before had any European power shipped such a large military force across such a vast distance. Those six thousand men, moreover, were meant to serve as the kernel of what would eventually become a much larger host. Sande expected his invasion to trigger a general revolt against mandarin rule, swelling his forces with Chinese rebels. Presumably, he imagined himself marching on Beijing at the head of a largely Chinese army, just as Cortés had marched on Tenochtitlán at the head of an army composed largely of indigenous allies. This was not the crackpot scheme of a crazy dreamer, therefore, but the serious proposal of a man who understood China and Chinese affairs through the lens of Spain's prior experience in America. Furthermore, he was not alone in seeing things this way. As Ollé points out, Sande's plan was endorsed by other members of Manila's colonial elite and was echoed by one Diego García de Palacios, who wrote to Philip II from Guatemala two years later, proposing a very similar scheme (2002, 79–82). Not only did these men see China as another Mexico or Peru, but they also envisioned the South Sea as a thoroughly domesticated oceanic basin, across which Spain could readily project its considerable military might in order to fully incorporate China into its overseas empire.

The Dovish Cartography of Bernardino de Escalante

Sande's proposal was rejected by the Council of Indies, which may very well have been influenced by the direct riposte of Bernardino de Escalante, a soldier turned cleric from Laredo, Spain, who became a confidante of Philip II by the time he was in his twenties. He was the author of one other published work besides the *Discurso de la navegación*, a *Diálogo del arte militar* (Dialogue on military arts) published in Seville in 1583. He wrote an unpublished *Regimiento de navegación a la India Oriental* (Rule for navigating to oriental India), a rutter that grew out of efforts to consolidate Portuguese and Castilian cartography in the wake of the Portuguese Annexation of 1580 (B. de Escalante 1992). He also penned numerous analyses of political, economic, and military matters, ranging from the invasion of England to the potential expulsion of Spain's *moriscos* (B. de Escalante 1995). In the *Discurso de la navegación*, he applies his expertise to the question of Spain's future China policy, arguing against military action and in favor of a diplomatic approach. Although he never mentions Sande or his plan by name, his book has long been understood as a direct response to the governor's hawkish scheme (Ollé 2002, 80–81).

Escalante must have understood that a convincing response to Sande had to rebut the governor's Sinophobic account of China and the Chinese, not just the invasion plan itself, so he turned to the work of Gaspar

da Cruz, published almost a decade earlier, and cribbed from it whole-
sale, adding material drawn primarily from João de Barros. The result is
an account of China and the Chinese that stands solidly within the Sinophilic tradition that had been pioneered by the Portuguese and embraced
by the Jesuits, yet that nevertheless located China precisely where the
Castilian Sinophobes had placed it, in the Castilian west, on the far side
of the South Sea from the New World. Escalante maps that ocean as a
well-contained, thoroughly domesticated maritime basin, across which
Spain can engage with China efficiently and peaceably.

Escalante quietly grounds his description of China in the theory
of climates, pointing out early on that the country extends northward
from a latitude of about 19° N to about 50°. Most of it therefore lies "en la
región que los Geógrafos llaman templada" and enjoys "el mismo clima
que España, Francia, e Italia" (in the region that the Geographers call
temperate . . . the same climate as Spain, France, and Italy) (151).[12] Most of
its inhabitants, therefore, resemble Europeans in appearance. Although
the inhabitants of Guangzhou and the surrounding coastline are "bazos"
[yellowish brown] in color, "como los de Berbería" (like the people of
Barbary), he explains, most of the Chinese are actually "blancos y rubios,
como en Alemania" (white and blonde, as in Germany) (82). The text
does not go so far as to assert that China's location in the temperate zone
should make for a culture as civilized as that of Europe. It simply goes on
to reiterate all the common tropes of Portuguese and Jesuit Sinophilia,
and it allows the readers to draw their own conclusions. It makes it easy
for a reader to believe that the text's Sinophilic account must be true,
because that is what one would expect of a people who inhabit the temperate zone.

Like the rest of Escalante's description, his account of China's political system responds to the expectation that the country's location in the
temperate zone could create in the mind of the reader, that the Chinese
will be similar to Europeans in almost everything that matters, including
their ability to govern themselves. According to Escalante, the mandarins
are not at all tyrants, but wise and effective rulers chosen through a meritocratic system of examination and protected from the corrupting influence of power by a clever system of checks and balances. Yes, the Chinese
penal system is dreadful, and the *bastinado* gruesome, but the mandarins
are methodical in the application of justice, and they struggle mightily to
find a person innocent before condemning him or her to death (129–34).
Despite the terrors that await the guilty, or perhaps because of them, the
Chinese are convinced, we read, that their system of government is the
best in the world and that only they, among all the peoples of the world,
can be said to see with both eyes. Europeans, by comparison, see with

only one, and the rest of the world is blind (142). Escalante does not confirm the boast, but neither does he deny it.

By insisting that the mandarins are reasonable and just, rather than capricious and tyrannical, Escalante robs the Sinophobes of the centerpiece of their legal and practical rationale for invasion. He further erodes that rationale by emphasizing the enormous scale and considerable sophistication of China's defenses. For example, according to Escalante, China's cities are surrounded by formidable walls made from the same clay used to manufacture porcelain. Like the Great Wall, they are all very well maintained by special ministers who are provided with all the resources they need to do the job. Guangzhou, for one, is surrounded by a forbidding moat and by walls of stone, brick, and mortar so thick, straight, and sturdy that ten to twelve horsemen can ride abreast atop of them, from one of the many towers to the next. A chapter devoted entirely to China's military establishment insists that every city has a garrison that is ready to defend against attack until it receives reinforcements from the massive imperial army. Both garrisons and regular army can be expected to fight well, for the Chinese are "muy mañosos y astutos en todas las facciones que se ofrecen de guerra" (very skillful and astute in every aspect of warfare), and their soldiers are always ready for a fight, since they get paid regularly, and generously (140–41). The final point was clearly meant to raise doubts among Spanish officials, who were well aware of their own considerable inability to pay Spain's soldiers with any consistency.

Escalante must nevertheless have been aware that his rivals might try to dismiss his account as the naïve musings of a man who has never seen China with his own eyes and can only parrot Portuguese and Jesuit propaganda, so he backs up his claims with a letter to Philip II from one Diego de Artieda Chirino y Uclés, a soldier who had participated in the conquest of the Philippines. Artieda praises the sophistication of Chinese culture and provides crucial details about China's military forces. Their artillery, Artieda claims, is both plentiful and superior to that of Europeans. Their wonderfully effective government has assured that every city is properly garrisoned. Even though the Chinese are not as "bellicose" as the Spanish, Artieda claims, their army numbers 300,000 foot and 200,000 horse, not counting the garrisons that defend each city. The soldier claims to have seen images of Chinese knights armed "con arneses y celadas Borgoñonas y lanzas" (Burgundian tack, helmets, and lances) (151). Escalante can only shrug his shoulders at the size and sophistication of such an army. "No se yo que poder puede bastar para para tanto número de caballería, y también armada, siendoles la artillería tan común como a nosotros," Escalante wonders, "y habiéndose de pasar allá en navegaciones tan largas" (I have no idea what power is sufficient to meet China's massive army, since the use of artillery is as familiar to them as it is to us, and we can only reach them by making long journeys by sea) (152). Sande may have been thinking big, but according to Escalante, not big enough. The problem was that "big enough" was far beyond anything Spain could hope to muster, much less move across the South Sea.

In light of such grandeur and power, an invasion was out of the question, but not an embassy, whose primary purpose would be to convince China's government to allow Catholic missionaries to enter the country. That embassy, however, would sail for China by way of the New World and the Pacific, not because that was the only route open to Spain, but because it was the best way to get from Madrid to Beijing. Although Escalante doubts that Spain was capable of moving a vast army across the South Sea, he had no doubt that that the way to China was shorter and faster by way of the Pacific Ocean than the Indian Ocean. He makes this point at the end of the text, where he insists that the maritime route to China by way of the South Sea is "segura y breve" (safe and short), and describes how ships regularly sail for the Philippines from the Mexican ports of Acapulco and Navidad, enjoying "viento en popa" (following seas) across the whole breadth of the ocean. Along the way the only land they hit are the "islas de los barbudos" (Isles of the Bearded Ones, i.e., the Marshall Islands), inhabited by a primitive population that does not bear arms of any kind, and the Islas de los Ladrones (the Marianas), inhabited by people who fight only with simple slingshots. The island of Luzón, which the Spanish firmly hold, lies only one hundred leagues from Guangzhou and enjoys a brisk trade with the Chinese mainland (153–54). However much he may differ from Sande about the nature of Chinese culture and the best way to approach relations with the Ming, Escalante shares his belief that the South Sea has become a Spanish lake and that China is best understood as mission territory lying in the Castilian West.

His attempt to map China into transpacific space, however, is not limited to these brief observations at the end of his discussion. The whole text can be understood as an effort to shift the location of China in the mind of the reader from the transasiatic east to the transpacific west. To understand how it does so, we must consider the principal way that Escalante's text deviates from its principal model, Gaspar da Cruz's *Tractado da as cousas da China*. Escalante does not just crib a description of China from his Portuguese predecessor, but also affixes an extensive introduction to that description, in the form of a history of Portuguese expansion from the medieval foundation of that kingdom through about 1515, the *Discurso de la navegación que hacen los Portugueses a los reinos y provincias de Oriente* referred to in the title of the work. The material is drawn from the work of João de Barros and does nothing to tone down its heroic register.[13] It even praises Portugal explicitly for "las grandes navegaciones y comercios que han hecho y hacen sus súbditos por todo el Oriente, con gran loor y estimación de la nación portuguesa" (the great sea journeys and commercial ventures that the subjects of the King of Portugal have made and are making throughout the entire Orient, earning praise and esteem for the Portuguese nation) (60). Curiously, this treatise written by an adviser to Philip II, before the annexation of Portugal by Spain, celebrates Portugal's imperial accomplishments much more loudly and explicitly than the Portuguese source from which it was derived.

The compliment, I argue, is thoroughly backhanded. While Escalante does indeed celebrate the Portuguese imperial enterprise, in doing so he also traces the progress of Portuguese imperialism from Lisbon, down the coast of Africa and around the Cape of Good Hope, to Calicut, Goa, and Malacca but he fails to mention Portuguese contacts with China and says nothing about Macau. He follows a series of explorers and conquerors as they face storms, navigational obstacles, and the opposition of native populations, especially Muslims. Although the Portuguese succeed in overcoming those obstacles, the reader's attention is drawn to the length, complexity, and perhaps even the vulnerability of the Portuguese route to the Indies. He or she is thereby set up to marvel at the brevity and safety of the Spanish route, which is nothing but smooth sailing across an empty ocean, save for a few islands inhabited by populations that should be easy to subdue. The overall effect is to lengthen the perceived distance of the Portuguese route while shortening the perceived distance of its Spanish counterpart, thereby pulling China away from Portugal's effective grasp and approximating it to Spain's effective sphere of influence. What matters most in Escalante, therefore, is not so much the geographical location of China, but its hodological location, that is, its location on a route of travel whose length is calculated subjectively, according to the ease with which it is traversed.

Clearly, the rivalry between hawks and doves, Sinophobes and Sinophiles, did not map neatly onto competing imperial projects and their respective cartographies. As far as Escalante was concerned, no matter how much he took from his Portuguese sources about the location of China and the Chinese in the hierarchy of human cultures, he did not need to concede anything to the Portuguese about its location in global geopolitical space. He did not need to map China as part of an Asian continent separated by the Pacific from Spanish America, or part of the Portuguese Indies, separated by the line of demarcation from the Castilian Indies. On the contrary, he could agree with Sande about the location of the Middle Kingdom in the transpacific west of the Castilian Indies, not the transasiatic east of the Portuguese Indies. In fact, he could insist on this point, much more deliberately and effectively than his hawkish rival, by constructing his text in such a way as to manipulate the reader's perception of the practical distances involved and thereby fix China's location in the reader's imagination to Spain's advantage.

Although Escalante's book was nowhere near as influential as that of his successor in Spanish Sinography, Mendoza, it did find readers, among them the Flemish printer Abraham Ortelius and his collaborators. The 1584 edition of the *Theatrum orbis terrarum* included the first map of European manufacture devoted entirely to China (fig. 28). It was based on a prototype by one Luiz Jorge de Barbuda, also known as Ludovicus Giorgius, a Portuguese mapmaker who worked for Spain after being smuggled out of Lisbon at the initiative of Giovanni Battista Gesio. He constructed it using source materials of Portuguese and perhaps even Chinese origin. After its publication by Ortelius, it became the prototype for a long series of European maps of the Middle Kingdom (Roque de

FIGURE 28. Abraham Ortelius, *Chinae olim Sinarum regionis, nova descriptio* (1584). From *Theatrum orbis terrarum* (Antwerp, 1584). Photograph: Courtesy of the Folger Digital Image Collection, Folger Shakespeare Library (call no. G1015.O6 1595 Cage [folio]; file 31794).

Oliveira 2003, 830–31). It follows Escalante and his Portuguese predecessors in their mistaken tendency to exaggerate the extent of China's network of navigable rivers, and it features a description of China on the reverse that cites Escalante as its source (112–13).

The Barbuda/Ortelius map may also answer to Escalante's effort to set China in the Far West rather than the Far East. Previous maps that represented the Middle Kingdom depict it as part of a larger Asian landmass, thereby acknowledging its status as *pars Asiae* and making it easy to visualize its proximity to the Portuguese Indies. Barbuda, by contrast, isolates China from the rest of Asia by depicting a wall of mountains to the west and north, as well as the Great Wall that separates China from Tartary. He crops the Malay Peninsula, thereby cutting the city if not the province of Malacca out of the picture, deemphasizing the Portuguese presence in the region and making it impossible to follow the Portuguese maritime route to the Middle Kingdom. Macau is conspicuous in its absence. He pivots the country ninety degrees clockwise, orienting the map toward the west. The reader approaches China by sea, not through the Indian Ocean by way of Malacca, but rather by way of Japan and the Philippines. A sailing vessel heads from the Philippines toward Guangzhou, recalling the trade between Manila and mainland China. All of this serves to depict China in a manner friendly to Castilian interests, without even mentioning the line of demarcation or saying a word about the territorial pretensions of the Iberian kingdoms. In this way, the map becomes a fitting illustration for Escalante's *Discurso de la navegación*, with its deliberate attempt to map the Middle Kingdom, not as the easternmost part of Asia, but as the westernmost province of the Spanish lake.

Mendoza's Spiritual Empire

Mendoza's *Historia de la gran China* is even more adamantly Sinophilic than Escalante's *Discurso*. Escalante, for example, tells us that the Chinese have their own system of writing and manufacture printed books, but he says nothing about the depth and breadth of Chinese learning, preferring instead to quote a few contradictory remarks from Barros and Cruz (115–16). Mendoza, by contrast, flatly affirms that the Chinese are people of "claros ingenios y buenos entendimientos" (lucid wits and strong understanding), and he affirms that they study both moral and natural philosophy (66).[14] Not only does he praise Chinese writing and printing, but he even catalogs the topics covered by Chinese books: geography, cartography, history, law, naval architecture, theology, medicine, astronomy, music, mathematics, architecture, divination, and military science (128–30). In Mendoza, the Chinese emerge even more clearly than they do in Escalante or the Portuguese sources as a fully rational people whose learning is commensurate with that of Christian Europe.

Mendoza also develops Escalante's effort to resolve some of the contradictions inherent in the Sinophilic tradition inherited from Galiote Pereira and Gaspar da Cruz. As we saw above, these Portuguese authors provide graphic accounts of the cruelty of China's judicial and penal systems, yet they often make no attempt to reconcile this material with their assertions that China's governors are wise and just. More aware than they, most likely, of the use the Manila Sinophobes were making of accounts of Chinese cruelty, Mendoza goes beyond his Portuguese sources, assuring the reader that the mandarins are assiduous in their efforts to distinguish the guilty from the innocent. He provides a detailed description of the work carried out by visiting inspectors, who assure that the judgments rendered by the mandarins are evenhanded. In doing so, he turns the tables on the Sinophobes, converting the *bastinado* and the hellish prisons from signs of mandarin tyranny into the very instruments of mandarin justice. China is one of the most orderly countries in the world, Mendoza argues, precisely because of the public's certainty that these punishments will always be visited on the guilty, but never on the innocent (116). Tyranny in China, according to Mendoza, is a thing of political past and the religious present (86–87). The real tyranny is that of the devil, who holds malevolent sway over the souls of the idolatrous Chinese.

Like Escalante and the other Sinophiles, Mendoza is a dove when it comes to Chinese affairs. He altogether refuses to entertain the possibility of invasion, insisting that his duty, as a man of the cloth, is to promote peace (99). Nevertheless, he seems eager to dissuade potential hawks by emphasizing China's robust defenses. He follows Escalante in marveling at the Great Wall, in arguing that Chinese cities are well fortified and properly garrisoned, and in providing jaw-dropping numbers for the size of China's army, and he even asserts that China's military establishment is larger than that of the rest of the known world combined and that its artillery is just as good as that of Europe, despite the poor examples that some of his coreligionists have seen in Guangzhou (95, 125–26). According to Mendoza, Chinese soldiers, "si igualaran en valor y valentía a las Naciones de la Europa, eran bastantes para conquistar el mundo" (if in bravery and valor they were equal the Nations of Europe, they would be enough to conquer the whole world) (99). The message is clear. Just as Chinese print culture can be favorably compared to that of Europe, so its military establishment bears comparison with any European army. To invade China would be foolhardy. Any future conquest of China had to be strictly spiritual.

Yet while the description of China that appears in the first of the book's two parts locates China at the apex of the hierarchy of non-European peoples, describing the Middle Kingdom in terms that seem almost utopian, it is the travel narratives that appear in the book's second part that locate the country in transpacific space.[15] These narratives tell of three separate visits to China. The first is the fateful 1574 embassy mentioned at the beginning of the previous section, whose failure convinced Sande of the necessity of the military option. The embassy was composed of two Augustinian friars, Martín de Rada and Jerónimo de Marín, and two *encomenderos,* Miguel de Loarca and Pedro Sarmiento. The second narrative involves a group of four Franciscans, led by Fray Pedro de Alfaro, who left Manila for China without official authorization in 1579, in a failed effort to establish a Franciscan mission in the country. The third narrative tells the story of another group of Franciscans, led by Fray Martín Ignacio de Loyola, whose failed attempt to establish a mission in China ended with an unplanned circumnavigation of the world between 1581 and 1584. A single narrator assumes responsibility for everything we read, using the third person to refer to the travelers in part 2, periodically referring the reader to material found in other parts of the text, and admitting to the addition of material based on Mendoza's own experience in Mexico (163, 320). Like the texts that frame Escalante's description, these narratives set China into a global space that is structured, not by the architecture of the continents or the theory of climates, but by the hodological space of maritime empire. They serve to promote the spiritual conquest of China by missionaries operating under Spanish, not Portuguese auspices.

Before turning to the travel narratives, however, we must touch on why Mendoza should take a side in the rivalry between the Iberian empires, given that he was writing during years immediately following the 1580 annexation of Portugal. As various scholars have demonstrated, the so-called Union of the Crowns (1580–1640) rendered the boundaries between Portugal and Castile in Iberia, America, and Asia quite porous, allowing subjects of the two kingdoms to migrate, interact, and even collaborate in ways that were impossible before.[16] This did not mean, however, that the old rivalry between the two kingdoms disappeared altogether. Under the terms of the annexation, the Hapsburg Monarchy had to keep Portugal and its empire separate from Castile and its overseas possessions, treating the two as separate possessions. So, even though the boundaries became porous, they still mattered, particularly when interests clashed across the lines of demarcation. The Union of the Crowns, therefore, did not make the old rivalry between Portugal and Castile go away. It simply converted it into a dispute internal to the workings of the composite monarchy of the Spanish Hapsburgs.

The rivalry extended to ecclesiastical matters, thanks to the considerable degree of control that the papacy had granted each of the Iberian kingdoms over the operations of the Church in its overseas possessions. At issue in the territorial disputes between Castile and Portugal was not just the scope of their prerogatives to conquer, settle, and trade, but also the scope of the Castilian *Patronato* and the Portuguese *Padroado*.[17] Each empire had its ecclesiastical allies, particularly within the missionary orders, whose ambitions were inextricably caught up with the interests of the crown that sponsored their work. The presence of missionaries in a given territory was often the only thing that defined the *de facto* boundary between Spanish and Portuguese territory, and the lines of demarcation could come into play in the definition of a diocese. Was China to be administered by the archbishop of Manila or that of Macau? Who would win the authority and the resources to evangelize the Chinese, the Jesuits in Macau or the mendicants in Manila? These were only some of the questions that emerged along the frontier of the *Patronato* and the *Padroado* in East and Southeast Asia.

For these reasons, the politics of Luso-Castilian rivalry still mattered when Mendoza was writing, and Mendoza clearly took the Castilian side. This should come as no surprise, given his background. At the age of seventeen, Mendoza left his native land of La Rioja for New Spain, where he spent the next twelve years of his life, joining the Augustinians along the way. In 1574, at the age of twenty-nine, he returned to Spain, where he studied at the University of Salamanca and eventually served as confessor to Antonio de Padilla Meneses, the president of the Council of Indies. At Padilla's request, Mendoza returned to Mexico as part of a diplomatic mission to China that turned back before ever crossing the Pacific, for reasons that remain unclear.[18] He spent the rest of his life in Spain, Italy, and Mexico, eventually serving as bishop of Chiapas. He was thus very much involved in the Spanish colonial project, even at the highest levels, and though he never made it to Asia, a part of his life was dedicated to the challenge of evangelizing the Orient. So, too, was the life of the order of which he formed a part. The Augustinians first reached East and Southeast Asia with Legazpi, and played an important role in the evangelization and governance of the Philippine Islands. Just as Francisco de Sande and others saw in their Philippine base a springboard for secular conquests on the mainland, so the Augustinians, along with the other religious orders in Manila, saw the islands as a launching pad for spiritual conquests.

The *Historia del gran reino de la China* thus downplays the precedence of the Jesuits and the Portuguese in matters Chinese. Living in Rome, the headquarters not only of the universal church but also of the Society of

Jesus, Mendoza must have been aware of the embryonic Jesuit mission established in China in 1583. Nancy Vogeley goes so far as to suggest that the accommodationist approach favored by the Jesuits may have even influenced Mendoza's commitment to peace with China (1997, 173–74). Nevertheless, Mendoza's text makes no mention of the Jesuit mission. Instead, it urges Philip II to dispatch another embassy to China with the hope of opening the kingdom's doors to the Augustinians, presumably from their base in the Philippines (244–45). His own book serves as a testament to the order's capacity to take on the challenge. Mendoza admits that he is not the first to write about China, but he also suggests that everything that has been published to date represents knowledge acquired "por relaciones" (by relation), suggesting that it is somehow second-hand, and not to be trusted. His book, by contrast, reflects the "clara y verdadera noticia" (clear and true account) that has emerged since the Spanish reached the Philippines (33). The 1575 embassy to China led by Mendoza's fellow Augustinian Martín de Rada, among others, becomes in Mendoza's *Historia* the first moment of Euro-Chinese contact that really matters, just as it had in Escalante's *Discurso*. Mendoza advertises his reliance on the eye-witness authority of Rada and other Castilian visitors to China, such as Alfaro and Loyola, in forging his description of China and thereby claims to replace supposedly dubious Portuguese knowledge of the Middle Kingdom with reliable, first-hand information of Castilian origin. This is probably why he never acknowledges his profound debt to Escalante, who had never been to China, and who relied so heavily on the Portuguese Sinography that Mendoza disavows.[19] It is the Spanish mendicants, Mendoza wants his reader to believe, who have truly discovered China, who understand the Chinese best, and who should therefore be sent to save their souls.

The travel narratives, therefore, play a crucial role in Mendoza's effort to replace his Portuguese and Jesuit predecessors as the true origin of European Sinology and thereby advance the cause of an Augustinian mission to China, launched from Manila under the auspices of the Castilian *Patronato*. They do so by lending their authority as eyewitness testimonial to underwrite the description in part 1. Yet they also do more than that. They provide readers with vicarious experiences of China meant to teach them how they should react to what they have learned about the country from Mendoza's description. In the three travel narratives, the reader accompanies people who have actually "been there," and learn lessons about the Middle Kingdom that Mendoza's description cannot teach by itself.[20]

Take, for example, the reader's experience of Chinese cities in the description and the travel narratives. In part 1, Mendoza devotes a chapter

to the cities of China that generalizes about their great size, their placement along rivers, and their sturdy fortifications (44–46). A subsequent
chapter describes Chinese domestic architecture, assuring the reader
that the houses of the great are suitably magnificent, while those of the
commoners are "muy buenas y muy bien edificadas" (very fine and very
well constructed) with whitewashed interiors, sturdy roofs, pleasant gar-
dens, and comfortable furnishings (47). Other chapters deal with dress,
manners, markets, and so forth. The procedure is analytical, and the
tone, admiring but declarative. In the account of the Rada embassy, by
contrast, the same details come together as the lived experience of a particular city, Quanzhou, or "Chincheo" in the text. We walk through the
city gates in the company of the ambassadors as the stalwart guards give
their salute and the masses gather to gawk. The streets are so crowded, we
read, that if one dropped a grain of wheat, it would not touch the ground.
We witness the surprise of the ambassadors at the pomp that accompanies an approaching mandarin, and we learn about the official's house
as the ambassadors enter it. We are invited to share in the astonishment
of the ambassadors as they walk through the streets, "tan suspensos que
iban como fuera de sí, pareciéndoles cosa de sueño" (so astonished that
it was as if they were traveling outside their own bodies, as if in a dream)
(204). Something similar happens in the Alfaro narrative, which quotes
the leader of the Franciscans as they tour the city of Fuzhou, or "Aucheo,"
saying that he had been in the principal cities of Flanders and Italy, and
that in all of them together he had not seen as many curiosities and as
much wealth as in a single street of that city (291). Although we learn
nothing about these cities that we could not have anticipated from the
generalizations in part 1, we learn how Castilian travelers reacted to what
they saw and thus learn something about how we should react to what we
have read. The purpose of the travel narratives is not just to demonstrate
that Mendoza's description is based on eyewitness accounts of China,
but to provide an experiential perspective upon the encounter with the
Middle Kingdom, one that aims at educating the reader's sensibilities.
Wide-mouthed astonishment, it seems, is the only possible reaction to
the wonders of China's urban life.

The first two travel narratives also teach the reader to love rather than
fear the members of China's governing class. The process is particularly
evident in the Rada narrative, which explains early on that Chinese vag-
abondage and mandarin "tiranía" (tyranny) are the reasons for the existence of so many Chinese pirates in the South China Sea (163). Upon
their arrival in China, the ambassadors are suddenly enclosed in the hold
of a Chinese naval vessel, and they fear that the audible commotion on
deck is that of preparations for their beheading. Later, when news of Lin

Feng's escape from Spanish encirclement reaches China and the ambassadors are accused of being nothing but spies, the Spaniards pay for the banquets they have enjoyed with sheer terror at what they believe to be their impending execution (228). On every occasion, however, their fears prove to be ill-founded. The first time, they find out that their sudden enclosure on the ship was intended to protect them from a surprise attack by pirates. The second time, the ire of the mandarins falls on a group of their colleagues rather than on the ambassadors. The fear of the Spanish travelers dissolves before the hospitality of the mandarins, and even morphs into fondness. On one occasion, a mandarin who has made a ceremonial appearance accompanied by his cane-brandishing servants proves to be a gracious host who treats his guests to food, drink, and conversation (194).[21] By the time the Spaniards depart, they have come to admire their hosts as fellow human beings. The official charged with seeing them off treats them with such "amistad y amor" (friendship and love) that the ambassadors sail away convinced that he and his colleagues are "hombres humanísimos y amorosos" (humane and loving men). The account ends in Manila, where everyone is optimistic about the receptivity of the mandarins to a future diplomatic mission, and even to the truth of the gospel (243–45).

Something similar happens in the Alfaro narrative. Fear of mandarin cruelty dissipates in scenes of cross-cultural respect and even fondness. At one point, the friars intervene as a man is condemned to suffer the *bastinado*, successfully convincing the official to spare the victim from punishment (269). Fault for the failure of the friars falls upon their faithless interpreter, who misrepresents the embassy to the mandarins for his own purposes. The friars, however, discover the double dealing much too late and must leave China having accomplished nothing. Their voyage nevertheless culminates in a tearful farewell from a mandarin whom the text describes as "apacible y humano" (gentle and humane), who feels sorry for the failure of the missionaries, and who recognizes that the friars, too, are "hombres de bien" (good men) (294). Like the Rada narrative, the Alfaro text tells us that, once you get to know them, the mandarins are not at all the cruel tyrants that the Sinophobes believe them to be, but rather measured, civilized, even likable people who are certain to respond to the reasonable arguments of a future diplomatic mission meant to open China's doors to the Gospel and its emissaries.

None of this quite tallies with the surviving accounts of the 1574 embassy written by its leaders, Fray Martín de Rada and Miguel de Loarca. Rada was the first person on record to have suggested that Spain should invade China from the Philippines, in a letter to the Viceroy of New Spain written in 1569, and his account of the embassy is redolent with Sinophobia (Rada 1569; also Ollé 2002, 41–42). Like his contemporary Sande, Rada admits that China is large, urbanized, and prosperous, but he calls the Chinese language "la mas bárbara y dificil que se a descubierto" (the most barbarous and difficult that has been discovered), because it uses a different character to represent each word, making it almost impossible to learn. Hence, anyone who can read is held in high esteem as a scholar, even though Chinese learning is never more than superficial. Although the Chinese make books, and Rada has acquired an impressive collection of them, he is sure that they contain only "el olor o nombre" (the scent or name) of most of the sciences. The Chinese are particularly deficient in mathematics, especially geometry, Rada complains. He finds Chinese sources to be almost useless in his efforts to determine the geographical dimensions of the Middle Kingdom. The friar never calls the mandarins tyrants, but neither does he say anything positive about China's government, and he is sure to include a grisly account of the *bastinado*, as well as remarks about the extreme servility of the Chinese people. On two occasions, he notes that the whippings are applied to everyone, regardless of station, and are even administered to women, suggesting that torture is applied indiscriminately, as it would be in any tyrannical regime (Rada 1575).

Loarca's account is more neutral in tone, neither praising nor disparaging Chinese culture and government and providing no description of the *bastinado* or the prisons, while at the same time including plenty of details that could lend themselves to a Sinophilic account of the country. Loarca, for example, provides a generous catalog of the range of subjects covered by Chinese books that could very well have served as the source of Mendoza's own list (Luarca [Loarca] 2002, 127–29). Where Loarca stands out, however, is in his eagerness to downplay China's military might. Like Rada, he provides eye-popping numbers for the size of China's army, but he also insists that the force is small for the size of the country and that therefore individual cities and provinces are only lightly garrisoned. China's fighting men, moreover, are "la gente mas vil que hay" (the most vile people there are), and the country's artillery is poorly made by European standards. The tack for the horses is poor, and the cavalry ride as badly as "los indios de la Nueva España" (the Indians of New Spain) (103–4). Loarca may have found in the Chinese an impressively politic people, but not a formidable foe.

As we have already seen, neither Rada's Sinophobia nor Loarca's effort to imagine China as militarily vulnerable come across in the *Historia del gran reino de la China*. Mendoza, of course, may have had access to different versions of the participants' reports, as well to other documentary evidence. We know, for example, that he interviewed Fray Jerónimo de

Marín, one of the 1574 ambassadors, while he was in Mexico City on his own abortive mission to China, but have no idea what he learned from those conversations. His handling of certain events described by Rada and Loarca nevertheless suggests at least some conscious manipulation of the sources. Both Rada and Loarca, for example, report witnessing a formal review of Chinese troops. Loarca puts their number at one thousand foot and horse, and he describes how they were drilled in attack and defense with the various weapons at their disposal, but then he goes on to make the disparaging remarks just cited (66). Rada puts the number at only six hundred and makes the same mixture of flattering observations about the drill and disparaging ones about the quality of the men and artillery (Rada 1575). In Mendoza, the number of men grows to twenty thousand pikemen and harquebusiers, and the drill becomes a spectacle of military precision that leaves the ambassadors dumbstruck. It is here that we read the remark that China's army is so vast and well-trained that it could conquer the world if only its soldiers were more spirited (231). The remark does not appear in the surviving copies of Loarca and Rada.

The Augustinian certainly seems to have touched up his primary source for the Alfaro narrative, the 1578 *relación* written by one of Alfaro's companions, Fray Agustín de Tordesillas. Blatant deviations from the source text are few but telling, and all of them serve to enhance the Sinophilic tone of the narrative. In the first instance, Mendoza appends a comment to an anecdote about the *bastinado* and the jails, claiming it demonstrates that "la justicia de este Reino es tanta como pueda ser en cualquier parte del mundo" (justice in this Kingdom is as plentiful as it can be in any part of the world) (267). Tordesillas never editorializes in this way about China's government. At another juncture, Mendoza claims to exclude a whole series of details about the wonders of a Chinese city because they are supposedly identical to what the reader has already seen in the Rada narrative (278). The Tordesillas narrative, however, includes no such details. Alfaro's praise for the street in Guangzhou at the expense of Flemish and Italian cities, cited above, does not appear in the Tordesillas document, nor does Mendoza's characterization of that Chinese official as "apacible y humano" (pleasant and humane). Mendoza's version of the narrative ends on a note of optimism regarding future mission work in Cochinchina, and the possibility that it will lead to the conversion of China itself (294, 302–4). The Tordesillas text does not include this coda.

If, then, the purpose of including the travel narratives is to provide the readers with a vicarious experience of China and the Chinese, then that experience is the one that Mendoza wants them to have, not necessarily the one that emerges from the record left by the travelers themselves.

The relationship between the description in the book's first part and the travel narratives in the second, it turns out, is actually the opposite of what it pretends to be. Rather than derive his Sinophilic description of China form the included travel narratives, Mendoza seems to have drawn his Sinophilic tendencies almost entirely from the bookish Sinography of the Portuguese, the Jesuits, and his fellow Castilian Escalante, and used it to modify the accounts left by the Spanish travelers. The readers are supposed to believe that they are seeing China through the eyes of Rada, Loarca, Alfaro, and the rest, and never realize that Mendoza has them looking through rose-colored glasses of his own fabrication. In doing so, the reader unlearns whatever Sinophobic bias he or she might possess and learns to embrace a Sinophilic perspective.

Among the experiences that Mendoza wants the reader to have is that of traveling westward to China, of experiencing the Middle Kingdom as the final destination of a Spanish evangelical project that has reached the East by sailing west. His version of the Rada narrative does this by beginning with Legazpi's expedition to the Philippines and briefly relating the successive arrival of the various missionary orders to the new colony (161–81). Mendoza's narrative of Alfaro's journey, meanwhile, begins with the arrival of the Franciscans in Manila from Spain and promptly turns to their quest for authorization to continue their work in China (249–53). Both narratives, therefore, figure missionaries on the move, stopping in the Philippines on their westward journey from the New World to China. In both narratives, moreover, divine providence guides and protects the traveling missionaries. In the Rada narrative, the dramatic defense of Manila from attack by the pirate Lin Feng becomes an opportunity to celebrate the heroism of the city's scant defenders, as well as a sign that God wants to safeguard the nascent Christianity of the Philippines so that it can eventually spread to China (170). In the Alfaro narrative, the Franciscans unexpectedly manage to convert a *sangley* priest to Christianity and so acquire the guide they need for their unauthorized voyage (250–51). While the Rada narrative sees the hand of providence in the calm seas and following winds that carry the embassy across the South China Sea, the Alfaro text finds it in the very fact that the Franciscans survive their stormy crossing without the help of a qualified pilot (185, 254–57). Upon reaching the Chinese coast, the Franciscans even benefit from a genuine miracle, when they pass through the cordon of the Chinese coast guard undetected (254–57). So, while Escalante insisted that the route to China by way of the Pacific was preferable to the alternative around the Cape because it was shorter and more practical, Mendoza portrays the westward route as the way that God himself has opened for Christianity's journey to the Middle Kingdom.

Mendoza's treatment of the myth that the Apostle Thomas preached the gospel throughout Asia during the first century of Christianity supports this project. The friar cites the supposed evidence, the existence of an Indian god with three heads, of paintings depicting twelve holy men, each of them of them supposedly relics of ancient belief in the Trinity and the Twelve Apostles. He even mentions Gaspar da Cruz's account of his visit to a shrine dedicated to the Buddhist holy woman Guanyin, who is often depicted with a child and was sometimes identified as a relic of belief in the Virgin Mary (57–59).[22] Yet while he admits that it is "verosímil" (verisimilar) that Saint Thomas preached in China in ancient times, he does not make much of the possibility, and, as Nancy Vogeley notes, he certainly does not take advantage of the opportunity to construct a providential view of history in which the imminent evangelization of China answers to the initiative of one of the Twelve Apostles (1997, 169). The reason for doing so, however, is not the one suggested by Vogeley, that Mendoza's history is secular in nature and does not truck in providentialist arguments. Rather, it is that the myth of Saint Thomas has the potential to undermine his attempt to map China into the Spanish west. It is the Portuguese and the Jesuits, after all, who are following in the footsteps of Saint Thomas, by approaching China from the west. If the Apostle did indeed make it to China as well as India, then providence would seem to be on their side.

Nowhere, however, is Mendoza's effort to locate China in the Castilian transpacific west more in evidence than in the third travel narrative, which tells how the Franciscan friar Martín Ignacio de Loyola sailed from Spain to China by way of the New World, and then from China to Spain by the Cape of Good Hope, between 1581 and 1584. Although Loyola's *Itinerario y compendio de las cosas notables que hay desde España, hasta el reino dela China, y de la China a España* (Itinerary and compendium of all the notable things That lie between Spain and the Kingdom of China, and from China to Spain) is sometimes called an "appendix" to Mendoza's text, it is no more or less supplemental than the other travel narratives. Alone among the three texts, however, it was published as a separate title, in Lisbon in 1586, by a publisher tied to the Discalced Carmelites of the Convent of São Filipe, founded in 1581 and named in honor of Philip II.[23] The very existence of the Lisbon edition, and its publication by a religious establishment associated with the assertion of Hapsburg sovereignty over Portugal, suggests that Loyola's text was perceived as a useful weapon in the propaganda war over the location of China.[24]

Loyola's circumnavigation of the globe was entirely accidental. By his own account, the friar had intended to establish a Franciscan mission in China, but he was expelled from the country under suspicion of espionage. Finding himself in Macau, he made his way back to Spain on Portuguese ships. By tracing the friar's route around the world, the Loyola narrative sets China into a global space, but one that is skewed toward Spanish interests even more explicitly than in the Rada and Alfaro narratives. It also includes an extensive description of New Mexico that makes unique contributions to Mendoza's effort to map China into the Spanish hemisphere. Not only does it help Mendoza figure China in relation to Spain's experience in America, just as Sande had, but it also helps him to paint an image of Spain's missionary future in both the New World and the Middle Kingdom.

The partisan character of Mendoza's world-making is evident in the remarks he makes about the dimensions of the Pacific and the inhabitants of the Ladrones. His account of Loyola's Pacific crossing is a textbook example of the rhetoric of smooth sailing. According to the text, the winds that carry the galleons westward from Acapulco are so steady and favorable that the crews barely need to touch the rigging and need not fear the rare storm. So easy is the way westward that the sailors have taken to calling the South Sea the "Mar de Damas" (the Sea of Checkers), after the notoriously simple board game (338).[25] There is some controversy, the text admits, as to the distance one covers sailing from Acapulco to the Ladrones, but the text favors the shorter estimate (339). Just as in Escalante, moreover, there is no one along the way who can effectively oppose Spanish intentions. There are only the inhabitants of the Ladrones, whom the text depicts as savages worthy of the name, but who could be converted to Christianity quite easily if only the Manila Galleons would stop to drop off a few missionaries with the necessary guards (340). Indeed, the way across the Pacific is even easier than the voyage across the Atlantic, which passes by the island of Dominica, where fearsome man-eaters wait to devour hapless Spaniards. While the text calls for missionaries to be dropped off in the Ladrones, in Dominica it calls for nothing but soldiers who can eradicate the cannibals (311–12).

The "Archipiélago" on the far side of the South Sea, with its capital city of Manila, is not limited to the islands we know today as the Philippines, but rather encompasses all of what we call insular Southeast Asia (342). The Strait of Malacca itself, moreover, is said to be "un mal estrecho y muy peligroso para las naos que van por el" (a very bad strait, very dangerous for the ships that pass through it) (383). It is the only place where the text reports trouble at sea, the shipwreck of a large commercial vessel that goes down with all hands and three thousand ducats of merchandise, right before the eyes of Fray Ignacio (383). In this way, the Strait of Malacca becomes a bookend to the Island of Dominica, as the only other place on Fray Ignacio's maritime tour of the world that functions as a site of resistance to the westward progress of maritime empire. Strait and island take the place of the lines of demarcation, which are never explicitly mentioned, marking the entrance to the Spanish Indies from the Atlantic, and the exit from those same Indies into the Indian Ocean beyond. What the text repeatedly calls "este Nuevo Mundo" (this New World) is clearly divided into two separate halves, the Indian Ocean Basin, which belongs to Portugal, and the transpacific Indies that belong to Spain (383, 385).

The Loyola narrative lavishes attention onto both sides of the "Mar de Damas," devoting five chapters each to North America and China. Two of the five chapters that deal with North America describe Mexico, including Mexico City, while the other three relate the discoveries made in New Mexico between 1582 and 1583 by Antonio de Espejo. The Mexico in question is not the Mexica polity of the past, governed by the tyrant Moctezuma and ripe for conquest by Cortés, but the current-day Viceroyalty of New Spain, an expansive, prosperous, bountiful, well-organized, Christian kingdom governed from an impressive metropolis that is home to the viceroy, the archbishop, the Inquisition, and the *audiencia*. Its native inhabitants, like the Chinese, are "gente dócil y de buenos ingenios y entendimientos" (a docile people with good wits and understanding), who take to Christianity quite readily, and who have never been found to backslide after baptism (318–24). Mendoza even goes so far as to claim that Mexico can be favorably compared with the largest and wealthiest of the kingdoms of the world, with the exception of China (324).

The three chapters about the Espejo expedition appeared for the first time in the definitive 1586 edition of Mendoza's text and provided the first account of Espejo's New Mexico to appear in print. Drawn from the conquistador's own account of his *entrada*, and possibly oral accounts of the expedition that Mendoza collected while in Mexico City on his abortive trip to China, the chapters devoted to New Mexico describe the region and its people quite favorably. They praise the province's natural fertility, its healthy, temperate climate, its agriculture, cattle-raising and mining industries, and they repeatedly note that the region's native inhabitants wear finely made cotton clothes, live in houses of several stories, and enjoy the benefits of good government (327–37; also Vogeley 1997, 170). At the end of the episode, Mendoza remarks that the "buenos ingenios" (strong wits) of the Indians of New Mexico exceed even those of the Mexicans and Peruvians. So reasonable a people, he argues, are sure to embrace Christianity quite readily (337). Mendoza leaves it for the reader to conclude that, in this respect, they are just like the Chinese.

According to Nancy Vogeley, Mendoza's favorable treatment of this region exemplifies the way that the early modern encounter with East Asia inflected the Spanish perceptions of the native inhabitants of America.

That encounter, she argues, taught Europeans to admire a non-Christian society as they never had before, breeding tolerance for other non-Christian peoples whose lifeways smacked of civility. Mendoza's admiration for China, according to Vogeley, led him to portray the religious practices of the New Mexican natives with gentle condescension rather than vitriol, to emphasize the courtesy that the inhabitants of New Mexico extended to the Espejo expedition, and to underscore the importance of the role played by the expedition's interpreter. All of this, in turn, served to suggest that the New Mexicans could be shepherded into the Christian fold through diplomacy and persuasion, without any need of military action, just like China (Vogeley 1997, 172).

Yet this very favorable portrayal of New Mexico could just as well have answered to expectations that were already in place long before the encounter with the Middle Kingdom. As we have seen in previous chapters, Europeans had long expected that the temperate regions of North America would be home to politic societies, and they had repeatedly marched or sailed into what we call the southwestern United States in search of them. It could very well have been these expectations, and not the encounter with China, that led Mendoza to see civility rather than savagery in the news from New Mexico, and to be optimistic about the possibilities for peace in future dealing with the region's inhabitants.

His treatment of New Mexico's human geography suggests as much. The text arranges the indigenous societies encountered by Espejo in ascending order of civility along a route of travel from south to north. The people Espejo first encounters, the *conchos*, *pasaguates*, and *tobosos*, are naked, fight with bows and arrows, live in homes made of thatch, and have no observable religion (326–27). As the expedition presses northward, however, it encounters increasingly more civilized people. First come the *yumanos* or *patarabueyes*, whose homes are made of brick and mortar and whose towns are well ordered, and who are in general people "de más policía que los que hasta allí habían visto" (of greater civility than others they had seen up to that point). Then comes a people whose name Espejo never learns, who possess blankets of blue and white "como las que traen de la China" (like those they bring from China). Their clothing is well made. They speak of a nearby people who live in three-story houses (329). As the expedition reaches a land with pine trees like those of Castile, a clearly temperate region, it encounters people who live in four-story homes, wear finely crafted leather boots, are governed by *caciques* (chiefs) and *alguaciles* (sheriffs) just like those of Mexico, and build temples to the devil. These people own parasols of Chinese style, decorated with the sun, the moon, and the stars (330). By the time Espejo reaches the Tiguas, the Cunamos, the Amejes, the Zuny, the Zaguato, and the Hubates, the towns are as large as Madrid, with as many as forty or fifty thousand inhabitants, and as many as eight plazas (332). The Amejes are singled out for their exceptionally good government, while the Zaguato are said to know of yet other people, just over the mountains, whose cities are so large as to make their own pueblo of fifty thousand people look like mere "barrios" (outskirts) (336). At the northern end of Espejo's itinerary, therefore, Mendoza maps another Cíbola, another phantom of civility equal to that of Europe. None of this is really new. Mendoza's itinerary map of New Mexico is just another version of the notional maps we have seen before, which imagine that the people of North America will display greater degrees of civility the farther north one marches, into the temperate latitudes.

The China of the Alfaro narrative, in turn, looks much as it does elsewhere in Mendoza's text. In fact, Mendoza insists that the only reason he includes a description of China based on Loyola's experience is to demonstrate how much agreement exists among his various eyewitnesses (358). Over the course of two chapters, Mendoza revisits earlier themes, like the country's astounding fertility and productivity, its heavily urbanized settlement patterns, its navigable rivers dotted with countless watercraft, its wealth of natural resources, its sophisticated governing class, its learning, its mighty army, and its false religion (358–67). He often refers the reader to other parts of the *Historia* for corroboration and further detail. The chapters end by insisting that the Chinese respond well to religious instruction and this makes them easy to convert. This is why so many of the resident Chinese of Macau and the Philippines have become good Christians (365–67). In this way, the description of China in the Loyola narrative provides a fitting coda to Mendoza's overall argument in favor of a spiritual conquest of China.

The narrative of Loyola's time in China that occupies the next three chapters begins by echoing the themes of the Rada and Alfaro narratives. Loyola's voyage from the Philippines to China, like the previous crossings, enjoys divine favor. Fair weather carries their frigate across the South China Sea, and providence preserves it from destruction by the guns of a Chinese naval vessel. God then gives the friars the strength they need to spit on a Chinese "idol" when it is expected that they will bow before it (348–52). Loyola and his companions, however, are wracked by fear of imprisonment, torture, and execution even more intensely than their predecessors. When a Chinese vessel carries the party toward a coastal prison, the Filipinos weep bitterly, one of the friars loses his mind, and a second becomes so deeply melancholic that he falls fatally ill (351). At no point, however, does the fear felt by this group of friars morph into friendship with the mandarins, as it does in the other accounts. With the

Loyola narrative, the pattern breaks. There are no banquets, no moments of cross-cultural sympathy. Loyola's party is never treated as anything but a group of criminals, and the travelers never doubt that imprisonment and execution will be their fate. They only escape death when the captain-general of Macau, Arias Gonzalo de Miranda, who happens to be in Guangzhou when the friars are brought there for sentencing, intervenes on their behalf, and negotiates their safe passage to the Portuguese *entrepôt* from which they sail for Spain by the Cape route.

Here, then, at the culmination of Mendoza's stridently Sinophilic account of the Middle Kingdom, the discourse of Sinophobia finally rears its head, apparently undaunted and undefeated. However sophisticated and wealthy the mandarins may be, the text admits, they are also to be feared. In addition they may not be receptive to the word of God, for their lives are too comfortable, their power too complete to make them open to change. Mendoza even goes so far as to call them tyrants, saying that they treat the people as if they were slaves (366). Nevertheless, in the very same utterance in which he makes this aspersion, Mendoza also turns the tables on the discourse of Sinophobia, by suggesting that the very cruelty of the mandarins will make for success among the common people. "Antes abrazarán con gran contento nuestra Santa Ley," the text suggests, "porque será causa de libertarlos de la tiranía del demonio y de los jueces y señores que les tratan como a sus esclavos" (They will sooner embrace our Holy Law, and will do so with glee, because it will free them from the tyranny of the devil and of the judges and lords who treat them as if they were their slaves) (366). Here, the characterization of mandarin rule as tyrannical does not serve to legitimate a military conquest, but to generate hope for the success of a spiritual conquest. Mendoza's Sinophilia finally co-opts the notion of mandarin tyranny in order to argue that the word, not the sword, is what will free the Chinese from the tyranny of their rulers, and of the devil himself.

In this way, Mendoza frames the Mar de Damas with two politic societies, China and New Mexico, both in the temperate zone, both on the northern frontier of Spain's current colonial possessions. Gone is that emphasis on westward movement that had always been a part of the Spanish imperial imagination. The new vector of imperial progress is now south to north, not east to west, from the subjugated tropics of New Spain and the Philippines to the temperate latitudes above them. The nature of that progress, moreover, is peaceful, not military, and its purpose spiritual, not political or economic. Spain's transpacific empire, in Mendoza's *Historia del gran reino de la China*, has become the sort of Christian polity that Las Casas had imagined, eschewing the sword and spreading the word. It is not just China that looks like a utopia in Mendoza's writing, but the whole Spanish Pacific.

One cannot help but wonder, however, how we should map it. The *Historia* remains agnostic about the geography of the northern Pacific, insisting that no one knows for certain how far New Spain extended beyond the lands discovered by Espejo, and the book says nothing about what happens to the coast of China as it trends northeastward (34, 318). The text nevertheless provides evidence that can be used to argue in favor of the possibility that the New World and Asia might be connected in the northern latitudes of the South Sea. Mendoza mentions the presence of Chinese trade goods—textiles and parasols—among the people of New Mexico, thereby connecting the region's economy with that of China, but saying nothing about the nature of the connection. Could the New Mexicans be involved in maritime trade across the South Sea? This was unlikely, since they are nowhere near the coast, and show no signs of possessing a seafaring culture. It is more likely, therefore, that they are trading with China overland, and that therefore North America is indeed continuous with Asia, just as so many Europeans had imagined it be during the course of the sixteenth century. New Mexico, therefore, is not just the northern frontier of the Viceroyalty of New Spain, but a borderland between Spanish America and China.

We can find a fitting image for Mendoza's Pacific world in a map made by another missionary, the Dominican Juan Cobo. In 1593, Cobo published one of Manila's first printed books, the *Bian zheng jiao zhen chuan shi lu* 辨正教真传实录 (Apology for the true religion), a philosophical propaedeutic based on Fray Luis de Granada's *Introducción al símbolo de la fe* (1584) and meant to establish the credibility of the Christian religion in the mind of an educated Chinese reader.[26] Commonly known as the *Shi lu*, the text includes a zonal map of the world oriented toward the south, centered on the Pacific, depicting China and Mexico on opposite sides of that ocean (figs. 29 and 30). The map emphasizes the location of both countries in the northern temperate zone and thereby tries to convince its Chinese readers that the Christians are just as civilized as the Chinese themselves, hailing as they do from a place whose climate was just as moderate as that of the Middle Kingdom. Yet the map also places the two countries on a single continuous Amerasian landmass that contains the intervening ocean on several sides. It was not enough, it seems, to demonstrate that Mexico was in the same climatic zone as China. It was also necessary to insist that the two countries formed part of the same continental landmass, the same geographical world. It is tempting to see in Mendoza's unexplained account of Chinese trade goods among

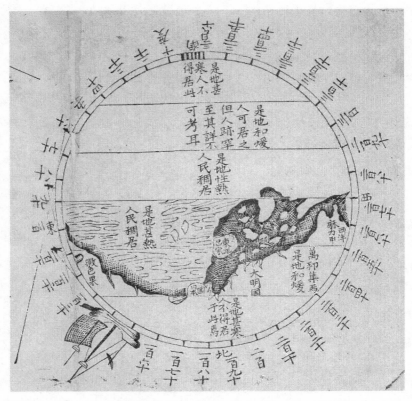

FIGURE 29. This zonal map of the world, oriented toward the south, depicts a continuous coastline from Mexico in the four o'clock position to China in the seven o'clock position. From Juan Cobo, 辨正教真传实录 (*Bian zheng jiao zhen chuan shi lu*) (1593). Property of the Biblioteca Nacional de España. Used with permission. Photograph: Courtesy of the Biblioteca Digital Hispánica (R/33396; bdh0000165702).

the New Mexicans a shadow of that same necessity, to see in the apparent similarities between the Middle Kingdom and Espejo's New Mexico a sign that both places formed part of a single circumpacific world that was only then coming into view.

The Return of the Hawks

Nowhere, however, does the vision of a fully realized transpacific empire appear more clearly than in the next prominent wave of hawkish Sinophobia that emanated from Manila, ironically in the same year that the

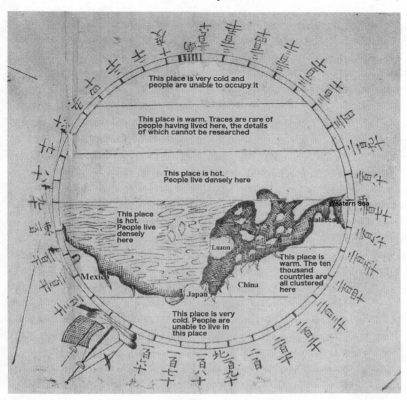

FIGURE 30. Juan Cobo's zonal map (fig. 29) with place names translated into English by Timothy Brook.

definitive edition of Mendoza's history came out in Madrid. In 1586, the so-called Synod of Manila, a gathering of the secular and ecclesiastical authorities that governed the Spanish Philippines, developed a plan to invade China and dispatched the Spanish Jesuit Alonso Sánchez to Spain to present it to the king. Like the Sande plan before it, the Sánchez plan grew out of frustrated diplomatic efforts. A few years earlier, Sánchez had traveled to China as head of an embassy charged with securing the loyalty of the Portuguese in Macau to Philip II and the permission of the Ming to establish a Spanish trading post on the Chinese mainland. Although he successfully achieved the first objective, he never came close to achieving the second. Returning to Manila after a frustrating voyage marked by constant fear of summary execution and the humiliating experience of repeated exhibition as a foreign curiosity, Sánchez argued before the Synod that the Chinese were so thoroughly convinced of their own superiority that they would never allow missionaries to enter their country unless they were compelled to do so militarily. He convinced the Synod to endorse an invasion plan, the most elaborate and ambitious one yet.

The story of Alonso Sánchez and the plan he placed before Philip II has been told many times, so I will not examine it in detail here.[27] Instead, I will limit myself to some remarks about its vision of transpacific empire, by way of providing a coda to this chapter. Sánchez and his collaborators placed even higher hopes than Sande did on Spain's ability to project power across the South Sea. The plan calls for an expeditionary force of over 20,000 men, including 5,000 or 6,000 Japanese and as many native Filipinos to be recruited locally in Manila, and 12,000 Europeans, including Spaniards and Italians, to be shipped across the South Sea from Europe and the New World, along with technical personnel of all kinds (artillery founders, machinists, etc.) from Spain (Vera 1859, 26). Like the Sande plan, it envisions the quick capture of one of the maritime provinces, followed by political chaos that will play to the invader's advantage, but adds a pincer maneuver on a global scale, supplementing the Castilian assault from Manila with a coordinated attack by the Portuguese from the *Estado da India*. These troops would march behind missionaries preaching the Gospel, persuading the Chinese to become Christians by both word and sword.

The document, however, does not limit itself to outlining a plan for military action. It also entices its reader by painting a picture of the transpacific empire that would emerge from the conquest of China. As lord of China, the document claims, the Hapsburg monarch would automatically enjoy suzerainty over nearby countries, such as Cochinchina, Cambodia, and Siam. Not only the Middle Kingdom, but all of the Indies of the West would come under Spanish control. Spain could then carve up the entire region into a series of highly civilized viceroyalties, stalwart Catholic bulwarks against Protestant incursion in the region. Countless churches and monasteries would be founded, bringing salvation to innumerable Chinese souls. *Encomiendas* and titles of nobility would be granted, enriching and ennobling many a Spaniard, and providing sure revenue for their king. Schools and universities would crop up all over, so the Chinese could learn Latin. A hundred galleons a year would sail between Spain and China, bringing untold wealth to the mother country. The Chinese themselves would make the voyage across the South Sea to Mexico and Peru, thereby participating fully in the life of the Spain's overseas empire (Vera 1859, 34–48).

In all these ways, the transpacific empire that would emerge from the conquest and conversion of China, argue the authors of the 1586 invasion plan, would solve all of Spain's most pressing problems, economic and political, and would bring unprecedented glory to church and crown. In doing so, moreover, it would compensate for what the document describes as the failure of the Spanish enterprise in the New World. The soldiers would be selected for their upstanding character and would be supervised by ecclesiastics, thereby preventing the wanton destruction characteristic of the conquest of America. Unlike the native inhabitants of the Americas, moreover, the Chinese were white and civilized, their women beautiful, chaste, and sophisticated. Spanish men would be able to marry Chinese women without reservation, just as the Portuguese already did in Macau, producing a mixed lineage that would readily adopt Christianity and European lifeways. Thus, while the conquest of the New World had led to nothing but "barbaridad y perdición y depravación temporal y espiritual de las leyes, gobierno, haciendas, pueblos y personas de todas las tierras" (barbarity and perdition and depravation, temporal and spiritual, of the laws, government, fortunes, peoples and persons of all the lands), the conquest of China would produce a fully realized Christian empire, one that would bring glory rather than shame to Catholic Spain (Vera 1859, 41). Ironically, however much the Sinophiles idealized China and the Chinese, it was the Sinophobes who ended up proposing the most utopian vision of all, locating China in an idealized Spanish Pacific world that promised not only the salvation of countless Asian souls, but the salvation of Spain itself.

7

The Kingdom of
the Setting Sun

As Mendoza's history of China painted a rosy picture for the future of Christianity in China, the established Christian mission in Japan reached an ominous turning point. In 1587, the Japanese strongman Toyotomi Hideyoshi (1536–1598) banished Christian missionaries from the island of Kyushu. These were members of the Society of Jesus, whose work in Japan went back to 1549, when Francis Xavier arrived on the shores of a Japan wracked by civil war. By their own accounts, the Jesuits had succeeded in attracting converts by the thousands, notably among the ruling elite.[1] Their work survived the 1587 ban only because Hideyoshi himself was reluctant to enforce it, aware as he was of the inextricable link between the Jesuit enterprise and Nagasaki's lucrative trade with Portuguese Macau. The official ban clearly did not dissuade other missionary orders from seeking Hideyoshi's permission to preach in Japan, or from obtaining it. During the 1590s, the Jesuits were joined by Franciscans hailing from Manila, who brought with them the promise of improved commercial ties between Japan and the Spanish Philippines.[2]

The approach that the friars took to their work, however, was diametrically opposed to the one adopted by the Society of Jesus. While the Jesuits conformed to Japanese lifeways as much as they could, taking pains to learn the language, to adopt local dress and customs, and to integrate themselves into elite society, the Franciscans chose to live in their own religious communities, wear their distinctive brown habits, say mass publicly, distribute rosaries, and preach to the lower classes by way of interpreters. Their very public approach to evangelization and their refusal to adapt to Japanese ways alarmed members of the Buddhist clergy and the Japanese ruling class, who had always harbored suspicions about the real intentions of the *namban* or "Southern Barbarians," as Europeans were known. Tensions rose, and when the galleon *San Felipe* foundered off the coast of Japan in 1596 and its captain boasted that the friars were the vanguard of a military invasion, those tensions finally flared.

Hideyoshi chose to enforce his edict in exemplary fashion, ordering the crucifixion of twenty-six Christians, including six Spanish Franciscan friars, seventeen Japanese Franciscan Tertiaries, and three Japanese Jesuit lay brothers, on the outskirts of Nagasaki on February 5, 1597.

News of the Nagasaki martyrdom traveled throughout the Catholic world as quickly as it could and inspired devotion to the martyrs, particularly in Spanish America, where they became the subject of a substantial iconographic tradition. The Franciscans were beatified in 1627 by Urban VIII and canonized in 1862 by Pius XI. Nevertheless, in the immediate aftermath of the event, controversy swirled around their actions and their fate. The Jesuits feared that that Nagasaki crucifixion spelled the beginning of the end for the Christian mission in Japan, and they accused the Franciscans of spoiling everything through imprudent behavior. The Franciscans, for their part, accused the Jesuits of playing a hand in the persecution that had led up to the execution, in an attempt to eliminate the spiritual and commercial competition represented by them and their ties to Manila.[3] Much of the confrontation took place behind the scenes as the two orders jockeyed for favor among the powerful at the Hapsburg and papal courts, but it also played out in print, as each order published its own accounts of the Nagasaki crucifixion. Franciscan histories glorified the victims, and emphasized the miracles and marvels that supposedly accompanied the event, while Jesuits accounts denied that any miracles had taken place and questioned whether or not the executions constituted true martyrdoms.[4]

One of the principal Franciscan contributions to the polemic was written by Marcelo de Ribadeneira, a Discalced Franciscan who had witnessed the crucifixions from a ship in Nagasaki harbor, and who had sworn an oath in Manila stating that Jesuit schemes to have the friars expelled from Japan had been decisive in turning the Japanese leadership against his order (Boxer 1974, 421 n. 7). Assuming the role of official advocate for the beatification of his executed brethren, Ribadeneira made for Rome, writing along the way. By the time he reached Europe, he had completed the better part of his *Historia de las islas del Archipiélago Filipino y reinos de la Gran China, Tartaria, Cochinchina, Malaca, Siam, Cambodge y Japón* (History of the islands of the Philippine Archipelago and the kingdoms of Greater China, Tartary, Cochinchina, Malacca, Siam, Cambodia, and Japan), which was published in Madrid in 1601.[5] The text relates the story of the executions in vivid detail, but also places them in the larger context of the history of the Franciscan Province of Saint Gregory, the administrative unit of the Franciscan order under which the mission in Japan had operated. The province had been established in Manila in 1586 by members of a Spanish branch of the Order of the Friars Minor and was authorized to preach throughout East and Southeast Asia. The first half of the text tells how the province came to be founded and relates its activities in the Philippines and on the Asian mainland. The second half covers the mission in Japan, plotting the Nagasaki martyrdom as the climax of the entire story.

Ribadeneira's text thus engages in a public debate over one of the most significant pieces of news to reach Europe from East Asia during the late sixteenth century. In so doing, it also fights a turf war with rival religious establishments in East and Southeast Asia, as Liam Brockey has demonstrated. The text denies that any religious establishment operating out of the *Estado da India*, including the Franciscans in Goa and Macau, enjoyed exclusive access to the region and its countless souls, and it takes specific aim at the Jesuits by including a Spanish translation of the bull granted to the Province of Saint Gregory by Pope Sixtus V, who happened to be a Franciscan himself (Brockey 2016, 8). Countermanding Gregory XIII's 1585 bull granting the Jesuits exclusive access to Japan, the new document authorized the members of the Province of Saint Gregory to preach in "las mismas islas Filipinas, como en otras cualesquier tierras y lugares de las sobredichas Indias y de los reinos llamados de la China" (in those same Philippine Islands, as in any other places and lands of the said Indies and the kingdoms known as China) (Ribadeneira 1970, 1.94). In this way, Pope Sixtus's bull mapped East and Southeast Asia as the mission territory of the Franciscans in Manila, an ecclesiastical counterpart of the secular Indies of the West. Yet Ribadeneira's narrative does not just bicker over the ecclesiastical boundary in Asia: it also delinks East and Southeast Asia from the cultural and religious history of adjacent South Asia and attaches the region to the history of Spanish America as plotted by members of the Franciscan order itself. As it turns out, Ribadeneira's attempt to define the historical meaning of the Nagasaki martyrdom was deeply rooted in the Spanish tradition of mapping the Far East as the transpacific West.

The *Historia del Archipiélago Filipino* even makes significant contributions to the way that could be accomplished. Unlike other attempts to map the Far East as the Far West, Ribadeneira's text hardly engages in geographical description at all. It makes only the barest mention of distances and longitudes, never traces a coastline in either word or image, and ignores the debates over the position of the antimeridian. In fact, the text even suggests that its author may have been astonishingly ignorant about some basic geographical details. As a result, the cartography of containment, which invokes specific configurations of land and water on the surface of the terraqueous globe to divide the Ocean Sea into a series of relatively small maritime basins, shines by its absence. So, too, does the rhetoric of smooth sailing, at least in the form that we have seen it so far. Rather than silence the difficulties involved in crossing the Pacific, the text's account of the journey made by the founding friars of the province from Spain to the Philippines via Mexico actually elaborates upon the suffering endured while traveling by sea. The text thus merits attention as an important counterpoint to the examples we have seen so far.

It also allows us to explore the particular role played by Japan in the Spanish geopolitical imagination. Scholarly interest in early modern writing about Japan has naturally focused on the massive output of the Jesuits, some of whom were keen and sensitive observers of Japanese culture. Yet Japan was not just an object of ethnographic inquiry. It was also a key location in many attempts to imagine oceanic space, from the fifteenth century through the sixteenth. Before Japan was Japan, it was Zipangu, an oceanic island known exclusively through the description that appeared in Marco Polo, but one that figured prominently in the imagination of anyone who wanted to imagine the Ocean Sea as a basin that could be traversed, rather than as an impassable expanse (Gillis 2009, 45–64). It continued to serve as a way station of the mind, moreover, after the New World appeared on European maps, and even after medieval Zipangu morphed into early modern Japan. In Ribadeneira's text, Japan becomes a way station on the westward pilgrimage of Christianity from its Mediterranean place of birth to East Asia, where the conversion of countless souls to Christianity will usher in the millennium of peace that is supposed to precede the end of the world.

This chapter begins by elaborating on the role played first by Zipangu and then by Japan as a significant way station on the way west. It then examines how Ribadeneira grafts this way of mapping Japan onto the spatial structure of Franciscan millenarian historiography, as it was developed over the course of the sixteenth century by his coreligionists in New Spain. As we shall see, the Nagasaki martyrdom, by Ribadeneira's reckoning, was not an event in the history of "the Orient," but a milestone in the history of an apocalyptic Occident under construction. A final section turns to Ribadeneira's candor regarding the miseries of travel by sea. As we shall see, the friar's anecdotes of the horrors of shipboard life have more to do with the celebration of a particularly Franciscan style of piety than with the nature of the oceans or ocean travel as such, yet they nevertheless provide a glimpse into how much Spain's relationship with the South Sea had changed since Magellan's Pacific crossing.

Zipangu, Japan, and Cartographic Liminality

Ribadeneira's text draws on an established vision of Japan, not only as a location to the west, but as a place of passage, originally from the New

World to the Indies of Marco Polo, and later from the New World to the Indies of the West. According to Marco Polo, Zipangu was the island home of a civilized people who enjoyed access to spectacular wealth, particularly in gold, that lay 1,500 miles east of the Asian mainland, at the easternmost edge of the known world. Its position gave it a prominent role in the designs of anyone who, toward the end of the fifteenth century, thought it was possible to reach the Indies by sailing west. Toscanelli identified Zipangu as a crucial way station on the optimal route from the Canary Islands to Cathay, one that made it possible to imagine how one could survive the otherwise impossibly long voyage across the Ocean Sea (Fernández-Armesto 1987, 250). Martin Behaim gave visual form to the idea on his globe, where Zipangu forms part of a chain of islands, some known and some hypothetical, that span the Ocean Sea from Europe to Asia (see fig. 16). As John Gillis argues, that chain of islands does not just answer the practical questions of the navigator pondering a voyage west, but speaks of the transformation of the Ocean Sea in the European imagination, from a menacing expanse at the edge of the world into a navigable basin (2009, 45–64).[6] In this way, Janus-faced Zipangu was crucial to the origins of what would become the cartography of containment and the rhetoric of smooth sailing.

Schooled by Toscanelli on the importance of Zipangu as a potential stopping place, and well aware of the appeal of the rich island as a destination in its own right, Columbus kept a keen eye out for it on his 1492 voyage and even convinced himself that he had found it, in the island he christened Hispaniola. That identification became the subject of some controversy over the course of the next half-century. The Italian mapmakers Giovanni Matteo Contarini and Francesco Roselli disagreed with the admiral and placed Zipangu west of the islands he had discovered on their 1506 world maps, while Johannes Ruysch went along with the explorer, identifying Hispaniola as Zipangu on his 1507 map of the world (see fig. 17). Oronce Finé first rejected the identification of Hispaniola with Zipangu on his double-cordiform map of the world (1531), but then accepted it on his single-cordiform (1534–36). (See fig. 13.) Caspar Vopel accepted it as well, as we can see on Vavassore's copy of his world map (1558). (See fig. 6.) Waldseemüller (1507), Apian (1524), Münster (1540), and Mercator (1544), by contrast, were among those that rejected the idea by clearly distinguishing between Hispaniola as an island in the North Sea and Zipangu as an island in the South Sea (see figs. 3 and 24). Others, like the crown's cosmographers in Seville, chose to avoid the issue altogether, leaving Zipangu off the map in recognition of the fact that no modern explorer had discovered an island answering to Marco Polo's description of civilized people and spectacular wealth.

Lawrence Wroth helps us understand what was at stake in all this. Since Zipangu marked the easternmost edge of the geography of Marco Polo, its location was intimately attached to ideas about the landmass we call North America. If one believed that landmass was where one could find Marco Polo's Cathay, then it made sense to keep Zipangu in what we call the Caribbean. If, by contrast, one believed that landmass was not part of Marco Polo's geography, but something new to European knowledge, then it made sense to place Zipangu west of it, in the South Sea (Wroth 1944, 201–4). This was true whether one believed North America was part of an insular America, the fourth part of the world, as it is on the 1507 Waldseemüller and other maps inspired by it; a large Indian island, as it appears on the 1540 Münster; or a portion of the fourth part of the world connected by a land bridge to Asia, as on the 1565 Gastaldi (see fig. 19). It was also true if one remained agnostic about the relationship between the New World and Asia, as on the 1544 Cabot (see fig. 22). In any case, when the island was placed in the South Sea, its location helped identify North America as something new and different, whether it was an extension of Asia or not. In those cases, the island could appear close to Asia, close to America, or somewhere in the middle, but it always marked the location where the world revealed by Spanish exploration gave way to the world described by Marco Polo and the Portuguese.

This game of cartographic shuttlecock, as Wroth calls it, came to an end as Zipangu morphed into "Japan" around midcentury. Knowledge of a place that went by that name passed from Malay informants to Portuguese knowledge networks, although it took some time to determine that Japan was the same place that Marco Polo had called "Zipangu." The name appeared for the first time in European writing as "Jampon" in the *Suma oriental* of Tomé Pires (ca. 1512–13), and as the name of a single oceanic island called "Giapam" on Giacomo Gastaldi's 1550 map of the western hemisphere (see fig. 23). In 1563, Antonio de Galvão made an important case for identifying Japan as Marco Polo's Zipangu, arguing that Japan's wealth in silver gave it away as the Venetian's rich island (Lach 1965, 1.2: 653). As time went on, the oceanic island of medieval fable became an Asian archipelago that hugged the mainland, morphing from one island into several and acquiring more realistic contours. In 1568, "Japam" appeared on a European map of its own for the first time, courtesy of the Portuguese mapmaker Fernão Vaz Dourado (Lach 1965, 1.2: 710). The cartographic model he pioneered developed as European mapmakers gained access to Japanese maps, particularly those of the *gyōgi* type, which depicted the general layout of the country's provinces and roads in diagrammatic fashion (Unno 1994, 366–71). The *gyōgi* tradition is clearly at work in the map of Japan by Luis Teixeira that Ortelius printed in 1595, which became the model for European maps of the country for the next hundred years (Lach 1965, 1.2: 709).

While Portuguese mapmakers transformed Zipangu into Japan, letters from the Jesuits replaced Marco Polo's fables with knowledge derived from direct and prolonged contact with the Japanese. Collectively, the Jesuits figured the Japanese as white and highly civilized, "the best people who have yet been discovered," in Francis Xavier's oft-cited phrase. They marveled at the warlike nature of the Japanese and, having arrived during Japan's Sengoku Period (1467–1568), when the country was deeply divided and at war with itself, they remarked upon its political volatility. Although they admitted that much of the populace lived in dignified poverty, they also heaped praise upon the size and orderliness of Japanese cities and testified to the literacy and learning of the Japanese elites. Not only did the Japanese have a written language, they noted, but various universities as well. The missionaries generally emphasized the reasonability, strong character, meticulous manners, and fierce sense of honor of the Japanese people, arguing that they could be brought to Christianity by means of rational persuasion alone. Most of their censure fell upon the Buddhist clergy, or "bonzos," who were generally characterized as morally degenerate but intellectually subtle, the only real obstacle to the triumph of the faith in Japan.

This image was forged primarily in the letters that Jesuits in Japan sent to their coreligionists back home, and which were then published in Lisbon, Madrid, and Rome. So firm and steady was the flow of letters that Japan came to figure more prominently than any other Asian country in the Jesuit compilations published between 1552 and 1600 (Lach 1965, 1.2: 674). The early contributions of Francis Xavier made their way into Ramusio's collections, converting the saint's initial impressions of the country into a standard account of Japan and the Japanese, but more detailed and more intimate accounts could be found in the powerful syntheses created by other Jesuits, such as Alessandro Valignano, Giovanni Pietro Maffei, and Luis de Guzmán. Valignano's efforts were particularly valuable, insofar as they were enriched by the insights of Luis Frois, a keen observer of Japanese culture, but they did not circulate in print during the early modern period.[7] The work of Maffei did, though, as did an important history of the work of the Jesuit order in Japan by Guzmán (1601). Donald Lach singles it out as the best synthesis of the Jesuit letters up to that point and claims that it "gives the reader the feeling that the Japanese are a real and plausible people" (1965, 1.2: 711).

In retrospect, we know that no one from the period even approached the depth and breadth of what the Jesuits had to say about Japan, but if we are to understand how Ribadeneira could continue to map the island as a stop on the way west, in other words, as a new and improved Zipangu, we have to appreciate that in Spain, at the beginning of the seventeenth century, many people simply did not read what the Jesuits wrote, and when they did, they did not always believe it. Despite the power and prestige of the Society of Jesus in Hapsburg Spain and Spanish America, its publications about the world beyond Europe did not always reach readers outside ecclesiastical circles, and its vision of Japan as a civilized country on the verge of embracing Christianity did not always take hold. As Christina Lee argues, this continued to be true even after the 1584 visit of four Japanese ambassadors to the courts of Spain and Italy as part of an elaborate publicity stunt meant to sell the Jesuit vision of Japan to early modern Europe and curry favor from the powerful. Although the Jesuits achieved many of the ends they sought, including the bull from Gregory XIII giving them exclusive access to Japan, the impact of the embassy does not seem to have endured, at least not in Spain. When Luis Sotelo tried to repeat the Jesuit trick in 1614, by bringing his own Japanese ambassadors to visit king and pope, Lee points out, the Spanish court failed to receive the Japanese visitors as ambassadors from a near-peer society. Some Spaniards, moreover, had no trouble referring to the visitors as "gente bárvara" (barbarous people), and one of the envoys even fell into slavery after his companions had decamped for home (C. H. Lee 2008, 357, 367). The conviction that non-Europeans lacked real civility ran deep, and it could endure whatever the Jesuits had to say about the Japanese, particularly if no one read it. Evidence for the durability of the notion that the Japanese were just as barbarous as any other "Indians" can be found as late as 1621, when Lope de Vega Carpio brought the Japanese to the Spanish stage in his play *Los mártires del Japón* (The martyrs of Japan). He had no trouble dressing his Japanese "indios" in the same feathered headdresses and equipping them with the same bows and arrows that a theater company might have used to stage Amerindians, or any other savage people of the Indies for that matter (C. H. J. Lee 2006, 24–27).

Lope's play also maps Japan as an oceanic island in the west, attesting to the durability of old habits of mind associated with Zipangu as well as to notions of Japanese barbarity. In the opening lines of the play, one of the courtiers of the Japanese ruler addresses him as, "Emperador invicto del poniente" (Unconquered Emperor of the west) (Vega Carpio 2006, 46). The tendency to imagine the island as a way station on the westward route had survived its transposition from the North Sea to the South and was alive and well in the early seventeenth century. As Europeans began to cross the Pacific, they kept an eye out for Zipangu just as Columbus had, and for precisely the same reasons. Pigafetta claims that Magellan's fleet must have passed the island without actually sighting it on its long and arduous Pacific crossing, quietly lamenting the fleet's failure to make landfall when it so urgently needed to do so (1969, 27). A few years later, Loaysa and del Cano made for Zipangu after sailing through the Strait of Magellan, desperate as they were for shelter from the storms that battered their fleet. Twenty years later, Alonso de Santa Cruz continued to think of Zipangu as an oceanic island lying somewhere in the South Sea, insisting that it lay so far to the east that it should be counted among the islands of the occident rather than those of the orient (2003, 291v). For the royal cosmographer, Zipangu was a border marker, just has it had been for Toscanelli, Behaim, and Columbus, although he thought of it as marking the end of the West rather than the beginning of the East. His contemporary Sebastian Cabot disagreed about the island's position, placing it much closer to Asia, but still in the west, and still inside the Castilian demarcation (see fig. 22). He also located it close to the Marianas, which were then the only confirmed way station on the route from New Spain to the Islands of the West, as if to suggest that Zipangu would make a much better stopping place. Cabot also spans the empty space between Asia and America with a description of Zipangu clearly derived from Marco Polo, thereby using conventional knowledge of the island to mediate between the known parts of the Spanish demarcation in a very concrete, material way.

Zipangu's function as an intellectual way station across the South Sea even survived its transformation into Japan. One of the earliest surviving texts describing a place called "Japan" is a 1548 account of the Villalobos expedition written by one of its survivors, García de Escalante Alvarado. Near the end of his account of this disastrous attempt to conquer and settle the Islands of the West, Escalante Alvarado shares what he knows about East and Southeast Asia, as garnered from Iberians in the Moluccas. One of them is a Galician merchant who claimed to have been to Japan, although his contention that the Japanese knew nothing of the use of swords sheds doubt on his veracity. Nevertheless, García Escalante's account of the country includes a number of the stereotypes that would form part of the standard image of Japan cultivated by the Portuguese, including remarks about the use of *kanji* characters to write with, as well as the claim that the Japanese language resembles German. The text even locates the country with reasonable accuracy at 32° N latitude, and 150 leagues from the coast of China, at roughly the position of the island of Kyushu (Varela 1983, 177).

Yet although Escalante Alvarado places Japan off the coast of China, he does not divorce it from the New World. On the contrary, he assimilates it to what others called "Greater India" or "the Provinces and Islands of the West," by noting that the clothing worn by Japanese workers resembles the kind that Coronado had seen "en la tierra dó fue" (in the land where he went) (2015, 178). He may have relied on others, including the Portuguese, for knowledge of East and Southeast Asia, but he mapped what he learned into a geopolitical imaginary that was shaped by Spain's territorial claims and the experiences of Spaniards looking out at the South Sea from New Spain. For him, Japan provided the missing link between what we call North America, as it was understood in New Spain, and East and Southeast Asia, as he learned about it from informants in the Moluccas. His conception seems to have been shared by Giacomo Gastaldi, the first mapmaker to use the name "Giapam" on a printed map. On his map of the western hemisphere, Giapam appears as a single oceanic island a mere 65° west of Cíbola, just over half the longitudinal distance separating Los Angeles from Tokyo.[8] (See fig. 23.)

Spaniards continued to map Japan as a landmark on the way west forty years later, after they had established themselves in the Philippines. In a 1587 letter addressing the need to send more missionaries to Japan, Domingo de Salazar mentions the recent arrival in Manila of a group of Dominicans, noting that the members of that order (his own) are reputed for their learning, and that this was precisely what one needed in order to spar with the clever bonzos. Clearly, the bishop of Manila is not interested in enhancing the Jesuit mission that had reached Japan under Portuguese auspices. He wants to develop a new missionary effort under the auspices of the Castilian *Patronato*. In order to get the resources he needs, he reminds the king that it is his duty to send preachers to Japan, just as he has done for "los *demas* Reynos de las indias occidentales" (the *other* kingdoms of the West Indies [emphasis added]) (Ruiz Vega 1587, 13r). For Salazar, Japan is one of the kingdoms of the West Indies, a part of what López de Velasco had christened the Indies of the West.

Of course, this statement could be understood as nothing more than a rhetorical ploy aimed at a monarch or minister who mapped Japan in this way, rather than as an indication of how Salazar himself imagined East and Southeast Asia in relation to America, yet there are other passages from his letter that suggest the contrary. The letter was meant to comment upon a report, signed by eleven Japanese Christians but written in Castilian, that had arrived in Manila from Japan, outlining the state of the Japanese church and including a crude sketch of a *gyōgi*-style map of Japan.[9] Salazar must have been quite satisfied to have had access to historical and geographical knowledge of the country that had not passed through Jesuit hands. Twice he mentions that certain aspects of Japanese life remained uncertain, because the only source of information was the Society of Jesus. Clearly, he was skeptical about the veracity of the Jesuits regarding Japanese civility and the success of their mission. Why would he not be skeptical as well of the way they mapped East and Southeast Asia as part of a Portuguese and Jesuit Orient? Like his countrymen, Salazar had arrived in the Philippines by way of Mexico and the South Sea, that is, by constantly traveling westward. Why would he, like his countrymen, not allow his experience of East and Southeast Asia as the terminus of his westward vector of travel to influence the way he mapped it into his world?

His fellow Dominican Juan Cobo not only maps Japan into transpacific space, but precisely as the sort of liminal location that it had represented for Escalante Alvarado. As we saw in the previous chapter, the map that appears in his *Shilu* locates East and Southeast Asia at the western edge of transpacific space, placing China and Mexico at either end of a single Amerasian landmass (see figs. 29 and 30). It also depicts Japan as a place of transition, between Mexico on one end and China at the other, along the line that divides the intervening ocean into two distinct spaces. Japan sits just left of the map's central vertical axis, jutting out from the Amerasian coastline as if it were a peninsula rather than an island or archipelago, along the border between the more darkly shaded waters to the east and the more lightly shaded ones to the west. According to Robert Batchelor, the clear contrast created by the differential shading reflects the Chinese distinction between *hai* 海 (sea) and *yang* 洋 (ocean). (E-mail message to author, October 12, 2018.) If he is right, then the treatment of the seas might serve as evidence of the active involvement of Cobo's Chinese collaborators in the production of the map. Yet by making that distinction, Cobo also converts Japan, along with Luzon to the south, into a place of passage from one oceanic space to another. In other words, he assigns it much the same role it had played in the Spanish geographical imagination since the Magellan expedition. On Cobo's map, Japan articulates the world that the Spanish have created in the eastern Pacific with China's sphere of influence in the western Pacific.

Back in Spain, meanwhile, Antonio de Herrera y Tordesillas was busy writing Spain's official history of its empire in the Indies, equipping it with a geography and cartography that insisted upon Castile's claim to the Indies of the West. The first installment of his history, *Historia general de los hechos de los castellanos en las islas y Tierra Firme del Mar Océano* (General history of the deeds of the Castilians in the islands and mainland of the Ocean Sea), appeared in Madrid the same year as Ribadeneira's text. It included the map that we saw at the beginning of the book, which maps Spain's overseas empire as everything that lay between the lines of demarcation, including East and Southeast Asia (see fig. 2). I shall have more to say about this map and the book in which it appeared in the next chapter. For now, I only note that the attempt to map the Far East as the Far West was alive and well in the Hispanic world at the turn of the sixteenth century, both on the front lines in Manila and back home in Madrid. Spaniards continued to map Japan into the west and even continued to think about it as a stopping place on the way from Mexico to China.

Luis de Guzmán took direct aim at this way of imagining not only Japan, but all of East and Southeast Asia. He was not just a sensitive commentator on the culture of the Japanese, but a warrior in the battle over how the Nagasaki martyrdom was to be remembered. Written in Spanish and published in Alcalá de Henares the same year that the histories of Ribadeneira and Herrera appeared in nearby Madrid, Guzmán's text stands as a direct rival of the friar's account of the event and everything that led up to it. In an appendix to the history proper, Guzmán defends the Society's conduct in Japan and absolves it of any responsibility for the executions at Nagasaki. The blame for the executions, he argues, falls squarely on the friars themselves, for failing to follow either the guidelines that Hideyoshi had established for the proper conduct of Christian evangelization or the advice that the Jesuits offered about how to properly comport themselves (1601, 2.645–712). Most significant in this context, however, is Guzmán's recognition that the battle over the meaning of Nagasaki cannot be separated from the larger war between Jesuits and mendicants over territorial prerogatives. Just as Ribadeneira provides a Spanish translation of the bull authorizing Franciscan activity in Japan and elsewhere, so Guzmán provides a Spanish translation of the 1585 bull by which Pope Gregory XIII granted the Society of Jesus exclusive rights to preach Christianity in Japan and defends the pope's reasons for issuing it and the Society's reasons for seeking it in the first place. He also points out that King Philip II took actions to make sure the pope's conditions were followed, and that Pope Clement VIII reaffirmed the Society's privileges in 1597. Japan, Guzmán insists, is Jesuit territory, and the Franciscans had no right to be there in the first place, nor do they have any right to return.

Yet Guzmán also recognized that the turf war among the missionary orders formed part of a larger struggle to define the location of what we call East and Southeast Asia. He makes his own contribution to that struggle at the outset of his book, in a pair of chapters describing "Oriental India." Guzmán carries out his description by tracing a discursive itinerary along the shores of Africa and the Indian Ocean Basin, reaching a crucial point of inflection when he reaches the mouth of the Ganges. There he acknowledges that there has been some debate about whether Oriental India ends at the mouth of the Ganges River or extends all the way to Malacca. Rather than take a position in the debate, however, he dismisses the controversy altogether by appealing to prevailing linguistic usage: "Se suele llamar con nombre de India Oriental (tomado ampliamente), toda la tierra firme que ay desde Malaca a la China" (It is common to call 'Oriental India' (taken broadly) all of the mainland from Malacca to China) (1601, 1.4). According to Guzmán, linguistic common sense tells us that everything east of the Ganges and of Malacca is part of "Oriental India" because it is part of the Asian mainland.

This statement merits close attention. Underneath its explicit appeal to common linguistic usage lies an implicit appeal to the architecture of the continents and its supposition that the parts of the world are the fundamental units of the world's physical and cultural geography, whose integrity had to be respected. Because "tierra firme" (the mainland) stretches from India to China, Guzmán suggests, so does "Oriental India." With a stroke of the pen, Guzmán discards the age-old belief that India *extra gangem* might be very different from India *intra gangem*, yet he also engages in a bit of political sleight-of-hand. Without any explanation, he substitutes the ancient boundary between the two Indias, the Ganges River, with the city that the Spanish geopolitical imagination identifies as the boundary between Portugal's empire in the East and Spain's empire in the West, Malacca. The obvious obsolescence of the old debate over the fundamental differences between the two Indias provides the impetus for casually discarding Castilian claims that the lands east of Malacca constitute a distinct region that can be considered separately from the rest of Asia and can be mapped into transpacific space. The goal of this gesture is obvious. If East and Southeast Asia form a natural part of Oriental India, then it makes sense that they should form part of the Jesuit enterprise in the East as well.

As we have seen, many of Guzmán's Spanish compatriots would not have understood this as a common-sense matter at all. Steeped in an established tendency to think of East and Southeast Asia as a region to the west, accustomed to the idea that the best way to get there was by crossing the Atlantic and the Pacific, reading Herrera's history and its maps hot off the press in Madrid, those Spaniards would have recognized Guzmán's statement for what it was, an affront to Spanish claims over transpacific Asia and to the prerogatives of the Castilian *Patronato* in the region. They would have turned to Ribadeneira's history and seen it for what it was as well, a determined attempt to map Japan as a stepping stone on the way west to a region that had always and continued to fall in the Castilian hemisphere, whose conversion to Christianity was to be accomplished by a Spanish branch of the Order of the Friars Minor and thereby bring about what had never even been on the radar of the Jesuit mission enterprise, the thousand-year reign of Christ on Earth.

A Franciscan Journey to the West

Ribadeneira maps Japan as a way station on the way to the rest of the transpacific West without providing very much mappable geographical information, such as distances, geographical coordinates, relative locations, and estimates of geographical area. He does not engage in a general description of the territory of the Province of Saint Gregory along the lines of Guzmán's description of the Jesuit Orient. He mentions the location of the antimeridian in order to defend the right of the Spanish Franciscans to preach in East Asia, but does so much as hint at the controversies surrounding it, much less try to mediate them by making geographical arguments of his own. Not only is he silent about such things, but he even makes the astonishing mistake of confusing the Philippine island of Cebu, where Magellan and Legazpi first landed, with the island of Luzón, where Manila is located (1970, 1.31). The only time he breaks this pattern is when it comes to Japan: "El reino de Japón son muchas islas, que están a un lado de la Gran China, de la cual distan doscientas leguas hacia la parte del Norte, en treinta y cuatro grados de altura" (The Kingdom of Japan is made up of many islands which lie to one side of greater China, two hundred leagues from it in the North, at 34° of latitude; 1970, 2.322). Even then, however, Ribadeneira's remarks are imprecise. He does not give the number of islands, says Japan lies "on one side" of China, but does not specify which, and then finally locates the islands to the north of the Middle Kingdom rather than to the west or northwest. Clearly, geography was not the friar's bailiwick.

Can this be explained as the simplicity of a Franciscan who lacked the intellectual sophistication of a Jesuit? Perhaps, but it can also be interpreted as the stamp of a uniquely Franciscan approach to the relationship between history and geography. As Zoltán Bidermann has argued in his analysis of Fray Paulo da Trindade's *Conquista Espiritual do Oriente* (Spiritual conquest of the Orient), published in Goa between 1630 and 1636, Franciscan historiography of "spiritual conquest" tended to eschew the practice of describing territory as precisely and concretely as possible, as had come to be expected of learned historiography. It did so, Biedermann argues, because "the stage for Franciscan conquest is the world as a whole, their mission a conquest undertaken in the name of God. In fact, it is even more than that: it is the Lord's own undertaking in the world He has created, and it must not be bogged down in the realities of geography" (2016). Trindade's text told the story of a cosmic struggle between God and the Devil as it played out in the efforts of the Seraphic Order to destroy idolatry in all its forms. It did not need to lay out the size and relative locations of the world's countries with geographical accuracy, but only to sketch "an unquestionable, God-given place in which the greatest of all battles would occur and the forces of the devil defeated" (Biedermann 2016). To describe geography with precision, it seems, would impoverish the tone that was sought, in which particular places were in effect interchangeable, each but an avatar of the cosmic battlefield upon which the real struggle played out.

Now, these generalizations do not apply to the entire body of Franciscan historiography. A subsequent Franciscan history of the Province of Saint Gregory, to cite but one example, engages in precisely the sort of geographical description that shines by its absence here (Antonio 1738). Nevertheless, what is true of Trindade's history is also true of Ribadeneira's text. Non-Christian religious images are "idols" by default, and the devil is to be found behind every idol, no matter which religious tradition is involved.[10] The friar's remarks about Japanese religion consistently attribute its beliefs and practices to the devil and to the powerful sway that those "siervos del demonio" (servants of the devil), the "bonzos" or monks of Japanese Buddhism, hold over the general populace (1970, 2.365). Like the writers of many Jesuit accounts of Japanese religion, Ribadeneira figures the bonzos as morally corrupt, hypocritical sodomites whose feigned piety and asceticism constitute a monstrous simulacrum of Christian religious life, taught to the bonzos by the devil himself (1970, 2.370). The Franciscans answered to the false asceticism of the bonzos, Ribadeneira claims, with the true life of penance at the heart of the Franciscan devotion, and the Japanese could see the difference. Their life in imitation of Christ attracted attention, provoked questions, and created opportunities for the friars to preach and attract converts (1970, 2.374–79). Far from disturbing the peace in Japan and provoking persecution, the friar argues, the very public way in which the Franciscans bore witness to Christ was the key to their effectiveness in the struggle against the devil and his idols. The Nagasaki martyrdom was nothing more, and nothing less, than the ultimate *imitatio christi*.

For Ribadeneira as for Trindade, geographical precision was not necessary to tell this tale of spiritual conquest, yet it would be a mistake to go so far as Biedermman does when he characterizes Trindade's text as "non-spatial." Ribadeneira's history is deeply spatial, despite the fact that it provides even less mappable geographical information than its Portuguese counterpart. We can begin to appreciate this spatial dimension by looking more closely at the way Ribadeneira figures Buddhism in East and Southeast Asia. Even though he clearly believes that all non-Christian religions are in essence forms of idolatry inspired by the devil, he also recognizes that those forms are unique and have unique histories. He explains, for example, that the people of East and Southeast Asia all practice the same form of idolatry, the religion we call Buddhism, and that Siam figures prominently as a center of Buddhist learning. The friar identifies Siam as "el seminario de la idolatría y adonde más estimados son los ídolos y los ministros de ellos" (the seminary of idolatry and the place where its idols and ministers are held in the highest esteem; 1970 1. 162).[11] In this way, East and Southeast Asia stand out from the rest of the devil's domain as a unique religious territory, even in the absence of explicit geographical description.

According to the friar, moreover, Siam is not just an important center of Buddhist belief and practice, but the very country where the religion had its origins. Ribadeneira locates the story of Siddharta Gautama's enlightenment there, rather than in India, where it actually took place.[12] According to learned men interviewed by Franciscans who have preached in Siam, the friar reports, the worship of idols began in that country when one of the kingdom's original rulers left his wife and children to become a hermit and then returned from the wilderness to impart the central tenets of the religion. Number one on the list was "honrar a los ídolos" (honor the idols; 1970, 1. 169). Having misplaced the origins of Buddhism, Ribadeneira cannot possibly tell the story of its dissemination correctly. Rather than have it migrate from India to other parts of Asia along different routes, at different times, suffering different transformations along the way, as we know Buddhism did, he has it spread from Siam, the "recámara de todos engaños" (the storehouse of all deceit), everywhere else, including Cambodia, China, and Japan (1970, 170). Buddhism thus becomes "la idolatría de Siam" (the idolatry of Siam), a single, homogeneous religious practice that is not only characteristic of the entire region but completely homegrown. One of the region's most significant cultural and historical ties with South Asia is severed, allowing East and Southeast Asia to emerge as an entirely independent empire of idols, available for incorporation into transpacific space by the missionary activity of the Seraphic Order, as interpreted by Marcelo de Ribadeneira.

That interpretation hinges upon the use the friar makes of an existing tradition of Franciscan historiography that had developed in New Spain over the course of the sixteenth century, itself based upon strands of millenarian thought deeply embedded in the intellectual and spiritual culture of the Franciscan order, going back almost to the time of Francis himself. Franciscan millenarianism often drew inspiration from the writings of the Calabrian monk and theologian Joachim of Fiore (ca. 1135 – 1202), particularly in the crown of Aragon, where various influential figures began to associate the dawn of the millennium with the conversion of "new worlds," first East Asia as it was opened up to evangelization by the fourteenth-century *Pax Mongolica*, and then the New World (Baudot

1990, 17).[13] Once they reached New Spain, the Franciscans developed a historiography steeped in this millenarian tradition. One of Ribadeneira's major accomplishments in the *Historia del Archipiélago* is to tie the history of the Province of Saint Gregory to the world-historical trajectory that his Novohispanic brethren had already charted and thereby map East and Southeast Asia as the western terminus of a Franciscan journey to the west, which was also the journey of Christianity toward the physical and historical end of the world.

The Franciscans in New Spain traced their origins to 1524, when the so-called "Twelve Apostles of New Spain" established the first Franciscan mission on the American mainland. That mission eventually became the Province of the Holy Gospel, headquartered in Mexico City.[14] Led by one Martín de Valencia (ca. 1474–1534), the Twelve Apostles held heady ideas about the significance of the work before them. Whether these ideas stemmed from Joachim of Fiore, as George Baudot and John Leddy Phelan have argued, or from a more orthodox Christian eschatology drawn from Augustine, as Elsa Frost has insisted, the Twelve Apostles believed that the conversion of the native inhabitants of the Indies would bring about the thousand years of peace that were supposed to precede the Second Coming of Christ.[15] By bearing the Faith across the Ocean Sea and founding a purified, evangelical version of the Catholic Church in the New World, they were helping to usher in the end of time. One of the clearest examples of this sort of thinking can be found in the unpublished *Historia de los indios de la Nueva España* (History of the Indians of New Spain) by Fray Toribio Benavente de Motolinía (1482–1568). Motolinía argued that the Christian church moved from east to west over time, and that by reaching the New World, it had finally arrived at the end of its geographical and historical trajectory (Rubial García 1996, 130).[16]

By the last quarter of the sixteenth century, however, it had become clear to everyone that the church in New Spain was far from the reformed, evangelical utopia that the Twelve Apostles had hoped to create. The indigenous people had not been quick to embrace the faith, and the Franciscan mission had failed to reproduce the purity and fervor of the primitive church in the New World. Yet just as this depressing realization was dawning on the next generation of Franciscan historians, the conquest of the Pacific by Legazpi and Urdaneta opened up transpacific space, and news began to circulate about China and the Chinese. This led Bernardino de Sahagún (1499–1590) to reinterpret the millenarian vision that had been bequeathed to him by his teacher, Motolinía. As James Phelan explains, Sahagún stripped Mexico of the role it had been assigned by the Twelve Apostles as the spatial and temporal "end of the world" and converted it into a bridge that mediated between Christianity's place of origin in the east and its next destination in the west, China (1970, 27). His friend Gerónimo de Mendieta (1525–1604) insisted that China could not be converted until the church in New Spain had been reformed, but he nevertheless concurred with Sahagún that the true *telos* of Christianity's westward pilgrimage was not the New World, but East Asia, and that the millennium would not begin until China was Christian (1870, 590).

Nevertheless, even as he disagreed with Motolinía about the role of the Spanish American church in the unfolding of universal history, Mendieta attempted to construct a sense of continuity with the past, and particularly with the authoritative reputation of the Twelve Apostles. One of the ways he did this was by retelling a story from the life of Fray Martín de Valencia, that we first find in Motolinía (1985, 304–10). Late in life, Fray Toribio writes, the leader of the Twelve Apostles made his way to Tehuantepec, where Hernán Cortés was building ships to reconnoiter the shores of the South Sea. He was eager to accompany the conquistador on his voyage, sharing as he did Cortés's belief that, somewhere along Mexico's Pacific shores, there were civilized people waiting to be discovered and converted to Christianity. As the ships were being built, Fray Martín traveled the region, and he stumbled upon an indigenous temple that impressed him with both its architectural sophistication and, as he saw it, its demonic purpose. The experience led him to prophesy that "se descubrirían en aquella costa gentes mas hermosas y de mas habilidad que éstas de la Nueva España" (people who are more beautiful and more capable than those of New Spain will be discovered along that coast) and to pray that he might live long enough to see them (1985, 306). He did not see his hopes fulfilled. After waiting for eight months for the ships to be finished, Fray Martín returned to his hermitage near Mexico City, and died shortly afterward.

Georges Baudot asserts that Fray Martín was trying to reach China on Cortés's ships, but in so doing he oversimplifies what was going on in Franciscan efforts to construct a world-historical destiny for the missionary work of the Order of the Friars Minor (Motolinía 1985, 305 n. 18). Motolinía himself never attaches Fray Martín's prophecy to an Asian location, not even Cathay, but instead focuses on the immediate opportunities that were beginning to arise along the frontiers of New Spain. First, he speculates that the people prophesied by Fray Martín might be the Yucatec Maya, whom the Franciscans first encountered in 1537, only three years after Fray Martín's death, and who seemed to be both culturally sophisticated and relatively receptive to Christian preaching. Then, however, Motolinía lavishes attention on the northwestern frontier, boasting that a fellow Franciscan, Fray Marcos de Niza, has spotted the populous cities that everyone had been hoping to find in that region (1985, 307–8).

Motolinía even goes so far as to mention that Cíbola and its Seven Cities were thought to be a "gran puerta para adelante" (a great door to what is beyond), but never specifies what he means by this (1985, 103). If he believed that the road to Cíbola might lead to Cathay along a continuous Amerasian coastline, then he chose to keep that belief to himself. If Fray Martín was heading anywhere that one could name, in Motolinía's view, it was to the ill-defined "Islands and Provinces of the West" that inspired the efforts of Cortés, Alvarado, Mendoza, Coronado, and other would-be conquerors among the friar's near contemporaries in colonial New Spain.

It was the next generation of Franciscan historians that reinterpreted this anecdote in order to convert Fray Martín's abortive trip to unknown parts west into a voyage to China. Mendieta makes a powerful contribution to the process in his *Historia ecclesiástica*, which provides the fullest account of this episode anywhere in sixteenth-century Franciscan historiography. In the following passage, Mendieta tells how Fray Martín left for Tehuantepec in the company of one Fray Domingo de Betanzos:

> Intentaron de embarcarse y entrar en la mar en busca de las gentes de la gran China, antes que hubiera la noticia que ahora hay de ellas, ni de la navegación, si se podía hacer o no. El primero que esto intentó fue el santo Fr. Martín, porque tuvo revelación que había otras muchas gentes hacia la parte del poniente, de mas entendimiento y capacidad que estas de la Nueva España. Y anhelaba su espíritu por ir a ellas y verlas en sus días, y convertirlas á su Dios. (Mendieta 1870, 587)

> [They tried to ship out and sail the sea in search of the people of Greater China, before there was knowledge we have today of them, the route, or whether it was possible or not to reach them. The first one to attempt this was the saintly Fray Martín, to whom it had been revealed that there were other people in parts west, of greater understanding and ability than those of New Spain. His spirit yearned to go to them, to see them before he died, and to convert them to his God.]

The passage exhibits a two-tiered temporality. On the one hand, it recreates the perspective of Fray Martín, who is eager to sail to "parte del poniente" (parts west) in search of "otras muchas gentes" (other people) whose name and exact location he does not know, but whose existence has revealed to him by God.[17] On the other hand, it speaks from Mendieta's own point of view, from which it is obvious that the people in question are the Chinese, not the inhabitants of some civilized country yet to be discovered.[18] So, just as Motolinía had interpreted Fray Martín's words and actions in light of the exciting opportunities that seemed to be opening up on the northwestern frontier in his day, so Mendieta interpreted the same episode from the founder's life in light of the equally exciting possibilities that had been opened up by the conquest of the Pacific.[19] Vague and ambiguous in its original form, the story of Fray Valencia's prophecy and abortive trip west became the perfect site upon which Franciscan historiography could reinterpret its self-assigned role in the unfolding of the millennium to suit changing conditions and emerging opportunities.

In Ribadeneira's text, the story of Fray Martín's prophecy and his abortive trip west serves to link the history of the Province of Saint Gregory with the work of the Franciscans in New Spain and the world-historical significance that Motolinía, Sahagún, and Mendieta had assigned to it. This involved overcoming a minor inconvenience, the fact that the two missionary enterprises had nothing to do with each other in institutional terms. The Province of the Holy Gospel in Mexico City had begun its career as a "custody," an official spin-off, of the Franciscan Province of Saint Gabriel in Extremadura, itself the product of a movement to reform the Observant branch of the Franciscan Order that unfolded between the end of the fifteenth and the beginning of the sixteenth centuries (Baudot 1995, 81–83). The Province of Saint Gregory, in turn, began as a custody of the Province of Saint Joseph in Galicia, composed of Discalced Franciscans who had split off from the Observant branch around the middle of the sixteenth century (Currier 1898, 240). So while both the mission to Mexico and the mission to the Philippines had bubbled out of the religious ferment of the early modern Hispanic world, the one was not an institutional child of the other. Each had emerged separately from different branches of the Franciscan Order that were sometimes at odds with each other. Ribadeneira himself had nothing to do with the work of the Franciscans in New Spain, and he had no particular intellectual ties to Motolinía, Sahagún, or Mendieta.

His text, however, says nothing about these distinctions. Instead, it stages a scene of recognition between the Spanish friars who stop in Mexico City on their way to the Philippines and the Novohispanic friars who serve as their hosts. According to Ribadeneira's text, the two groups, Discalced and Observant, immediately recognized each other as spiritual brethren. The Spanish friars, the text claims, admired the strictness with which their Novohispanic counterparts observed the Rule of Saint Francis. Whatever criticisms the Discalced Franciscans may have had of the Observants in Iberia did not apply here. The Novohispanic friars, in turn, interpret the voyage that the Spanish were undertaking as the fulfillment of the prophecy of their beloved Fray Martín de Valencia

(1970, 1.41). Apparently, it did not matter to them if Discalced Franciscans from Spain took over the work that might have been done by Observants from Mexico. If this actually happened, then the friars in New Spain actually handed Ribadeneira the narrative material that he needed in order to overcome the institutional and spiritual divisions of his order and present the mission in East Asia as a continuation of the work in New Spain; yet even if nothing of the sort ever occurred, that is the effect that is achieved by relating the scene. Institutionally, the Province of St. Gregory may have nothing to do with the Franciscan mission to the New World, but spiritually, providentially, it is an extension of its work across the Pacific, to the lands whose conversion must be achieved if the millennium is ever to begin.

At this point, Ribadeneira ascribes the identification and the millenarian implications that go with it to the friars in Mexico City, keeping them at a distance from his own perspective as narrator, but by the time he reaches the events leading up to the Nagasaki martyrdom, he embraces that millenarianism as his own.[20] The events surrounding the crucifixions come couched in a series of biblical-style portents and miracles, just as they do in other Franciscan chroniclers of the event. A comet appears. Ash rains on Kyoto, red earth on Osaka, and worms on other places. The seas become especially rough, producing waves strong enough to wipe out an entire town. Powerful earthquakes rock Kyoto, Osaka, Fushimi, and elsewhere. A man chopping wood finds a perfectly formed cross in the center of a log he had split open (1970, 2.415–18). The night of the martyrdom, three rays of light illuminate the hill where the crucifixion takes place. A cross is sighted over the Japanese sky. A column of fire and a host of new stars appear over a shrine to the Virgin Mary. After the execution, the crucified bodies of the martyrs remain incorrupt, and the crows refuse to pick out their eyes. The body of the lead Franciscan, Pedro Bautista, continues to give blood days after the event, while the blood of two other martyrs remains liquid nine months later in the porcelain vessel in which it has been collected. Every Friday, the dead Bautista is seen saying mass at the Nagasaki leper's hospital founded by the friars (1970, 2.498–501). All of this serves to tie the Nagasaki martyrdom to the crucifixion of Christ, and to the martyrdoms of the early church, suggesting that the Franciscan mission in Japan had successfully borne witness to the purified evangelical Christianity that was the hallmark of Franciscan observance.

The account also associates the Nagasaki martyrdom with the dawn of the millennium by telling us that, upon hearing that the Japanese authorities were maneuvering to have them executed, the friars and their followers prepared "vestiduras blancas" (white garments) for the day of their martyrdom (1970, 2.429–30). Elena Isabel Estrada de Gerlero points out that the white robes recall various passages from the Book of Revelations that mention white robes as the garment of the martyr, specifically Revelations 3:5, 6:11 and 7:14 (2000, 75). Specifically, they recall the white robes of the just on the Day of Judgment. The portents and miracles, therefore, do not just look back toward the crucifixion of Christ and the martyrs of the early Church, but forward to the Apocalypse. And if the millenarian implications of all this imagery are not enough, Ribadeneira finally comes clean about his interpretation of the world-historical significance of everyting that has happened, in his final chapter, where he asserts that "ya estamos en el fin de los tiempos y en la última edad del mundo" (We are already in the end times and the final age of the world; 1970, 2.636). Not only have the Franciscans of the Province of Saint Gregory picked up where Martín de Valencia left off, embarking on the final stage of Christianity's momentous march west, therefore, but in the portentous martyrdom of their brethren at Nagasaki they have actually turned the page on universal history, announcing the beginning of the end of its eschatological unfolding.

Japan, in this vision of Christianity's westward march across the South Sea to East and Southeast Asia, is not a final destination, but a place of passage. The Nagasaki martyrdom has not jeopardized the future of the Japanese church, as the Jesuits claimed, but has rather set the cornerstone of a solid religious enterprise, one that will now spread to the rest of the region:

> Aquel reino . . . que aunque hasta ahora ha sido seminario de idólatras, de aquí adelante ha se ser escuela de la verdad evangélica, y olvidando los principios temporales que tuvo, vendrá tiempo que principalmente se celebre la memoria de Cristo crucificado, representada muy al vivo en cada uno de estos dichosísimos crucificados. Y placerá al Señor que aquella gentilidad, que está entenebrecida con sus errores, por este medio está alumbrada en el camino del cielo como con soberana luz. (1970, 2.632–33)

> [That kingdom . . . which until now had been a seminary of idolators, henceforth shall be a school of the gospel truth, forgetful of its origins. There will come a time when the memory of Christ crucified, as exemplified so vividly in each of these men so fortunate to be crucified, will be celebrated. And it will please the Lord to enlighten that gentility that now lies in the shadow of error, showing it the way to heaven by means of his sovereign light.]

The friar imagines a future for Japan in which idols will give way to the cross, error to truth, darkness to light, yet he also alludes to the destiny of the entire region. By referring to Japan as a "seminario de idólatras" (seminary of idolators), Ribadeneira recalls the language he used to figure Siam as the place that gave birth to the region's unique version of idolatry. Japan, he asserts, will be transformed from a "seminario de idólatras" into an "escuela de la verdad evangélica" (school of the gospel truth), and he thereby not only suggests that the country's conversion will be deep and complete, but that it will take over Siam's role as the center of cultural diffusion in the region. Japan will cease to be a recipient of Siamese error and will become a teacher of Christian truth. The people, "Aquella gentilidad" (that gentility), that will pass from darkness into light are not just the inhabitants of Japan itself, therefore, but everyone in all of East and Southeast Asia.

The millennium has not yet dawned, but is right around the corner, as it always seems to be in the minds of anyone who prophesies its coming. More importantly, the history of the Province of St. Gregory has assumed its place, fully and completely, in the particular vision of Franciscan millenarianism that had been pioneered by the Twelve Apostles and significantly revised by Sahagún and Mendieta. The Franciscan mission to East and Southeast Asia is playing its prophesied role in the westward march of Christianity toward the historical and geographical end of the world. Japan, moreover, has assumed the same role in that trajectory that it had played in the geographical imagination of Spaniards interested in expansion across the South Sea. It is not a final destination, but a way station along the route to bigger and greater things. Here, Japan does not offer rest to weary mariners, but hope to Franciscan missionaries who want to see their work throughout Spain's Indies as the vehicle of Christianity's pilgrimage westward to the physical and historical end of the world, despite the failure of Spanish colonialism in America, and perhaps also in the Philippines, to produce the New Jerusalem.

Stinking Ships and Endless Oceans

One might think that the way west, in this heady eschatological vision of the Franciscan missionary enterprise, would be couched in the rhetoric of smooth sailing and would appeal to the cartography of containment, insisting on the ease of the westbound voyage across the South Sea and the lake-like character of the world's oceans. A world in which the oceans served to connect rather than separate would serve as clear evidence of God's will that Christianity should travel across the face of the globe, reaching all those lands where the devil still reigned. Nevertheless, both tropes shine by their absence in Ribadeneira's *Historia del Archipiélago Filipino*. There is no cartography of containment because there is no cartography, no attempt to map physical geographies in either word or image. More surpisingly, however, there is also no attempt to portray Franciscan voyages across the Atlantic and Pacific Oceans as unremarkable transits across perfectly domesticated bodies of water. On the contrary, Ribadeneira relates the sufferings of the Franciscans aboard ship, even revels in them. His candor about the subject provides an opportunity to reflect upon how much things had changed since von Sevenborgen and Oviedo silenced the rigors of Magellan's Pacific crossing only eighty years before.

Like Pigafetta, Ribadeneira figures the suffering of human bodies at sea. A passage early on about the first leg of the journey from Spain to the New World sets the tone:

> Y quien por experiencia sabe cuán trabajoso es el mal olor del navío y la estrechez de el, y las indisposiciones penosas que los primeros días causa el mar, podrá estimar el trabajo que estos siervos de Dios tendrían en su viaje, juntándose a esto la mucha clausura que guardaban, llevando portero en un estrecho lugar de la popa que les dieron, y haciendo parte de la popa iglesia y no saliendo de allí sin licencia y necesidad, ocupándose las horas acostumbradas en rezar el oficio divino, y en hacer disciplinas y tener oración, con tanto concierto como si estuvieran en un muy religioso convento. (1970, 1.39–40)

> [Anyone who has experienced the trying stench and confinement of a ship, and knows how woeful the indisposition caused by the first days at sea can be, will be able to appreciate the travails that these servants of God would endure on their journey. Add to that their decision to remain strictly cloistered, shut up in the tiny corner of the stern of the ship that had been assigned to them, converting that stern into a church, and never leaving it without permission or out of some necessity, occupying themselves in saying the divine office, disciplining themselves, and praying with as much accord as if they had been in a convent.]

Ships stink. Quarters are cramped. Seasickness is all too real. Life aboard is so monotonous that the activities of a month-long crossing can be adequately summarized in a sentence. Elsewhere we read about more extreme forms of suffering. One friar is so tormented by the devil on his ocean crossing that he spends the whole voyage flagellating himself (1970, 1.192). Another works in the ship's galley, so that his brothers will not lack for food, and ends up dying from the heat (1970, 1.217–18).

Yet another endures two separate attacks by French corsairs (1.235–37. A particularly learned friar is wracked by sea sickness, and is tempted by the devil himself to turn back (1970 1.257–58). Others drown in a shipwreck (1970, 1.575). Ocean travel in Ribadeneira's text is hardly the jump across the pond that Acosta makes it out to be in his remarks about its ease and regularity. It is instead a drawn out and miserable business, taxing at its best and calamitous at its worst.

These accounts of suffering at sea, however, actually point in an entirely different rhetorical direction from those we have seen before. Pigafetta was trying to portray the fleet's crossing as a singular event, unprecedented and perhaps unrepeatable, by suggesting that the Pacific was so broad that it was practically impossible to traverse. Ribadeneira, by contrast, is not trying to tell the reader anything about travel by ship or about the oceans that the reader does not already know. As we can see from the passage above, he assumes that the miseries of shipboard life are well known. The primary point he is trying to make is about the friars. The group that finds itself confined to cramped quarters in the stern actually intensifies its suffering by treating their confinement as a type of cloister, thereby converting the necessary travails of shipboard life into spiritual opportunity. When the same group finds out that the Pacific crossing is even longer and more miserable than the voyage across the Atlantic, they look forward to the chance it offers to "crucificarse con Cristo" (crucify themselves with Christ; 1970, 1.43). This is typical of the friars in Ribadeneira's text. They embrace suffering as an opportunity to imitate Christ and thereby heave ever more closely to the central principles of a Franciscan vocation. The seasick friar tempted by the devil to abandon his mission, for example, nevertheless manages to apply himself to the study of Japanese. Another sees in the widespread illness aboard ship an opportunity to care for the sick, giving his provisions to the hungry to the point of neglecting his own health (1970, 1.288). This was the way that the Franciscan order, Ribadeneira explains, had introduced the faith to far-off lands:

> Cristo nuestro Maestro, conquistó el mundo y quitó al demonio el señorío tiránico que tenia, con pobreza, humildad, descalcez y hambre, afrentas y deshonras, hasta ser crucificado como ladrón entre ladrones . . . y de esta manera en nuestros días plantaron la fe en la Nueva España, Perú y Filipinas muchos varones franciscanos verdaderamente evangélicos. (1970, 2.407)

> [Christ our Teacher conquered the world and deprived the devil of the tyrannical lordship he enjoyed with poverty, humility, barefootedness

and hunger, insults and affronts, up to the point of being crucified as a thief among thieves . . . and this is how in our own days so many truly evangelical Franciscans have planted the faith in New Spain, Peru, and the Philippines.]

By embracing the suffering of shipboard life, the friars were not just making the best of a necessary evil, but were living the evangelical life at the heart of their mission and its success. The fact that the suffering takes place at sea is in fact incidental, as we can see from the fact that the friar also mentions the hardships of life in the tropics, with its with its heat, its humidity, and its "mosquitos y sanguijuelas . . . que hacen carnicería en los pies descalzos" (mosquitoes and leaches . . . which butcher one's bare feet; 1970, 1.57. See also 1.69, 90).

The real interest of this aspect of Ribadeneira's text, therefore, lies not in its candor about all the misery, but in the conditions of possibility toward which that candor gestures. Underneath all the friar's rhetoric about the devil and his idols, not to mention his grand claims about the impending millennium and the role of the Province of St. Gregory in helping to bring it about, the *Historia del Archipiélago Filipino* is a travel narrative, one that tells the story of how a group of Spaniards, all members of the Franciscan order, made their way from Spain to Mexico to Manila, and from there to the various countries of East and Southeast Asia. The scale of the journeys, individually and collectively, is quite impressive. All of the friars could be said to form part of the new global elite that Serge Gruzinski identifies as one of the principal social groups to emerge from the project of Iberian globalization. These were people whose lives played out across vast distances and diverse geographies, on a level that would have been impossible to imagine just a century before, and who learned to think about human affairs on a global scale (Gruzinski 2004, 259–76). Ribadeneira certainly did, as evidenced by his effort to map the history of the Province of St. Gregory into the historical and geographical displacement of Christianity across the Atlantic and the Pacific, and to see in the far-away Nagasaki crucifixion an event of world-historical significance.

Nevertheless, just as one can identify a certain grandeur in such displacements, and in such breadth of vision, one can also recognize a certain banality in them. None of the voyages Ribadeneira describes are anything like those of Magellan. While they are clearly unsafe and obviously uncomfortable in the extreme, they are not daring ventures into the unknown, in search of undiscovered islands or of novel routes across uncharted oceans. They are typical examples of travel by sea, miserable and risky but nevertheless ordinary, insofar as they involve established infrastructure, the regular shipping routes of Spain's maritime empire. However much Ribadeneira's friars try to convert their misery into opportunities for sanctity, they are not heroes, but commonplace travelers, enduring the same agonies that awaited anyone who climbed aboard a ship that was headed for distant lands across a broad ocean. The risks such people assumed were quite real, requiring them to put their affairs in order before they walked up the gangway, but they were not singular, unique, or mysterious. They were both common and familiar, even among people who never set foot on an ocean-going vessel.

Ironically, Ribadeneira's tales of suffering at sea, seen in this light, begin to look like literary kin to the rhetoric of smooth sailing, insofar as they gesture toward the regularity of ocean travel circa 1600. The rhetorical trope of smooth sailing says nothing about such suffering, preferring instead to emphasize the relative ease and frequency with which moderns cross the oceans. Ribadeneira's candor about shipboard life exposes that rhetoric as the hyperbole that we already knew it to be, but it does nothing to undermine its basic assertion, that travel across the oceans has become a regular feature of early modern life, even if few people, in absolute terms, actually engage in it. In its commonplace predictability, the shipboard suffering of Ribadeneira's friars actually attests to the predictable regularity of the galleon sailings that knit the disparate places of the Hapsburg empire into a global web of maritime activity.

8

The Anxieties of
a Paper Empire

While Ribadeneira dreamt of the conversion of Asia and the ensuing millennium, his secular counterparts fretted about the challenge posed by England and Holland to Spain's secular ambitions in the transpacific West. These two countries, one a former Hapsburg ally and the other a rebel province, had begun during the late sixteenth century to flex their muscles as maritime powers in their own right, preying upon Iberian shipping, assaulting Spanish and Portuguese outposts, and embarking upon colonial projects of their own. The assault was ideological as well as military. While English privateers and Dutch fleets nibbled away at Iberian strongholds in the Atlantic, Indian, and Pacific Oceans, writers and printers in both countries fostered the growth and development of the Black Legend, that infamous set of stereotypes that characterized Spanish rule in the Old and New Worlds as an exercise in cruelty and tyranny. Spain responded to the challenge on the ground and on the page, spending money on soldiers, ships, and fortifications and devoting resources to the writing of history in an imperial register. As Richard Kagan argues, official history was just the thing to counter the Black Legend, by portraying Spain's overseas empire as a model of good governance and Christian piety, thereby defending the monarchy's battered reputation (Kagan 2010, 263–82). When paired with geography and cartography, I add, it would also assert the crown's territorial prerogatives in the face of incursions by England and Holland into territory that Spain had always seen as its own, thanks to the papal bulls and the Treaty of Tordesillas.

By 1600 at the latest, it was clear that the Moluccas were falling into Dutch hands, but when the Hapsburgs finally responded with military force, they did not do so from Macau, Malacca, or Goa, using the resources of the Portuguese *Estado da India*, but from Manila, with a force under the command of a Spaniard. The strategy was indicative of the extent to which the increasing Castilianization of the Hapsburg Monarchy extended to colonial affairs in East and Southeast Asia. Having become without question the major player in what was supposed to be a dual monarchy, Spain had begun to assert authority over Portugal's maritime empire, over and against the provisions that had governed the original Union of the Crowns established in 1580. Historian Kevin Sheehan explains that the duke of Lerma even went so far as to strip the Portuguese Council of State of its authority over the kingdom's possessions overseas, by creating a separate Council of India that would advise the crown on all matters having to do with the Portuguese Indies. Its first president was Francisco de Mascarenhas, a former governor of India who was thought to be loyal to the Hapsburgs (2008, 258). Although the new arrangement only lasted ten years, the very fact that it was attempted suggests just how far powerful Spaniards like Lerma were willing and able to undermine Portugal's traditional prerogatives.

It was during this period of emerging challenges from foreign powers and increasing Castilian hegemony over the dual monarchy that the crown issued the two official histories I examine in this chapter. The first is Antonio de Herrera's *Historia general de los hechos de los castellanos en las islas y Tierra Firme del Mar Océano* (General history of the deeds of the Castilians in the islands and Mainland of the Ocean Sea), better known as the *Décadas* (Decades), published in two installments (1601 and 1614), both in Madrid. The second history is Bartolomé Leonardo de Argensola's *Conquista de las Malucas* (Conquest of the Moluccas), also printed in Madrid, but in a single volume that appeared in 1609. Both were written by official historians. The first, Antonio de Herrera y Tordesillas, was the official chronicler of the Indies and thus a successor to Juan López de Velasco. The second was the Latin secretary of the president of the Council of Indies. Insofar as these two ponderous texts have been studied at all, they have usually been examined separately. Here, I analyze Argensola's history as the unofficial continuation of Herrera's incomplete chronicle, an attempt to fill in the significant gaps left by his project and to answer to some of its most significant silences. Together, Herrera's *Décadas* and Argensola's *Conquista* construct a cartographic and historiographic vision of the Spanish Indies that answers to the Protestant challenge in Southeast Asia by insisting upon the transpacific range of Spanish sovereignty. Yet while Herrera's project remained incomplete, Argensola's is haunted by the very real limitations of Spain's transpacific reach, and the insuperable challenge posed in the region by China. In the end, both historians map paper empires that betray the anxieties of a monarchy in crisis.

Herrera Sets the Stage

Herrera's *Historia general* does not provide as comprehensive an account of Spain's overseas adventures as its bombastic title promises. It is undoubtedly lengthy and detailed, regrettably dry and ponderous, but it is also curiously incomplete.[1] Beginning with the 1492 voyage of Christopher Columbus, it relates the history of Spain's overseas empire only as far as the year 1554. As a result, it never gets to the colonization of the Philippines by Legazpi, the subsequent attempts to evangelize China, or any other "deeds of the Castilians" in the Indies of the West. Such events receive only passing mention, in micronarratives that pepper the description of the Indies that serves as a general introduction to the *Décadas* as a whole (Herrera y Tordesillas 1991, 1.219, 235–39).[2] Herrera's *Historia general* nevertheless attempts to incorporate the Indies of the West into the Castilian geographical and imperial imaginary, perhaps more aggressively and certainly more publicly than any official history before it. It does so primarily through geography and cartography, but also through historical narrative, in the attention it gives to Spain's South Sea enterprises before 1554, and more importantly, in the way it creates an experience of reading that can also be understood as a form of cognitive mapping.

Antonio de Herrera assumed the job of Chronicler of the Indies in 1596, and five years later, published the first volume of the *Historia general,* which includes a general description of the Indies and the first four of Herrera's eight decades, covering the years from 1492 to 1531. The work went quickly because Herrera had plenty of unpublished material to adapt and assemble, sometimes with minimal alteration. These included the *Sumario* of Juan López de Velasco that I examined in chapter 5, and which served as the basis for Herrera's geographical introduction. That introduction bore its own title, *Descripción de las islas, y tierra firme de el mar oceano, que llaman Indias Ocidentales* (Description of the islands and mainland of the Ocean Sea, which are known as the West Indies), and appeared as a separate publication the same year as the first volume of the *Historia general.* It was greeted with enthusiasm, getting translated into Dutch (1622), French (1622), Latin (1622), and English (1625) in short order (Cuesta Domingo 2015, 27–28). It was even anthologized with other New World materials and equipped with illustrations by Theodor de Bry (1622). The original edition described the Indies in ninety-six pages of prose and in fourteen maps based on those that Velasco had prepared for the *Sumario,* and which included a general map of the West Indies and a regional map of the Indies of the West. (Figures 2 and 31.) This section

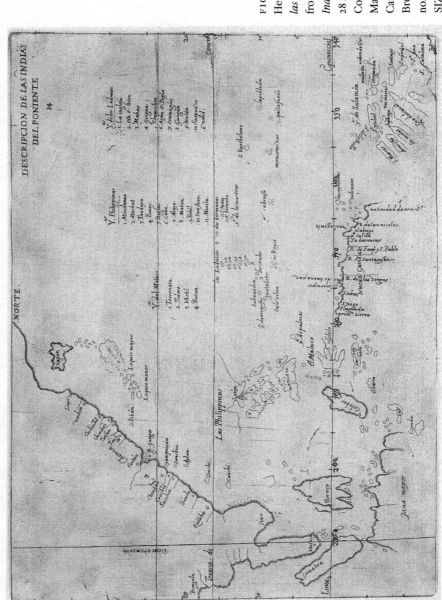

FIGURE 31. Antonio de Herrera, *Descripción de las Indias del Poniente* from *Descripción de las Indias* (Madrid, 1601). 28 x 35 cm. Photograph: Courtesy of the JCB Map Collection, John Carter Brown Library, Brown University (call no. B601 H564h/1-SIZE; file 01808-019).

examines the way Herrera adapts the geographical and cartographical material he inherited from López de Velasco into an ambitious vision of Spain's transpacific empire, and then it makes some observations about how the narrative of the subsequent *Décadas* attempts to populate Herrera's maps with an expansive vision of empire and imperial history.

Like other official Spanish maps of the Indies, Herrera's maps disregard Portugal's territorial claims in East and Southeast Asia, yet unlike most of the other maps we have seen, these were designed for circulation in print. This fact alone serves as a measure of the extent to which the Spanish Hapsburgs, by the dawn of the seventeenth century, were willing to favor the interests of their Castilian subjects over those of their Portuguese ones. Nevertheless, these maps were not specifically designed to slight Portugal, nor were the *Décadas* as a whole, despite their insistent usage of "Castilian" instead of "Spanish" in the title and elsewhere. While its composition in Spanish clearly suggested a Spanish reader, the *Historia general* was also intended for a European audience. One of the things it told that audience, the English and the Dutch in particular, was that there was a new sheriff in East and Southeast Asia, the kingdom of Castile, who would not be as easy to push around as Portugal had been.

In order to do so, Spain had to go public, for the first time, with the tripartite, transpacific geography that Velasco had forged for Spain's overseas empire, in order to assert Spanish sovereignty over the entirety of the demarcation, including East and Southeast Asia. Like Velasco's *Sumario*, Herrera's *Descripción* divides the "Indias occidentales" into three regions, the "Indias meridionales" (Indies of the South), the "Indias septentrionales" (Indies of the North), and the "Indias del poniente" (Indies of the West). The general map of the Indies gives visual form to the ensemble, while the regional maps provide some degree of detail. By the standards of Northern European print cartography, Herrera's maps are bereft of information and crudely engraved. His map of the Indies of the West looks like nothing more than a sketch when compared to its competitors by Ortelius (1570), Plancius (1594), Linschoten (1595), and Hondius (1606).[3] (Figure 32.) Yet it was the very existence of such maps, and the budding commercial enterprises they facilitated, that made Herrera's simple images so necessary. His "Descripción de las Indias de Poniente" may have represented a crude example of the engraver's art by comparison with Linschoten's "Exacta et accurate delineatio . . . China, Cauchinchina, Camboja sive Champa . . . ," but it put East and Southeast Asia where they belonged, in the Castilian West, and placed them out of bounds to Dutch merchants.

Unlike Velasco, however, Herrera abandons any hope of tying the Indies of the West to the Indies of the North and South by appealing to

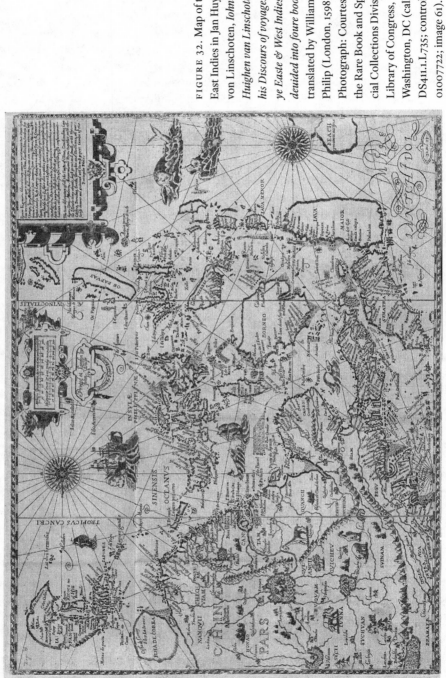

FIGURE 32. Map of the East Indies in Jan Huygen von Linschoten, *Iohn Huighen van Linschoten his Discours of voyages into ye Easte & West Indies: deuided into foure bookes,* translated by William Philip (London, 1598). Photograph: Courtesy of the Rare Book and Special Collections Division, Library of Congress, Washington, DC (call no. DS411.1.L735; control no. 01007722; image 61).

physical geographies, known or imagined. As we saw in chapter 5, Velasco had flirted with the theory of Amerasian continuity on both his chart of the Indies and the text of his *Geografía y descripción*. The *Geografía y descripción* rehearsed arguments about Amerindian origins that relied upon the existence of a physical connection, as yet undiscovered, between the Old World and the New, while Velasco's chart of the demarcation depicted the Pacific shores of Asia and America as converging lines, suggesting that they might meet and form a single continent somewhere in the undiscovered north (see fig. 11). Herrera's general map, by contrast, leaves little room for doubt that he believes the New World to be insular (see fig. 2). Rather than turn northwestward toward Asia, Herrera's American coastline in the North Pacific takes a sharp turn toward the northeast, just as it does in Gómara's description of the Indies insisting upon their insularity and, before that, in Sebastian Münster's map depicting the New World as an island (fig. 24). Herrera's text, meanwhile, clearly states that the world is divided into four parts, Europe, Asia, Africa, and a fourth part, the "islas y tierra firme del mar Océano" (islands and mainland of the Ocean Sea), which "comunmente son llamadas las Indias Occidentales y Nuevo Mundo" (are commonly called the West Indies and the New World). He refuses to call them "América," insisting that the name applies only to what we call South America (1.174–75).

Herrera thus hesitates to assign the fourth part of the world a proper name, limiting the purview of the toponym "America," reducing the obvious alternatives to nothing but common names, and proffering an alternative of his own that could not be more generic, "islas y tierra firme del mar Océano." While his map thus favors the theory of American insularity, his text fails to fully embrace the invention of America. Nevertheless, this does not stop him from mapping the lands in question into a political space and from assigning them a sovereign. Whatever one wants to call them, the islands and mainland of the Ocean Sea lie within "la demarcación de los Reyes de Castilla y de León, que es un hemisferio o medio mundo" (the demarcation of the Kings of Castile and León, which is a hemisphere or demi-world). (1.174–75). A framework at once political and geometrical suddenly displaces the putatively natural and geographical armature that had been guiding Herrera's world-making. The architecture of the continents gives way to the cartography of the demarcation. It is then and only then that what we call North and South America acquire proper names, as the Indies of the North and South. Those names, moreover, are highly relational. Not only do they tie the northern and southern landmasses to each other, but they also tie them to the part of Asia that also lies within the demarcation and belongs to the crown of Castile, the Indies of the West. As if to emphasize the purely political yet nevertheless proper nature of these names, Herrera explains of the Indies of the West that "aunque son parte de la India Oriental, se nombran de Poniente respeto de Castilla" (although they are part of Oriental India, they are called of the West with respect to Castile) (1.148).[4] Not "commonly called" the Indies of the West, but simply "called," because they are to the west with respect to Castile. In other words, their proper name stems from their relation to Castile, both in terms of direction and propriety. It is the proper names of political geography, not of natural geography, that count here.

Herrera's treatment of the South Sea pulls in a similar direction, abandoning the putatively natural geographies that Velasco uses to contain the Pacific. Herrera moves Japan, the Ladrones, the Philippines, and the Solomons westward with respect to their positions on the Velasco chart, making the North Pacific seem emptier, and the trip from Acapulco to Manila, longer. He abbreviates the length of the coastline of New Guinea and includes nothing to suggest the existence of a *Terra Australis* in the southwest, thereby eliminating the convenient southern boundary that helped define the North Pacific as Velasco's small, well-contained *Golfo de poniente* (see fig. 11). While later versions of the same map add a chart of the climatic zones to the empty space where many a mapmaker places the hypothetical southern continent, the original version dresses the emptiness in nothing but a skimpy cartouche explaining that "entre los meridianos señalados se contiene la navegación y descubrimiento que compete a los Castellanos" (the designated meridians contain between them the navigation and discovery that is incumbent upon the Castilians). It is as if Herrera were actively stripping the South Sea of all those natural geographical features, known and surmised, that Velasco and others had used to contain it and daring the reader to think of the Spanish Indies as a nakedly political entity, its existence grounded in nothing but the legal framework of the lines of demarcation.

Herrera nevertheless wants his reader to believe the lands between the lines hang together, and his map is designed to do just that. First and foremost, Herrera preserves Velasco's regional scale and framing, leaving the rest of Asia out of the picture. He also ignores period convention by failing to label the mainland of the Indies of the West as *pars Asiae*. These omissions liberate the Indies of the West from their supposedly natural home on the continent of Asia and make them available for remapping in relation to the Indies of the North and South. Second, he uses line and language to reinforce the relationship among the three parts that is inherent in their proper names. Although the telescoped coastline of New Guinea and the Solomon Islands can no longer carry the reader's eye across the South Sea to Peru, it now receives an important assist from the strategically placed toponym "Mar del Sur," which picks up where that coastline leaves off. Along with the names and lines indicating the location of the tropics and the equator, the toponym takes the place of the galleon's itinerary on the Velasco chart, substituting reading for imaginary sailings, but nevertheless carrying the reader across the South Sea in ways that help domesticate the ocean in his or her imagination and bring its opposing shores together. The South Sea itself, moreover, remains relatively narrow by comparison with what we find on the Seville charts. Its breadth of roughly 110° of longitude from Peru to Gilolo is closer to the estimates that came out of the Villalobos enterprise than those that emerged from the Magellan expedition. Although it may not be as well bounded as Velasco's "Golfo de poniente," it is small enough to look like a bridgeable expanse rather than an insuperable obstacle or an ontological barrier between the Indies of the West and those of the North and South. In all these ways, Herrera's cartography encourages the reader to imagine the East Indies of the Dutch as an extension of the West Indies of the Spanish.

The job of Herrera's historical narrative, as we have seen, is to populate this map with historical memories flattering to Spain and favorable to its interests. Presumably, if the narrative extended beyond 1554, it would insist that Spain's conquest of the Philippines was legal, minimally violent, and beneficial to the native inhabitants, and that its efforts to sponsor missionary work elsewhere in East and Southeast Asia were godly and good. The Spanish presence in the region would be slight, but its purpose noble, particularly in light of the dire need that the sophisticated but idolatrous people of China and Japan had for instruction in Christianity. Yet as we have also seen, Herrera never gets past 1554, so if we set aside the micronarratives that pepper Herrera's description and the occasional proleptic remarks that color his historical narrative, we have only his accounts of Balboa's discovery of the South Sea and of the various expeditions to the Spicery on which to fasten if we are to understand his approach to the history of the deeds of the Castilians in the Indies of the West.

A passage from Herrera's account of Balboa's discovery, however, helps us understand how difficult it can be to interpret Herrera's historiography by approaching it on a granular level. Herrera has the conquistador and his men marveling at the spaciousness of the ocean, when he suddenly abandons the scene to engage in theoretical speculation:

No se puede encarecer la admiración que los castellanos tenían oyendo que aquella mar (para ellos tan nueva) no tenía fin, y por la grandeza de ella, que después se ha descubierto, me ocurre tratar aquí cuál sea mayor, la tierra o la mar. (1.605)

[It is impossible to appreciate the wonder that the Castilians felt upon hearing that the sea (to them so new) was without end, and because of its great size, which has since come to be known, it occurs to me to ask here which is larger, the land or the sea.]

Speaking with the advantage of hindsight, from the perspective of someone who knows more about the actual size of the South Sea than did Balboa and his men, the narrator goes on to explore some of the cosmographical issues we touched upon in chapter 1, asking whether there is more land or water on the surface of the terraqueous globe, and concluding for theological reasons that land must predominate over water. This means that the South Sea cannot be as large as it seems to be, that somewhere beyond its known horizon there must be a great deal of land awaiting discovery (1.605). The text does not say what form that land might take, but the logic of the passage certainly allows for an undiscovered continent in the southern hemisphere. In this way, the text conjures the *Terra Australis*, or something like it, out of Balboa's very limited experience of the South Sea and manages to contain the Pacific.

Yet why would Herrera speculate in this way about the existence of undiscovered South Sea shores, when he says nothing about the matter in his *Descripción de las Indias* and fails to include any speculation about the *Terra Australis* in his map of the demarcation? The question, it turns out, is misguided, because Herrera was not the author of the passage in question. The chronicler lifted his account of Balboa's discovery almost verbatim from the unpublished *Historia de las Indias* (History of the Indies) written by Las Casas between 1527 and 1559.[5] In reflecting upon these cosmographical issues, Herrera is not so much speaking in his own voice as channeling that of the Dominican friar. Whether or not Herrera agreed with Las Casas is beside the point for now: what matters is the lesson that we learn about reading the *Historia general*. Although it presents itself as a coherent history of the Indies penned by a single author, it is actually something like an anthology of disparate materials that have not always been corrected for consistency. If we are to interpret the *Historia general* in ways that might answer to what Herrera himself intended, we need to be careful about hanging too much significance on particular passages like this one and rather keep our eye on the overall design, as well as on those elements that we know were created under Herrera's direction, like the maps and the next object of analysis, the title-page illustrations.

Each of Herrera's decades, as well as the *Descripción*, opens with an illustrated title page, most probably designed by Herrera himself.[6] The one of most interest to us graces the Third Decade (fig. 33). Like the other title-page images, it features a cartouche with the title, the name of the

FIGURE 33. Antonio de Herrera, title page illustration from the Third Decade of his *Historia general de los hechos de los castellanos* . . . (Madrid, 1601). Photograph: Courtesy of the JCB Map Collection, John Carter Brown Library, Brown University (call no. B601 H564h/1-SIZE; file 01808-004).

author, and the arms of the kingdom of Castile and León that serves to invest the text with the authority of the monarch himself, acting through his servant, the historian. In the corners we find portrait medallions of the principal players, including Hernán Cortés in the upper left, and Ferdinand Magellan in the upper right. Most important here, however, are the two columns of vignettes that appear on either side of the central cartouche. The series on the left depicts events from the conquest of Mexico, beginning in the lower left with a view of Mexico City before the conquest, ascending through three vignettes depicting the capture of the Mexica lord Cuauhtemoc, the initial encounter with the ruler of Michoacan, and a scene from Cortés's march to Honduras, and ending with a view of Mexico City as reconstructed after the conquest. On the right, we have scenes from the Magellan expedition, beginning in the upper right with the discovery of the strait and descending through the passage into the South Sea, the death of Magellan, and the arrival of the *Victoria* in Spain. The series ends with the 1524 *junta* in which Spanish and Portuguese cosmographers attempted unsuccessfully to settle the controversy surrounding the location of the antimeridian. The image puts them where they had planned to meet but did not, on the bridge spanning the Guadiana River, which defined the boundary between the two kingdoms.

Herrera's narratives of the conquest of Mexico and the first circumnavigation of the globe begin in the Second Decade and continue in the Third, so the title-page image actually interrupts the progress of the two stories, inviting readers to step back and reflect upon them. The two histories are interlaced with each other, as well as with other narrative threads involving events in Spain, Hispaniola, Darien, and elsewhere. Narrative interlace is characteristic of the *Décadas* as a whole, which recount the history of the Indies year by year, rather than place by place, as Oviedo and Gómara had done. This requires the narrator to jump from one location to the other, occasionally mapping his movements with remarks like this one: "Cuando Hernando de Magallanes iba navegando, y como se ha visto, en Barcelona, Tierra firme, y otras partes, sucedió lo que se ha dicho, Hernando Cortés que se hallaba en la isla de Cozumel, estaba muy contento con Gerónimo de Aguilar" (While Ferdinand Magellan was sailing, and the events narrated earlier unfolded in Barcelona, Tierra Firme, and other places, Hernán Cortés found himself on the island of Cozumel, happy to have found Jerónimo de Aguilar) (1.782). The remark assists the reader with the ongoing task of mapping the different narratives into their respective geographical settings, and helps him or her appreciate the fact that far-flung events narrated sequentially actually unfolded simultaneously.

The title-page illustration for the Third Decade does something similar. It disentangles the two interlaced story lines, Magellan and Cortés, allowing the reader to comprehend each of them as a single series of significant events that unfolds without interruption from another story line.[7] It assigns equal significance to both, thereby compensating for the fact that the text devotes much more attention to Cortés than it does to Magellan. The reader learns that the number of pages devoted to any particular narrative is not an accurate measure of its overall significance. More importantly, however, the reader also sees the two stories unfold simultaneously. Just as he or she can follow each story individually, in linear fashion, so he or she can also jump from one story to the other, in the same way the text does. This time, however, the reader can readily capture the simultaneity of mutually distant events. Her perspective is that of a divinity, or that of the man who discovers the aleph in the story by Jorge Luis Borges, that magical object in which one can see everything that everyone is doing all at once, no matter where they are doing it. In this case, however, the space under surveillance is not the world as a whole, but the Indies, imagined as a theater in which the heroic deeds of the Castilians unfold everywhere, all at once.[8]

In this way, the title-page illustration of the Third Decade captures the utopian aspiration of the text itself, to provide not only a comprehensive history of Spain's overseas empire, but a map of empire saturated with historical memory. The title page of the Third Decade does not just help the reader imagine the conquest of Mexico and the circumnavigation of the globe as a series of events unfolding simultaneously: it teaches the reader how to read the entire text of the *Décadas*, by imaginatively reconstructing the simultaneity of events that the text can only present sequentially, and by systematically plotting events into their respective locations, so as to fill the empty spaces of Herrera's understated maps with a plethora of places rendered meaningful by the "deeds of the Castilians," that is, by Spanish colonialism and imperialism. In this way, the title-page illustrations, this one in particular, become the mediating term between the cartography and the historiography. They allow us to imagine how Herrera might have handled the twin problems of distance and simultaneity had he ever extended his history beyond 1554. We can picture a title page for one of the decades that remained unwritten, perhaps one that interlaced the conquest of the Philippines and the birth of transpacific trade with the consolidation of Spanish authority in Peru under Viceroy Francisco de Toledo. Such an image, and such a narrative, would have taught the reader to move back and forth across the Pacific and to think of the Indies of the West as an integral part of that vast theater of Spanish colonialism that so prominently featured the New World.

But why did Herrera never complete his history? Kagan argues that he had every intention of doing so, but that his progress was stymied by a fall from official favor. Fed up with Herrera's arrogance, the duke of Lerma stripped him of his position, exiled him from the court, and appointed the humanist Pedro de Valencia to take his place, at twice Herrera's salary (Kagan 2010, 272–79). Nevertheless Valencia, as Kagan himself relates, made no progress on the official chronicle either, and his failure might point to a problem that was much larger than the personal fortunes of individual historians. Valencia was charged with producing an addendum that dealt with the war in Chile, a nasty little police action that had been rendered notorious by Alonso de Ercilla's wildly successful epic poem, *La araucana*. As far as we know, Valencia never wrote a word of the chronicle and defended his inactivity before the Council of Indies by explaining that no one could write truthfully about the Chilean war without including the "injusticias y crueldades" (injustices and cruelties) that the Spanish had committed there, and thereby defaming "la nación española" (the Spanish nation). Any history of the Chilean war, in other words, stood a good chance of becoming another *Brevíssima relación de las Indias*, grist for the mill of the Protestant Black Legend (Kagan 2010, 278). Perhaps this is also the reason that Herrera never extended his chronicle beyond the year 1554. He would not have been able to avoid the Chilean war, which Ercilla had rendered unavoidable but also unjustifiable, and perhaps he knew that not even he, with his ample capacity for spin control, could make that square peg of brutality fit the round hole of imperial history.

To continue the chronicle, it helped to strike out in another direction, quite literally. While the war against the Mapuche in southern Chile had devolved into an unending quagmire, Spain's efforts to assert its authority in the South Sea had produced a quick and apparently decisive victory. In 1606, Pedro Bravo de Acuña, the governor of the Philippines, had sailed to the Moluccas with an army of Spaniards, Portuguese, Tagalogs, and Pampangans to take the island of Ternate. The Portuguese had been expelled from the island by an indigenous revolt in 1575, and from neighboring Tidore in 1605, by the combined efforts of the Moluccans and the Dutch East India Company. Acuña scored a quick and decisive victory, kicking the Dutch out of the most important islands of the Spicery and establishing Castilian control over Ternate and Tidore for the very first time. The duke of Lemos, Pedro Fernández de Castro y Andrade (1560–1622), considered this to be his most important accomplishment as president of the Council of Indies, and he commissioned his Latin secretary, the historian and poet Bartolomé Leonardo de Argensola, to write a history celebrating Acuña's achievement. Argensola dutifully accepted the assignment, despite his stated distaste for the material, and abandoned the court for provincial life in Aragon as soon as he was done, yet the text he produced does not read like the job of someone eager to get the job over with so he could go home (Leonardo de Argensola and Colas Latorre 1996, 25 n. 56). Appearing in print in 1609, just as Herrera was falling from grace and Valencia was dithering over Chile, Argensola's *Conquista de las Malucas* does not just tell the story of the Acuña expedition, but ranges broadly over the natural and moral history of the Indies of the West, celebrating the "deeds of Castilians" in that part of the world. As such, it stands in for that part of Herrera's general history that never got written, albeit quite accidentally. It is to Argensola's history that I now turn.

Argensola Erases the Antimeridian

The *Conquista de las Malucas* was a success among Spanish readers, but there were those critics who carped that the text indulged too often in metaphor and included amorous intrigues more appropriate to poetry than to history. They also joked that Acuña's victory had been so quick and complete that its tale could have been told in a single folio, forcing Argensola to pad his text with all sorts of irrelevant material (Bartolomé Leonardo de Argensola 2009, 9).[9] Indeed, while Herrera's *Historia general* provides less than the title promises, Argensola's *Conquista de las Malucas* provides much more. The first nine of its ten books tell the whole history of the Moluccas during the sixteenth century, as Argensola understood it, but starting with the middle of the third book, the text interlaces events in the Moluccas with lengthy episodes set in other places, some of them only tangentially related to events in the islands themselves. These episodes include accounts of the circumnavigation of the world by Sir Francis Drake and his pursuit by the Peruvian colonial official Pedro de Sarmiento (books 3 and 4), the annexation of Portugal by Spain (4), the defeat of the Spanish Armada (5), an Iberian incursion into Cambodia (6), the history of Dutch expansion into the East Indies (7), and the so-called revolt of the Manila *sangleys* that took place in 1603 (9). Along the way, we get a wealth of geographical and ethnographical material, including descriptions of China, the Philippines, and of course the Moluccas, but also the Strait of Magellan, Borneo, Java, and numerous other South Sea locations.

The author's brother and a fellow writer, Lupercio Leonardo de Argensola, defended the book from its critics. Acuña's expedition, Lupercio explained, was an effort to recover territory that had once belonged to Portugal and had been usurped by Holland, so one could not truly appreciate the significance of the expedition if one did not first understand how the islands had been won and lost by the Portuguese. One also needed to remember, Lupercio added, that souls, not spices, were the real stakes in the struggle over the Moluccas. The history of Iberian involvement in the islands thus had to be placed in the context of the global struggle against Protestant (i.e., English and Dutch) heresy.

These arguments have not convinced all of Argensola's readers. John Stevens, the person who translated the *Conquista* into English during the eighteenth century, did not believe that there was any real unity to the text (Leonardo y Argensola 1708, Preface). The noted Hispanist Otis Green dismissed it as "diffuse," and full of "lamentable digressions" (Green 1952, 52). More recently, however, John Villiers has rescusitated Lupercio's argument, noting that several of the supposed digressions, most notably the ones about Francis Drake, Pedro Sarmiento de Gamboa, and the Dutch incursion into the East Indies, help provide the ideological rationale for Acuña's punitive action (2003, 471). Carmen Nocentelli has gone even farther in her defense of Argensola's text, celebrating its baggy inclusivity as "a synthesis of totalizing ambitions—a veritable summa of geographic, historical, and ethnographic information drawn from a vast array of sources" (2013, 91).

Like any history, however, the *Conquista de las Malucas* has a significant spatial dimension, one that has been neglected by its readers. The text does not just tell the story of European involvement in Southeast Asia over the course of the sixteenth century, but also maps the region into the world-making projects of Spanish imperialism. We can understand how it does so by picking up on the spatial dimensions of each of the two major themes identified by Lupercio. By examining Argensola's treatment of "Spain and the Spicery," we discover how his text rewrites the history of Iberian rivalry in Southeast Asia in ways that render the disputes over the antimeridian irrelevant, while continuing to favor the interests of Castile over Portugal. In the *Conquista de las Malucas,* the old frontier between the Spanish and Portuguese empires in Asia is erased, converting all of the Indies into a single imperial space rightfully dominated by the subjects of the Hapsburg monarch. Yet while Argensola erases the monarchy's most important internal boundary, he also defines its external boundaries against foreign incursion. We can discover how he does this by attending to Lupercio's second theme, "The True Faith against the Heretic." Argensola's narratives of Drake's circumnavigation, Sarmiento's voyage to the Strait of Magellan, and Dutch incursions into the Spicery do not just provide the ideological justification for the Acuña enterprise: they also serve to map the external boundaries of Argensola's unified imperial space.

The *Conquista* shows the same disregard for Portugal's territorial pretensions in East and Southeast Asia that we see in Herrera's *Décadas* and other texts of Castilian origins. It opens with a description of the islands and their inhabitants punctuated by tales of Iberian arrivals, first following the historical and geographical itinerary of Portuguese expansion into the Indian Ocean all the way to Francisco Serrano's 1511 encounter with Ternate, at the tail end of Portuguese expansion into the Indian Ocean basin, and later that of the Magellan expedition across the Atlantic and Pacific, until its 1521 arrival in Tidore. Along the way, Argensola rehearses the controversies over the antimeridian, only to insist that they were rendered moot when Philip II became king of Portugal (1–31). In this way, it sets the history of the Moluccas into the inescapable context of Iberian disputes over the division of the globe, only to eventually erase all traces of the antimeridian and construct the global space of the Hapsburg empire as a single continuous expanse.

That eraser works through a combined strategy of omission and commission. The text never mentions certain inconvenient details, such as the fact that Philip II had sworn to respect the rights and privileges of Portugal and its overseas empire, or that controversies over the lines of demarcation continued to rage during the so-called Union of the Crowns. Neither does it mention how these controversies came to a head with Acuña's conquest of the Moluccas. As Richard Sheehan points out, the Acuña expedition was a textbook example of Portugal's marginalization within a governing structure dominated by Castilians. The operation was planned and executed by members of the Council of Indies, all of them Spaniards, without any involvement from the Councils of Portugal and India. It involved the projection of military power from Manila rather than Goa or Malacca, under a Spanish commander, not a Portuguese one, and when all was said and done, it left the Moluccas under Castilian control, over and against Portuguese protest (Sheehan 2008, 309–39). Five years after the event, the eminent Portuguese cosmographer João Baptista Lavanha (ca. 1550–1624) continued to insist that the Moluccas were on the Portuguese side of the antimeridian by 3° of longitude, and that control over the islands should be remitted to the crown of Portugal (Sheehan 2008, 331–33).[10] Argensola remains silent about all of this, plotting the story of the Acuña expedition from an exclusively Castilian perspective, as if that were the only possible way of telling it.

Argensola even suggests that this Castilian takeover of Portuguese territory answered to the needs and desires of the Moluccans themselves. He cites a speech supposedly delivered in 1600 by an ambassador from Ternate before the governor of the Philippines:

No podemos dejar de admirarnos de ver con que alcanzando las fuerzas portuguesas tan insignes victorias como las de Calicut, contra los Turcos en Dío, contra los Egipcios, contra los de Cananor, de Zeilán, de las Javas, Sumatra, y contra tantas naciones *para aquella parte*; y los Castellanos por esta otra contra Camboja, Mindanao, Japón, Conchinchina, China; solo nosotros los Malucos, que *estamos en medio de dos poderíos de un solo Monarca* quedamos expuestos a las últimas armas de unas islas rebeldes. Si es así, que el Rey de España permite (antes manda) que seamos socorridos por Filipinas, ¿por qué no es obedecido? (239)

[We cannot help but wonder how Portuguese arms, having achieved such signal victories as those of Calicut, over the Turks at Díu, against the Egyptians, against the people of Cananor, that of Ceylon, that of the Javas, Sumatra and so many other nations *of that side*; and the Castilians on *this other side* against Cambodia, Mindanao, Japan, Cochinchina, and China; only we the Moluccans, *who are in the midst of the two dominions of one sole Monarch*, find ourselves exposed to the fury of rebellious islands [i.e. the Dutch]. If the King of Spain permits (no, commands) that we should be aided by the Philippines, why is he not obeyed?]

The speech, almost certainly an authorial fabrication, evokes a broad geography spanning the Indian Ocean, Southeast Asia, and East Asia, divided into symmetrical territories (*partes*), one Portuguese and the other Castilian, defined entirely by military victories, not by any treaty or line of longitude. The Moluccas lie between the two territories, the sole place where neither Castilian nor Portuguese arms have made themselves felt. The speaker clearly does not care which nation rules the Moluccas and does not understand why help cannot come from the Philippines. His only concern is that the monarch's will be done, that the Dutch be expelled from his islands. Clearly, the reader is supposed to sympathize with the ambassador and realize as he does that the king is obligated to help them with whatever resources he has available, whether Castilian or Portuguese. Acuña's expedition becomes the dutiful and merciful answer to the ambassador's plea for help.

The ambassador's point of view finds iconographic expression in the text's frontispiece illustration, which personifies the Spice Islands as a female figure, "Maluca," recalling the commonplace tendency to allegorize the four parts of the world as women, not to mention the more general tendency to depict territory as female (fig. 34).[11] The image mixes signs of civility (the town in the background, the clothing, the sword) with signs of exotic and even erotic tropicality (the crocodile, the volcano, the clove trees, the close-fitting top) in ways that reflect standard stereotypes

FIGURE 34. The Moluccas appear as an allegorical woman staring at the arms of Philip IV as they orbit the world. Title page from Bartolomé Leonardo de Argensola, *Conquista de las Malucas* (1609). Photograph: Courtesy of the Hans P. Kraus Collection of Sir Francis Drake, Rare Book and Special Collections Division, Library of Congress, Washington, DC (call no. DS646.6.L56; digital ID rbdk do33).

of the East Indies. The cloves in her cornucopia precisely echo the shape of her breasts, transferring her sex appeal to the land and its products, confounding sexual and colonial desire. The image, however, does not remind us in any way of the disputes between Spain and Portugal over the Moluccas. No hint of the rivalry between the Iberian kingdoms appears anywhere. On the contrary, the image indicates that the whole controversy has been transcended. Miss Maluca gazes heavenward, not at the arms of the kingdom of Castile and León that grace the title pages of Herrera's *Décadas*, but at the arms of King Philip III, with its multiple European and Iberian escutcheons, including those of both Castile and Portugal. Those arms follow the zodiac, the path of the sun through the heavens, reminding the reader that this is an empire on which the sun never sets. Philip rules the Moluccas by right, no matter which way you cut it.

According to Argensola, Spaniards do not have the right to rule the Moluccas because of the Treaty of Tordesillas, but rather by dint of their moral superiority vis-à-vis their Portuguese cousins. Just as there was a Black Legend of Spanish misrule in America, so there was a Black Legend of Portuguese misrule in Asia, and as Villiers points out, Argensola draws upon some of the very sources of that legend, including Diogo do Cuoto and Jan van Linschoten, to develop a highly critical account of Portuguese colonialism in the islands (Villiers 2003, 462 n. 42).[12] He dramatizes the exceedingly violent measures supposedly used by Portuguese governors to establish and maintain power over the Spicery. He condemns their Machiavellian machinations, labeling the viciousness of the Portuguese "insufrible" (insufferable) and "inhumana" (inhuman) on several occasions, and at one point he even goes so far as to call their brutality "indigna de Caribes, cuanto más de la hidalguía portuguesa" (beneath even that of Caribs, not to mention the Portuguese nobility) (49; also 35, 37).

Ironically, the *Conquista* ends up resembling the *Araucana* of Alonso de Ercilla, in broad but significant ways, despite the general effort of official historians like Argensola to avoid the poem's subject matter. Just as Chile was originally won by Valdivia, but lost to the cruelty that followed from his cupidity, so the Moluccas were won by Portugal, only to be lost to their own greed and tyranny. Just as Ercilla, moreover, invites the reader to sympathize with the Araucanians no matter how barbarous they might be, so Argensola invites the reader to sympathize with the indigenous inhabitants of the Moluccas. Like the *Araucana*, the *Conquista de las Malucas* has the reader witness a gathering of indigenous leaders in which they air their grievances against their would-be conquerors and organize their revolt. There, the king of Tidore mobilizes the rhetoric of the Black Legend, calling the Iberians "los ladrones del orbe" (the thieves of the world) who "roban, matan, avasallan, y con falso nombres nos privan de nuestro imperio, y hasta que convierten las provincias en soledades, no les parece que tienen introducida en ellas la paz" (steal, kill, enslave, and with false title deprive us of our rule, and do not believe they have pacified any place until they have converted into a barren wasteland) (63–64). The historian even follows the epic poet by deploying tales of suffering women to inspire sympathy for the plight of the oppressed indigenous population. After a Ternatan queen loses her son to a Portuguese assassination plot, the reader joins her at the graveside, as she weeps for her loss and rails against Portuguese "tiranía" (tyranny) (54). Argensola's general imitation of Ercilla lays the groundwork for his critique of the Portuguese. It is their fault, the text powerfully implies, that the Moluccas were lost.

The point, however, is not simply to beat up on the Portuguese and delegitimate their territorial claims. In fact, Argensola does whatever he can to qualify his argument in order to defend the Portuguese from the worst possible criticism. He insists, for example, that even the cruelest governors had the best of intentions and that they did what they did for the sake of the common good, not out of self-interest, adding that occasional cruelty is the inevitable fault of the bravest of men (98). Like other imperial apologists, Argensola emphasizes the contributions of the Portuguese to evangelizing the people of insular Southeast Asia (53). The real tragedy of the Moluccan rebellion of 1575, according to the historian, was the widespread destruction of Christian churches, the apostasy of countless Moluccans, and the murder of over sixty thousand Christians (96). Argensola even equips his text with a dashing and capable Portuguese hero, André Furtado de Mendoça (1558–1611). The text follows this scion of one of Iberia's most powerful aristocratic families as he puts down native uprisings across the Portuguese *Estado da India*, drives the Dutch from the island of Ambon, and invests the fortress that the Dutch had built for their Moluccan allies on Ternate (169–73, 254–75). By the standards of what had been said of Portuguese Molucca in print, this is not an outright attack upon Portuguese colonialism in the region, but rather a measured critique of it.

The point, therefore, is to use the established reputation of the Portuguese as failed colonialists to effect a *translatio imperii* of sorts with regard to the Moluccas. While the Tidoran ambassador argues that help should come from whichever imperial outpost was most capable of providing it, whether it was Portuguese or Castilian, the text as a whole argues that the military relief of the Moluccas had to come from the Philippines because the Castilians were the only ones in possession of the necessary resources, material and moral. One of the most famous episodes of the entire text, the story of the princess Quisaira of Tidore in book 4, paves the way for this argument. In this popular novelistic episode, which was later translated, anthologized, and adapted for the stage in both France and Spain, the Portuguese captain Ruy Dias Acuña loses the fair Quisaira and eventually his own life to a Moluccan rival, thanks primarily to his own indolence and prevarication (140–43). Nocentelli convincingly argues that the episode serves to satirize the ethos of Portuguese chivalry, indict Portuguese colonial rule, and pave the way for the future success of that other Acuña, the Castilian hero of Argensola's text (2010, 576–77). By this account, Quisaira represents the intradiegetic embodiment of the allegorical "Miss Molucca" who graces the title page of the *Conquista de las Malucas* (fig. 34). The whole text is the tale of how Portugal lost her, but Spain won her back.

The same could be said of the story of Furtado de Mendoça, or Hurtado de Mendoza, as Argensola refers to him. Mendoça emerges in the *Conquista de las Malucas* as an advocate of inter-Iberian cooperation. The Portuguese captain knows that he cannot take Ternate alone, but has given up on his unresponsive, uncooperative countrymen in Goa and Malacca, so he turns for aid to Acuña in Manila. Argensola quotes a letter from Mendoça to Acuña, reporting on the dire military situation in the Moluccas, suggesting that if the Dutch position in Ternate goes unchallenged, Manila will be the next place to fall, and arguing that "todos, como Cristianos y vasallos de su Majestad, debemos oponernos a ruina tan grande" (all of us, as Christians and vassals of His Majesty, should resist so great a ruin) (252). Just as the Tidoran ambassador had pleaded for help from whichever outpost of the dual monarchy could supply it, so Mendoça goes to whoever has the necessary resources in order to get the job done. If that means turning to the Castilians in Manila, then so be it.

Acuña sends men, guns, and supplies, but the battle narrative that ensues does not serve to celebrate the potential of military cooperation between Portugal and Castile. Instead, it serves to suggest that Portugal is exhausted, and Castile, vigorous. At a revue of the troops on a Ternatan beach, the commander of the Castilian reinforcements, Juan Juárez Gallinato, finds the Portuguese soldiers to be so sickly, poorly armed, and incompetent in the use of the harquebus that he begins to doubt their chances for success. The Castilians, by contrast, turn out in "tres cuerpos experimentados y lucidos en grande concierto, y del valor que luego se mostró" (three splendid, experienced, and orderly formations, of the valor that was later demonstrated) (266). When the Iberians begin to bombard the Ternatan fort, the Portuguese artillery turns out to be useless. The barrels are short and the stone shot shatters when it hits the walls. The Castilian artillery, by contrast, is longer barreled and more effective. Size matters, it turns out. Neither the big guns nor the bravery of the Spaniards, however, can compensate for the lack of powder, the ravages of tropical disease, the imminent threat of an approaching Dutch fleet, or the indifference of the Tidoran allies. Mendoça asks his officers what they should do. Gallinato responds for the majority, saying they should persist in their attack. Mendoça nevertheless decides to withdraw. Argensola does not directly accuse Mendoça of cowardice, but simply reports Acuña's efforts to defend him from criticisms of his decision and insists, in his own defense, "Este fue el suceso tan prevenido y amenazado, referido por mí con indiferente afecto" (This was the event that everyone had been waiting for, which I have related with indifferent affect). Any judgments of value, he adds, "no son para el escritor" (are not for the writer) (276). The message is nevertheless clear. Portuguese *fidalguía* has failed, and Castilian *hidalgía* must come to its relief, if Christian empire is to endure.

That relief comes from across the other side of the Pacific. Having traced the itinerary of Portuguese expansion eastward to its geographical extreme and its moral nadir, Argensola presents the westward progress of Spanish expansion as a new solution to old problems. Philip II, recognizing that the Portuguese in Goa have forgotten about the Moluccas, decides that the Philippines will serve as his new base in East and Southeast Asia (145–46). From early on, Manila appears in Argensola's text not just as a commercial center or a base for evangelical work but as a forward military position, strategically situated south of Japan, east of China, and north of the Spicery (154–55). By contrast with the Moluccas, where Portuguese cruelty and incompetence have reversed the initial good fortune of Christianity and civilization, the Philippines enjoy good governance and proper belief, "La Religión y la policía son las mismas que en España" (Religion and polish are the same as in Spain), except for those parts where Spanish rule has not yet made itself felt. "Los Españoles y criollos" who govern the Philippines, Argensola claims, "no desdicen de su buen origen" (The Spaniards and criollos do not belie their good origins; 156). Spanish *hidalguía*, unlike Portuguese *fidalguía*, has successfully transplanted itself to the tropics, successfully making the move that Oviedo had hoped it could make.

Pedro de Acuña becomes the exemplar of this brand of Castilian masculinity that manages to rule successfully while far from home. He makes his first appearance in book 5, long before Argensola gets to the Ternatan campaign, as a veteran of Spain's Mediterranean conflicts with the Ottoman empire who has accepted a position as governor of the crucial Caribbean port of Cartagena (182). The text then builds up Acuña's image and promise by interlacing brief snapshots of his achievements in Cartagena with accounts of various failed attempts to retake Ternate launched from Manila and Malacca (200–201). Finally, the text follows Acuña to the Philippines, where he exhibits unique foresight and sound judgment in matters military and diplomatic before setting off for the conquest of the Spice Islands (244–62). If Herrera's map of the Spanish Indies served to announce to all concerned that there was a new, Castilian sheriff in East and Southeast Asia, then Argensola's narrative gives that sheriff a name, Pedro Bravo de Acuña. He is there to assert Castile's rights to the Indies of the West, effectively taking over from the failed Portuguese and sending a clear message about Spanish capacity and resolve to the heretical Dutch and English. That *translatio* that Argensola is trying to effect is also a global remapping that converts the Moluccas from an eastern outpost of Portugal's flagging *Estado da India* into a western outpost of Spain's supposedly vigorous and emerging transpacific empire.

Do Not Enter

Argensola's preoccupation with the English and Dutch brings us to Lupercio's second theme, "The True Faith against Heresy." While Argensola's history of the Iberian contest for control of the Spicery leads him to erase the lines of demarcation and imagine the global empire of the Spanish Hapsburgs as a single, undivided space, the historian's concern with the Protestant challenge to Iberian control of the spice trade requires him to delimit that space against English and Dutch incursion. So, while Argensola erases the old boundaries with one hand, he draws new boundaries with the other. Those new boundaries, however, do not take the form of lines of longitude that define an exclusive space of settlement, trade, and conquest, but of strategic points of entry into an expanded version of the South Sea. This way of territorializing the space of Hapsburg sovereignty responds to Argensola's primary world-making strategy, which imagines a global network of maritime routes that converge on the Moluccas, rather than a framework of continental landmasses or a series of climatic zones. In the *Conquista de las Malucas*, the world is imagined through the seaways constructed by and for the European maritime empires. His boundaries appear at the choke points of the network.

The global dimension of the struggle over the spice trade, whether in its early phase involving only Castile and Portugal or its later phase involving England and Holland as well, requires Argensola to shift his focus away from the principal locus of his narrative, the Moluccas, to other places elsewhere in the world. Sometimes, he makes the transition abruptly, as he does when he turns his attention away from a military conflict in Borneo to tell of the events leading up to the annexation of Portugal, "En Europa no se sabía la nueva de ellas [las islas], adonde y en África se temían más altos peligros" (The news [from the islands] was unknown in Europe and Africa, where everyone dreaded greater dangers; 100). He does something similar when he turns to his account of the Spanish Armada and the emergence of Holland as an independent, Protestant, maritime power (158). At other times, however, he gets from one part of the world to another by following the geographical and historical itinerary of an actual long-distance voyage, whether it be that of da Gama, Serrano, Magellan, Loaysa, Drake, Sarmiento de Gamboa, Legazpi, Furtado Mendoça, or van Neck. He even follows the voyage of a Moluccan ambassador, Cachil Naique, from Ternate to Lisbon (132–34). Of course, he does this because the voyages themselves are part of the story he is trying to tell, but the result is that the reader learns to imagine the world of early modern global imperial conflict as a network of maritime routes across the Atlantic, Indian, and Pacific Oceans.

Three of these voyage narratives are particularly important to our understanding of Argensola's boundary-making efforts, the stories of Francis Drake, Pedro Sarmiento de Gamboa, and Jacob Corneliszoon van Neck. The interrelated stories of Drake and Sarmiento serve to mark the eastern point of entry to the South Sea. Argensola follows the English privateer on his journey around the world between 1577 and 1580, tracing his route from England across the Atlantic, through the Strait of Magellan, up the west coast of South America, before wintering somewhere in California and then heading to the Moluccas, where he becomes the first Englishman to trade in cloves at their source. At this point, Argensola interrupts the narrative to denounce the privateer. According to the historian, the primary threat Drake posed had nothing to do with the treasure ships he took on the Pacific, or the load of spices he carried away from Ternate, but with the pernicious ideas he helped introduce to the people of the South Sea: "Éste fue el primero que abrió el paso a los Sectarios Hugunotes, Luteranos y Calvinistas que después pasaron a aquellos mares con navíos cargados de textos pervertidos, Biblias heréticas y otros libros de inficionada doctrina" (He was the first to open the way for the Huguenot, Lutheran, and Calvinist sectarians who later passed into those waters aboard ships loaded with perverted texts, heretical Bibles and other books of infected doctrine; 105). The text then assures the reader that Providence is already staunching the infection, by sending "Religiosos Españoles" (Spanish religious orders) with an "Evangelio limpio . . . a los últimos Indios" (a clean Gospel . . . to the most distant Indians), supported by the "ejércitos" (armies) of the king, whose "más principal instituto es la causa de la Religión" (most important establishment is the cause of Religion; 105). God, acting through the Catholic monarchy, will respond—is responding—to the threat of heresy, and will assure that the South Sea becomes a purely Catholic space.

That response, the reader learns, began as soon as the authorities in Peru became aware of the privateer's presence in what had been until then exclusively Spanish waters. The text tells how the viceroy of Peru, Pedro de Toledo, dispatched Pedro Sarmiento de Gamboa, an accomplished soldier, naval captain, poet, scientist, antiquarian, and historian, to chart the southern coast of Chile and the Strait of Magellan, and if possible, to kill or capture Drake. The text follows Sarmiento to the south of Chile, through the Strait of Magellan, and across the Atlantic to San-lúcar de Barrameda, where Sarmiento goes to report his findings regarding the possibility of fortifying the strait to the king and the Council of Indies, but it devotes the greatest attention to Sarmiento's activities in and around the Strait of Magellan. Argensola mentions Sarmiento's encounter, not just with the giants that everyone else had reported, but with an actual cyclops, and he shares the rumor that Sarmiento glimpsed a "ciudad fantástica" (fantastical city) from afar (118–19). This would be the mythical "Ciudad de los Césares" (City of the Caesars) that had become part of the lore surrounding Patagonia. Argensola closes on this note, clearly eager to generate a sense of wonder around the region. That barely glimpsed center of civilization shines like the Seven Cities of Cíbola in the account of Fray Marcos de Niza, waiting for its own Cortés or Pizarro to come along.

Nevertheless, while the text invokes this sense of wonder surrounding the area of the strait, it focuses on the scientific and legal dimension of Sarmiento's work in the area, noting his exhaustive efforts to chart the shores of southern Chile and of the strait itself, and attending to his efforts to assert Spanish sovereignty over the whole region. He reports how Sarmiento renames the strait "Madre de Dios; para que por esta devoción alcance de su Hijo la salud de aquellas no numerables provincias" (Mother of God, so that through this devotion her Son's salvation might reach those lands without number), thereby restoring the sanctity (and virginity) of the channel that Drake had defiled (105–19). He reproduces the full text of the ceremony of possession that Sarmiento carried out at the strait's eastern mouth, and he tells of the cross Sarmiento erected there, with a bottle at its foot containing a message to all comers that "aquellas tierras y mares eran del Rey de España" (those lands and seas belonged to the King of Spain). The ceremony of possession and the message in a bottle alike appeal to papal authority to back up Sarmiento's claims, even though his actual geographical position lies to the west of the original line of demarcation (117). Sarmiento, in effect, transfers the function of the line of demarcation as the eastern boundary of the Spanish Indies to the Strait of Magellan, substituting that ill-defined meridian with a clear geographical point of entry. By reproducing Sarmiento's texts in full, Argensola does not just report Sarmiento's speech acts, but performs them anew for the benefit of his European readers.

He also broadens their significance. The viceroy's instructions to Sarmiento clearly state that the purpose of his mission is to provide for "la seguridad de todas las Indias del Mar del Sur" (the security of all of the Indies of the South Sea; Sarmiento de Gamboa 2000, 31). Although the document does not specify what it means by that capacious phrase, the Indies of the South Sea, it is more than likely that the viceroy was referring only to the viceroyalty of Peru. Toledo had been concerned for some time with the state of the viceroyalty's defenses, proposing in 1573 that the crown fortify its most important ports (Bradley 2009, 11–12). When he sent Sarmiento to the Strait of Magellan, in October of 1579, the viceroy had no idea that the corsair was sailing across the Pacific toward the Spice Islands, where he would arrive in November. He only knew that the privateer's attack had vindicated his assessment of Peru's vulnerability to maritime assault, and that the time was ripe to propose more aggressive measures to the crown. Sarmiento, in turn, was a cosmographer and a veteran of Álvaro de Mendaña's first South Pacific voyage (1567–69), who might have been expected to define the "the Indies of the South Sea" more broadly, but if he did, it does not come across in his account of the expedition or in his views about the fortification of the Strait of Magellan. His memorial outlining the advantages to be had by settling and fortifying the strait emphasizes the existence of local commercial opportunities rather than the potential colony's broad geopolitical value (Sarmiento de Gamboa 2015, 49). It is reasonable to suppose, therefore, that both men were primarily, if not exclusively, concerned with the security of the Pacific shores of South America and the opportunities posed by a new colony in the area of the strait, rather than with the security of what Schurz christened "the Spanish Lake."

Argensola, however, tells the story of Sarmiento's expedition from a metropolitan perspective, which did indeed appreciate the global geopolitical significance of the Strait of Magellan and its potential as a route of access to East and Southeast Asia for friend and foe alike (Sarmiento de Gamboa 2015, 48). He therefore manipulates the chronology of his narrative to make it seem that Toledo and Sarmiento are reacting to Drake's arrival in the Moluccas, not just to his raids along the coast of South America. The text relates the news of Drake's activities in the Spicery before getting to the story of Sarmiento's expedition, and it omits any mention of dates, allowing the reader to construct a relationship of cause and effect between the first event and the second that did not obtain in fact. When he reports that Sarmiento and the viceroy took action, "para la conservación de las Indias, de su paz y Religión" (to conserve the Indies, their peace and their Religion), it seems that they are concerned with the Indies of the West just as much as with the Indies of the South (105).[13] In this way, Argensola converts Toledo and Sarmiento into defenders of the Moluccas, not just of Peru. The Strait of Magellan, in turn, becomes the strategic point of access to the entire South Sea, including the Indies of the West. The unspecified "tierras y mares" that Sarmiento claims for the King of Spain are no longer just those in the vicinity of the strait, but everything from the strait to the west, all the way to the Moluccas, at the very least. Sarmiento, in effect, becomes something he never attempted to be, a second Balboa, claiming for Spain the whole of the Pacific Ocean.

The historian maps the western boundary of this imperial space by telling the story of the first Dutch expedition to reach the Moluccas in book 7.[14] Just as he begins his account of Drake's voyage in England, so he also begins his narrative of the voyage of Jacob Corneliszoon van Neck in Holland. He briefly describes the country, the character of its inhabitants, and the nature of its heresies, which according to Argensola are indistinguishable from atheism (212–15). He then follows the Dutch around the Cape of Good Hope, through the Indian Ocean and insular Southeast Asia, where they reach Ternate in 1598, following a novel route through the Strait of Sunda rather than the Portuguese-controlled Strait of Malacca. In Ternate, they offer support to the sultan in his struggle against Tidore, establish a trading post, and introduce "las pestilencias de Calvino" (the pestilence of Calvin; 234). Just like Drake, the Dutch are not just competitors for control of the spice trade, but carriers of a dangerous infection, Protestant Christianity.

The Dutch pause at a series of islands along the way, giving Argensola the opportunity to provide all sorts of ethnographic details about insular Southeast Asia, but their most significant stop comes much earlier along their route, at the uninhabited island of Mauritius, where part of the fleet stops to take on food and water. As Patricia Seed explains, the island was strategically situated along the new route from the Strait of Mozambique to the Moluccas that the Dutch were pioneering. It was thus likely that the Hapsburgs would try to challenge Holland's entrance into the region by contesting their claim to this particular island (Seed 1995, 166). In full awareness of this possibility, the vice-admiral of the fleet, as Argensola reports, named the island in honor of Maurice, Count of Naussau and Prince of Orange (1567–1625), and *stadtholder* of the Dutch Republic. To mark his claim, he hammered a sign to a tree with the arms of Holland, Zeeland, and Amsterdam, and the words, "LOS CRISTIANOS REFOR-MADOS" (THE REFORMED CHRISTIANS; 218).[15] Like Sarmiento's activities in the Strait of Magellan, the actions of the Dutch on Mauritius were purely symbolic and did nothing to establish effective territorial control, but they were nevertheless significant. Written in Spanish, the sign was there for the benefit of the Spanish and Portuguese and was meant to challenge the pretensions of the Hapsburg Monarchy to exclu-sive rights over commerce and colonialism in the Indian Ocean basin. A tyrannical parody in the west of Sarmiento's legitimate actions in the east, the Dutch ceremony of possession becomes for the patriotic Castilian reader an act of political, military, and religious provocation.

By reproducing this provocation within his text, Argensola transforms it into a call to action, implying that Acuña's success in Ternate must be seen as the first battle in a prospective campaign to drive the Dutch out of this maritime space, where only the subjects of the Spanish Hapsburgs can legally settle, conquer, and trade, and where only Catholicism can be allowed to take root. That space, moreover, includes everything along the network of maritime circulation that stretches from the Strait of Mozambique to the Strait of Magellan, crossing both the Indian and the Pacific Oceans, integrating Africa, Asia, and America. Ironically, this is precisely the space in which the Dutch East India Company had been entitled to operate. Argensola takes over this new and expanded South Sea, as the Dutch themselves have territorialized it, and converts it into a vast circulatory system that must be purged of Protestant contagion, so that its countless souls can brought to the true faith. Acuña's voyage from the Philippines to the Moluccas, therefore, must continue along the itinerary of the Dutch, following it in reverse and coming to the relief of a beleaguered Portugal.

This ambitious vision of the Hapsburg empire in the South Sea, however, is haunted by the shadow of yet another enemy, China, which does not figure in either of the themes that unite the *Conquista de las Malucas*, as Lupercio Leonardo would have us see it. Neither the story of "Spain and the Spicery" nor that of "The True Faith against the Heretic" needed to feature Chinese characters or Chinese settings in order to meet the expectations of the sixteenth-century European reader. The Middle Kingdom nevertheless merits a description in Argensola's text, at the end of book 4, right after Argensola's account of Philip II's decision to use the Philippines rather than the Portuguese *Estado de India* as the preferred base for an assault on the Moluccas. This description raises fears about future conflict between Spain and China over control of the Indies of the West, fears that find an ambivalent response in one of the text's most famous episodes, its account of the revolt of the *sangleys* in book 9.

The Chinese Threat and the Chimera of Empire

Argensola's description of the Middle Kingdom stands oddly apart from the rest of his narrative. It does not report any events that impinge upon the narrative's larger patterns of causality, nor does it follow any travelers on their way to or from the Middle Kingdom. Argensola says nothing, for example, about the various attempts to found Christian missions in China that we read about in Mendoza's *Historia del gran reino de la China*. The description nevertheless serves to inscribe a series of hopes and fears that reverberate throughout Argensola's vision of the Spanish South Sea and that inspire an indirect response in one of the text's most frequently cited episodes, its account of the 1603 revolt of the Manila *sangleys* in book 9. The *Conquista de las Malucas* is not just about the Spicery, the Philippines, or the Dutch menace. It is also about Spain's problematic future in a part of the world where China was always the elephant in the room.

Argensola's description of China rehearses some of the major tropes of early modern Iberian Sinophilia that we saw in chapter 6. The historian presents the Middle Kingdom as a large, very populous, highly productive country in the temperate zone, full of sizable and sophisticated cities that are linked by navigable rivers. The Chinese sail these rivers aboard sophisticated watercraft, engaging in a frenzy of industrial and commercial activity. Country and city alike produce so many desirable goods that China has no demand for anything that foreigners have to sell, except precious metals. Its people work hard, but have refined tastes, enjoying delicious banquets and elaborate theatrical representations. Sovereignty resides entirely in the country's monarch, although he distributes his authority to a complex hierarchy of officials, the mandarins. The text says nothing about the rigor of China's judicial or penal system, and never so much as intimates that the governing class might be willfully or indiscriminately cruel. It claims that the population holds the mandarins in such awe that they do not dare look them in the face but it does not suggest that the people fear their rulers or wish to be free of them (146–49). The claim that the mandarins were tyrants that was so crucial to the discourse of the hawkish Sinophobes finds no place on Argensola's pages.

The historian nevertheless manages to make a case for war against China through his treatment of Chinese religion. Like every other Iberian writer, Argensola presents it as a risibly false system of belief and practice inspired by the devil and ultimately devoted to his worship, but he never makes the common claim that the Chinese do not believe very fervently in their own gods and are thus generally amenable to Christian conversion, nor does he identify any aspects of Chinese religion as relics of a primitive Christianity introduced by Saint Thomas in antiquity (e.g., 58, 63, 148). On the contrary, he claims that many of the Chinese are actually atheists who reject religion altogether, seeing it as a threat to their liberty as human beings. "Con esta impía credulidad y pestilente ateísmo," Argensola explains, "resisten a la eterna salud con que nuestras armas y nuestros predicadores les convidan" (With this impious credulity and pestilential atheism, they resist the eternal health to which our arms and our preachers invite them; 148). He goes on to add that "cualquier guerra que por introducción del Evangelio se haga, es importantísima y de suma ganancia" (any war waged to introduce the Gospel is of utmost importance and profit; 150). Although he never comes out and says so explicitly, it would seem that war against China is not only justified, but necessary and urgent.

Thus, despite its apparent Sinophilia, and its eschewal of the most obvious trope of Iberian Sinophobia, Argensola's description of China bears an impressive burden of contempt and fear. By claiming that an unspecified number of Chinese are actually atheists, Argensola dramatically expands a trope that we find deployed in the writings of Alonso Sánchez. According to the Jesuit, the mandarins thought that religion was the opiate of the masses, that they themselves were the only real gods, their comfortable life the only real heaven, and the prisons they administered the only real hell. The mandarins, in other words, were atheists, but they were the only ones. The members of the general population were run-of-the-mill idolators (Sanchez 1583; Sánchez 1588). Argensola, by contrast, expands Chinese atheism to the general population, claiming that disdain for god and religion can be found in "muchas partes" (many places). Atheism, in Argensola, is not just a characteristic of China's arrogant rulers, but a phenomenon so widespread as to become characteristic of China as a whole (148). In this, China resembles Holland, whose heresy, according to Argensola, was indistinguishable from atheism, but China is undoubtedly larger and more powerful, and it poses a greater threat to Spain's ambitions in the Indies of the West. In the *Conquista de las Malucas*, therefore, it is actually China, not Holland, that emerges as Spain's ideological Other, and its most dangerous political and military rival.

In order to make this case, Argensola puts a new twist on yet another trope of early modern descriptions of China, one which we have not yet discussed. Ming China was undoubtedly large, but it was less extensive than the China of the previous dynasty, the Yuan, thanks in part to a deliberate strategy of withdrawal and consolidation (Brook 2010, 28). Although the Yongle emperor (r. 1403–1424) invaded Vietnam and launched an ambitious program of maritime exploration under the leadership of the eunuch Zheng He, the dynasty as a whole eschewed foreign military adventures. His successors scrapped the Yongle emperor's expansionist schemes, closed China's borders, and forbade contact with the outside world. Europeans who wrote about China were well aware of this history of territorial contraction and converted it into a *topos* of any consideration of China's borders or its history. In the hands of the dovish Sinophiles, the trope became an example of the wisdom of China's governing class. Escalante, for example, praises the Chinese because they "tuvieron mas prudencia que los griegos, cartagineses y romanos, los cuales por conquistar tierras ajenas se apartaron tanto de las propias que las vinieron a perder" (were more prudent than the Greeks, the Carthaginians and the Romans, whose foreign conquests took them so far from their own lands that they eventually lost them; 108). Unlike those ancient empires, the Chinese realized when they had become overextended, and wisely withdrew from their empire rather than allow it to exhaust their resources. If the China hawks found in such praise a cautionary tale aimed at their own imperialist ambitions, Escalante would probably not have minded.

Argensola rehearses this *topos* of Chinese history at the very beginning of his description of China, but upends its significance, by emphasizing the negative consequences that the withdrawal from empire had for the people that China left behind. According to Argensola, the inhabitants of China's former territories, which presumably included most of Southeast Asia, found themselves "expuestos a la tiranía y a la infidelidad" and "después de largas guerras cedieron a los mas poderosos. De aquí comenzaron los Reyes de la India, obligados a no dejar las armas, a no guardar fe para sustentarse unos contra otros" (exposed to tyranny and infidelity . . . after long wars, yielded to the most powerful. This was the origin of the Kings of India, who could never put down their arms, could never trust each other enough to make alliances among themselves; 146). While contraction may have been good for China, it was bad for the region as a whole, which fell into chaos, endemic warfare, and tyranny.

There is no doubt that Argensola counts the rulers of Tidore and Ternate among the tyrants in question. Not only does he use the term explicitly when he first mentions them, but he also equips his text with numerous episodes that depict the sultans of the Spicery as willful, deceitful, lascivious, and cruel (14, 22). As soon as the Ternatans cast the Portuguese out of their island, the men of the island's aristocracy descend into lust, betrayal, and murder as they vie for power over women and the throne (164). Elsewhere we read about jealousy among members of the royal family, double-dealings with Portugal, and insidious moves against Tidore, including a successful attempt to poison and kill their sultan. Sultan Babullah's death is said to have been the result of his frantic attempts to satisfy his disproportionate sexual appetite, or perhaps of witchcraft or poison, the favorite "tiranicida" (tyrannicide) of the region (138). When the Ternatans are finally defeated in book 10, the surrounding islands rejoice at their liberation from Ternatan tyranny, but Acuña must nevertheless remain circumspect because not even the Tidorans, who have been steadfastly loyal, can really be trusted (343).

As others have pointed out, this characterization of life in the Moluccas was typical of early modern European accounts of the Spicery, including

Argensola's documented sources (Lach 1965, 1.2: 606; also Villiers 2003, 136; Nocentelli 2013, 109). Galvão, Barros, Lopes de Castanheda, Cuoto, Resende, Bariel, and Botero all present the islands as a scene of endemic internecine conflict. According to Leonard and Barbara Andaya, this characterization stemmed from the failure of Europeans to understand the pattern of "complementary dualism" between Ternate and Tidore that was characteristic of Moluccan political life, and which combined friendship and enmity in ways that looked chaotic and capricious to European eyes (Andaya 1993, 47–112; Andaya 2015, 164). This allowed European observers to subsume the Moluccas, not under the category of "the Oriental" as Nocentelli (2013, 93) would have us believe, but under the category of the tropical savage. The Moluccans, in Argensola's text, count among those "indios asiáticos" (Asiatic Indians) who are inherently incapable of governing themselves, just like other inhabitants of the tropics (13).[16] In their constant quarreling, they resemble the New World Indians of Juan López de Velasco much more than they do the Chinese, whose government was famous for its rationality and effectiveness. By withdrawing from their Indian empire, the Chinese have deprived the people of the Indies of the guidance they so desperately needed, allowing them to revert to the chaos and violence typical of life in the tropics.

This state of endemic conflict and arbitrary rule, Argensola argues, is only a temporary affair, "hasta que otro mayor poderío los avasallase" (until another, greater power should make them [the people of the Indies] its subjects; 146). The power in question is obviously Spain, or the Hapsburg Monarchy, which has stepped into the chaos created by the Chinese withdrawal, restoring good government to the tropical Indies, or at least to the Philippines and the Moluccas. The trouble, however, is that China has come to regret its decision to abandon its former empire and is now ready to challenge Spanish rule. "No pasa ningún año sin amenazas de ejércitos Chinas" (Not a single year passes without threats from Chinese armies), Argensola warns. Supposedly, the Chinese are preparing for war, hoping for "victoria contra los Españoles que ocupan aquellas tierras que ellos dejaron por imprudencia" (victory against the Spanish, who occupy those lands that they [the Chinese] so imprudently abandoned; 150). In fact, Argensola would have us believe, war has already begun. He wraps up his diatribe by equating the Chinese with the *sangleys*, insisting that no one is to be trusted less than they and heralding his narrative of the 1603 revolt, whose history he tells in book 9 (150–51).

The so-called revolt of the *sangleys* is one of the most storied events in the history of the Philippines. Some historians do not hesitate to call the incident a pogrom, believing that the violence that began on October 3, 1603, and ended roughly three weeks later with 23,000 *sangleys* dead was instigated by the Spanish.[17] According to José Eugenio Borao, this is how the event comes across in early modern Chinese accounts, which were derived from the testimony of *sangley* survivors. The Ming chronicles tell of Spanish preparations to carry out a massacre and of lethal attacks upon groups of unarmed, defenseless *sangleys* (Borao 1998, 9–11). The Spanish sources, unsurprisingly, point the finger at the resident Chinese population, and they even identify a ringleader in a Christian *sangley* official, Juan Bautista de Vera.[18] All the sources, however, identify a strange event that took place four months earlier as the real beginning of the story. On May 23, a group of mandarins arrived in Manila claiming that they had come to investigate a rumor that the islands were home to an enormous mountain of gold called "Cabit." The governor escorted them to the port of Cavite so they could see for themselves that no such mountain existed, and then the officials made their way back to China.

The mysterious visit of the mandarins triggered a period of increased racial and ethnic tension in a city that was already marked by considerable unease. The Spanish in Manila numbered no more than seven or eight hundred, a tiny fraction of the city's primarily Chinese population, most of which had emigrated from the coastal province of Fukien during the previous thirty years. The Spanish were alarmed by the influx of so many exotic aliens and tried to contain them in a walled ghetto called the Parian, just as their ancestors had tried to contain the Muslim and Jewish populations of medieval Spain, yet the number of *sangleys* soon exceeded what the Parian could accommodate, and Chinese communities sprang up across the Passig River from Manila proper, particularly in the areas of Tondo and Binondo. Only a small number of these people were merchants involved in the galleon trade. The vast bulk were laborers and artisans who provided the community with all sorts of goods and services. In many ways, they *were* the city of Manila. The Spanish were the city's ruling class, their presence in the Philippines tolerated by China and Japan because it brought wealth to the region in the form of American silver.

Arriving in a city that must have seemed very familiar to them, the mandarins behaved in every way as if they were in a place under Chinese rule, proudly displaying their insignia of office and even administering justice among the *sangley* community. Governor Acuña eventually put an end to this behavior, but only after some prodding from members of the Audiencia who were distressed by this apparent challenge to Spanish sovereignty. The high-handed behavior of the Chinese officials must have lent credence to the rumors that flew after their departure, that the real purpose of the visit was to reconnoiter the city and to incite the *sangleys* to revolt against Spanish rule in preparation for an imminent invasion from the mainland. Manila's decidedly Sinophobic archbishop, Miguel de Benavides, fueled the flames from the pulpit of the cathedral, preaching conspiracy theories and impending doom.[19] Acuña, meanwhile, made improvements to Manila's fortifications, ordered the manufacture of arms, took a census of the *sangley* population and its armaments, and recruited the support of the resident Japanese population in case of an eventual armed confrontation with the *sangleys*. These are among the moves that the Chinese sources identify as preparations for a massacre. In the Spanish sources, they become wise precautions against a possible rebellion.

The violence broke out across the Passig River from the walled city of Manila, and it initially swung in favor of the *sangleys*. After successfully defending Tondo and Binondo from a *sangley* assault, Luis Pérez das Mariñas led roughly 135 men in pursuit of the enemy, only to get surrounded and slaughtered. The rebels put his head and those of other Spanish soldiers on pikes and displayed them from the walls of the Parián, which lay just outside the walled Spanish city. Over the course of the fighting, the Parián was burned to the ground, and many of its prosperous Chinese merchants, some of whom had converted to Christianity and had hesitated to take sides in the revolt, lost their property and even their lives. The *sangleys* assaulted the city walls, only to fall back under fire from the city's cannon. Eventually, the Spanish received sorely needed reinforcements from the nearby province of Pampanga. Armed bands of Pampangans and Japanese under Spanish command took the fight to the various strongholds that the *sangleys* had constructed outside the city, butchering the rebels as they went. One of the final encounters was in the port of Batangas, about fifty miles south of Manila, where a group of *sangleys* was trying to arrange for maritime transport to the mainland. With the slaughter of the *sangleys* in Batangas, the violence came to an end.

It is not my intention to reconstruct the events in any more detail, or to assign blame for the eruption of violence. Instead, I examine Argensola's account of the violence of 1603 as an ideologically loaded piece of official historiography that responds to some of the anxieties about China that emerge in the historian's description of that country. The account of the violence is based on the work of Diego de Guevara, the prior of the Convent of Saint Augustine in Manila, who was apparently charged by the governor and the Audiencia with delivering an officially approved narrative of the events to the court of the king. In general terms, Argensola's version of the uprising coincides with those of other Spaniards, including Antonio de Morga, a member of the Audiencia who published his own history of the Philippines in Mexico City the same year that Argensola's book appeared in Madrid, but it also stands out from the others in certain key ways (Morga 2007, 174–78, 187–95).[20] The poet-historian presents the incident in decidedly Sinophobic terms, casts a particularly favorable light on Pedro de Acuña, deflects responsibility for the number of *sangley* casualties onto the Japanese and Pampangans, and includes some unique and telling details about the misadventure of Pérez das Mariñas. It is these details in particular that allow us to understand Argensola's narrative of the 1603 revolt as a troubled attempt to mediate his fears of impending conflict with the Middle Kingdom.

Pedro de Acuña rejected the notion that the *sangley* revolt had been engineered by external actors, but others were quite willing to lay the blame on the mandarins.[21] An account of the revolt by Miguel Rodríguez Maldonado that appeared in print in Seville in 1606 stated quite plainly that "los Mandarines habían venido con cautela a ver la tierra, y era con su orden el levantamiento" (the Mandarins had come to inspect the country discreetly, and the revolt was undertaken by their orders; Rodríguez Maldonado 1606, 6). Argensola seems to have agreed, although he does not put it quite as directly. Instead, he provides evidence meant to steer the reader toward that conclusion. The first is the opinion of Archbishop Benavides. While Morga simply reports the prelate's conviction that invasion was imminent, Argensola grants the archbishop credibility by adding that Benavides spoke Chinese, had visited China, and had suffered "tormentos y crueldades" (torture and cruelty) there (Morga 2007, 177; 288–89). He then confirms Benavides's suspicions about the Chinese by citing a letter written from the vicinity of Guangzhou by the prominent soldier and cosmographer Hernando de los Ríos Coronel, calling China the "verdadero Reino del demonio" (true Kingdom of the devil), and like Alonso Sánchez before him, claiming that Chinese civility was nothing but a façade hiding a more sinister reality (289).[22] Finally, Argensola cites historical precedent, claiming without citing a source that the Chinese had seized the island of Hainan by first inciting a revolt there. All of this serves to mobilize established Sinophobic attitudes in order to add credibility to the theory that the visiting mandarins had arranged the revolt.

Yet while Argensola advances an interpretation of the event that had nothing to do with the way Pedro de Acuña understood it himself, the historian also transforms the governor into the hero of his tale. In his hands, the measures that the governor took during the months leading up to the violence become "sana diligencia" (discreet precautions), never clumsy provocations (291). In Morga, by contrast, the very public nature of Acuña's actions lend credence to rumors of all kinds (Morga 2007, 188). In Argensola, it is the gossip of the Japanese, not Acuña's public actions, that raises suspicions about an impending massacre (291). While Morga spreads responsibility for the Spanish victory among the governor, "sus oficiales" (his officials), and the various captains that venture out beyond the walls, Argensola puts the spotlight on Acuña's leadership, particularly when it comes to recruiting aid from the Pampangans in order to bring the revolt to a quick and decisive end (303). Argensola's Acuña also shows courage and prudence in the aftermath of the violence, when he decides to report the event to the authorities in China, despite fears that doing so will only bring the wrath of the Ming down upon the fragile colony. Morga, meanwhile, cites the supercilious response of a Chinese official explaining why the Chinese have decided to spare the Spanish the punishment they deserve for slaughtering their countrymen (Morga 2007, 197–200). Argensola cites the same letter, but also Acuña's response justifying his actions and warning the Chinese that the Spanish are perfectly capable of defending themselves (307–11). In Argensola's account, Pedro de Acuña emerges even more clearly than in any of the others as the man of the hour, precisely the person that Manila needed in the fall of 1603.

This marked bias in favor of Acuña could have stemmed from the treatment of the governor in Argensola's source, the chronicle written by Guevara, yet whatever its origins, there can be no doubt that it fit very well with Argensola's larger project of casting Acuña as the embodiment of a heroic Castilian masculinity that had arrived in the Southeast Asia to take over from the failing Portuguese, set the government of the region on a firm footing, advance the cause of religion, and respond to the challenge of the Protestant incursion. By depicting Acuña as the hero of the *sangley* revolt and suggesting quite powerfully that the revolt itself was the spearhead of a Chinese invasion that never materialized, Argensola also figures Acuña and the Castilian masculinity he represents as a bulwark against the looming menace of Ming China. While the description of China in book 4 had raised the specter of possible Sino-Hispanic conflict, the account of the *sangley* revolt in book 9 assures the reader that the Spanish are perfectly capable of holding their own against the Chinese, whether they rebel from below or invade from outside.

The whole narrative nevertheless reads like an anxious attempt at political damage control. All of the sources put the number of *sangley* dead somewhere in the low twenty thousands. Why, one might ask, was it so high? Rodríguez Maldonado provides one possible explanation. According to his chronicle, as the Spanish counteroffensive began to gather momentum, "el Señor Gobernador despachó luego a los pueblos de su Majestad, dándoles aviso que no se reservase ninguno, sino que todos los que se hallasen, los pasasen a cuchillo" (the Governor sent a dispatch to all of His Majesty's towns, advising them that no one should be spared, but rather that everyone who was found should be put to the knife; Rodríguez Maldonado 1606, 6). Acuña, in other words, decided to take no prisoners and grant no clemency, and he ordered the indiscriminate execution of *sangleys* throughout the Philippines. It is not clear from Maldonado's text whether he meant any and all *sangleys*, or just those who had taken up arms. In any case, nothing of the kind appears in Argensola, where responsibility for the carnage gets displaced onto other actors. In a key incident, we follow the forces of one Martín de Herrera as they pursue and cut down defeated *sangleys* who are fleeing for their lives. For once, the text admits that the violence got out of hand, but it places the blame on the Japanese and Pampangans, claiming that they "son tan carniceros" (are such butchers) that not even the firmest Spanish captains can hold them back (305). Readers familiar with the history of the conquest of Mexico might recall the controversies surrounding the massacre in Cholula, and the efforts of Hernán Cortés to explain away the violence of the event, so as to preserve his own image as a just and capable commander (Padrón 2004, 105–6). Argensola does something similar, preserving Acuña from any real responsibility for the carnage and thus inoculating him from possible accusations of tyrannical behavior. The true tyrants, after all, are the "Kings of India," not the Spanish.

The text nevertheless attests to the difficulty, perhaps even the impossibility, of resisting the Chinese menace in the long run. It does so in its account of the grim fate suffered by Luis Pérez das Mariñas and his men. Das Mariñas was the son of a former governor of the Philippines who had died at the hands of mutinous *sangley* galley slaves in 1593 and was thus uniquely motivated to exact revenge on the rebels of 1603. All of the accounts agree that he and his men were cut down when their pursuit of fleeing rebels led them into swampy country where their prey had supposedly taken refuge. Only Argensola, however, shares what Das Mariñas supposedly said to inspire his men to follow him into what turned out to be a trap. "Que le siguiesen," he yelled, "que con veinticinco soldados bastaba para toda la China" (That they should follow him, that twenty-five men were enough for all of China; 295). The captain's words evoke the old notion that Spain could conquer China with a handful of men. With them, the dream of conquering China suddenly irrupts into Argensola's text in its most hyperbolic form. Das Mariñas becomes, for a moment, Martín de Rada, Francisco Sande, Alonso Sánchez, or any of the other hawks who imagined themselves among a tiny group of conquistadors toppling the mighty empire of the Ming. The captain's path, however, leads to death and dismemberment, not victory. In the aftermath of the battle, the field is found littered with the dented helmets of the slaughtered soldiers (295). While dented armor often serves to figure the heroism of the epic warrior, here it becomes a poignant image of defeat and a tacit admission that fantasies of empire in the Indies of the West were nothing but quixotic chimeras.

Conclusion

It is ironic that the Spanish crown only went public with its ambitious vision of the Indies as a transpacific empire at a time when it had become clear that Spain would be incapable of turning that vision into a reality, that indeed it would be hard-pressed to hold on to the little it had won. Herrera's *Historia general* appeared in print after a long decade of military, political, and economic setbacks that had left Spain reeling. The *Conquista de las Malucas* appeared about a month after the signing of the humiliating Twelve Years Truce, in which the monarchy of Philip III grudgingly accepted the grim reality that its considerable resources had been exhausted by the effort to quell the apparently indomitable rebellion of the Netherlands. The Spanish Hapsburgs found themselves grossly overextended, barely capable of defending what they already possessed, much less of expanding their empire. There was even talk of abandoning the Philippines, which were thought to be too distant, too vulnerable, and too poor to be worth defending. Herrera says nothing about these discussions, while Argensola reports them only to celebrate the repeated decision to hold on to the Philippines for the sake of the Christian mission in the region (85–86). No moment in either of these texts more clearly reveals the enormity of the gap between the poverty of Spain's material circumstances in East and Southeast Asia and the counterfactual ambition expressed by the kingdom's official chroniclers.

Herrera's *Historia general* teaches the reader to range across the whole map of the demarcation, including the Indies of the West, only to have him or her run up against the text's silence regarding everything that occurred after 1554, which means everything of importance that occurred in the transpacific west. In the end, the text does so little to fill the Indies of the West with historical memory that it effectively underscores its emptiness of any real "deeds of the Castilians" in the region, signaling the inconsequential nature of Spain's presence in East and Southeast Asia in the very act of asserting Spain's prerogatives there. It is no wonder, then, that when Theodor de Bry published a Latin translation of Herrera's *Descripción de las Indias* as part of an anthology of texts about the Indies, he equipped the publication with a map that limited its geographical purview to the New World, tacitly rejecting the text's own territorial claims.

By 1601, Herrera's readers beyond the Pyrenees were so accustomed to mapping the world in terms of an American West and an Asian East, it seems, that they were not willing to buy what Herrera was selling.

Argensola, by contrast, chronicles the deeds of the Castilians in the Indies of the West, highlighting Acuña's recent victory over the Dutch and the Ternatans in Southeast Asia, yet his account of a transpacific empire revolves around anxieties and silences. As we have already seen, the impending conflict with China looms large, and the text's assurances that Spain will be able to hold its own against the Ming threat may not have looked credible to readers familiar with China's considerable military capacity. Argensola's readers may have also been aware of the failure of Sarmiento's effort to colonize and fortify the Strait of Magellan in 1581 or of the continued penetration of the strait by the English and eventually, the Dutch, from 1587 on. The *Conquista de las Malucas* says nothing about these events. It simply rehearses the story of how Sarmiento erected an elaborate "Do Not Enter" sign, without saying anything about how completely ineffective the gesture turned out to be.

The text also remains mum about the schemes of the China hawks to conquer the Middle Kingdom, despite its obvious sympathy with their view that war with China was inevitable, even desirable. Perhaps Argensola thought them too absurd to merit mention, or perhaps he realized that they might cast a shadow over Acuña's accomplishment. With the conquest of the Moluccas, Spain had finally managed to project power across the Pacific, just as Sande and Sánchez had hoped it would, but on a paltry scale by comparison with their ambitious designs. Spain, as it turned out, was indeed capable of moving men across the Pacific, but only in the hundreds, not the thousands or tens of thousands. It was capable of recovering an island, but not of conquering an empire.

Conclusion

While Argensola was priming the ideological pump of Pacific expansion, the Portuguese pilot Pedro Fernandes de Queirós (1565–1614) was campaigning for the resources he needed to launch a voyage to the South Pacific in search of the *Terra Australis Incognita*, writing one memorandum after the other to the Hapsburg monarch. One of these memoranda, known simply as the *Eighth Memorial*, appeared in print in Seville and Pamplona in 1610, and it was soon translated into English, French, German, and Latin, and anthologized by both Theodor de Bry and Samuel Purchas. It has come to be considered one of the most influential pieces of European travel literature of the seventeenth century (Kelly 1965, 235–50). The text is a fitting tribute to Spain's dreams of Pacific empire. It says nothing about the dimensions of the South Sea or the difficulties that Queirós and his companions had experienced sailing it, but instead provides a glowing description of the *Terra Australis*, idealizing the tropical portions that Quierós thought he knew first-hand and imagining a great Spanish port city on its shores that would facilitate communication between America and the Philippines. It even proposes that the city will help provide contact with the politic civilizations that are sure to exist in the continent's temperate interior, countries that were supposed to be even wealthier than China or Japan (Fernández de Quirós 1991, 171–80). The *Eighth Memorial* also serves as a fitting coda for the story this book has outlined. Although it captured the imagination of many a reader and helped assure that the *Terra Australis* would find a place on many European maps until the eighteenth century, it failed to clinch the deal that Queirós was trying to strike. The pilot would go to his grave without making the voyage he had planned, which would have been his fourth in the waters of the South Pacific. The problem was not the relative attractiveness of the potential prize. With its economy in tatters and its military establishment overextended, Spain was in no position to pursue an ambitious new imperial enterprise on the far side of the world.

The Spanish Pacific was settling into a new pattern, one that would characterize its existence for the next two hundred years or so. No longer would anyone dream of mainland conquests, whether in Asia or in the hypothetical continent of the southern hemisphere. From 1610 onward, the crown's efforts would necessarily concentrate upon defending what it already possessed, the Philippines and the Moluccas, with an occasional attempt to secure its power over Mindanao or extend it to the neighboring Palau Islands. The lucky few involved in the galleon trade, meanwhile, would concentrate on the profits to be had in the exchange of silver for silk and all the other goodies of Asia. As the horizon of possible conquests became more constricted, the geopolitical imaginary associated with this activity went through a significant transformation. While Spanish officialdom continued to map the Philippines as part of the West Indies, at least on certain occasions or in certain contexts, the notion of mapping the islands as part of a broad transpacific West spanning all of East and Southeast Asia seems to have faded away. Most seventeenth-century Spanish sources with which I am familiar seem to map the Pacific the way everyone else did, as a space of encounter between America and Asia, with no pretense to any kind of natural internal coherence, whether geographic or ethnographic.

Two examples will have to suffice. If anyone should have imagined the Spanish Pacific along the lines I have charted in this book, it was Antonio de Morga, a colonial official whose career began in Manila and then took him to Mexico City and Quito. Nevertheless, when Morga describes the Philippines at the outset of his *Sucesos de las Islas Filipinas* (Events of the Philippine Islands, 1609), the first history of the islands from a secular vantage point, he maps them quite unambiguously into the Orient:

Según los cosmógrafos antiguos y modernos, la parte del mundo llamada Asia tiene adyacentes grandísima copia de islas mayores y menores, habitadas por diversas naciones y gentes, enriquecidas así de piedras preciosas, oro, plata y otros minerales. . . . Llámanlas de ordinario en sus libros y descripciones y cartas de marear, el grande archipiélago de San Lázaro, que son el mar océano Oriental, de las cuales, entre otras mas famosas son las islas del Maluco, Céleves, Tendaya, Luzón, Mindanao y Borneo, que ahora se llaman las Filipinas. (Morga 2007, 13)

[According to ancient and modern cosmographers, the part of the world called Asia has adjacent to it a great abundance of large and small islands, inhabited by diverse nations and peoples, and enriched with precious gems, gold, silver, and other minerals. . . . Their books and nautical charts usually call them the great archipelago of Saint Lazarus, which is in the Eastern Ocean Sea, among which the most famous are the islands of the Moluccas, Celebes, Tendaya, Luzón, Mindanao and Borneo, and which are now called the Philippines.]

The passage appeals to the authority of both ancients and moderns, as expressed in books, maps, and nautical charts, to underwrite its central contention: that the Philippines are the same rich archipelago that the Western tradition had always located at the eastern end of Asia. Morga goes on to explain that they are called the "islas del poniente" (islands of the west) only because Spaniards reach them by sailing west (Morga 2007, 217). At the end of the book, he even describes the round-trip route taken by the Manila Galleons, in order to help the reader understand how this Asian archipelago is connected to Spanish America, but unlike Herrera, who mentioned only the westward journey and claimed that it was like sailing along a "apacible río" (a peacable river), Morga describes the much longer, stormier, perilous eastward journey (2007, 303). As a result, the maritime link between America and the Philippines does not look like a manifestation of the inexorable westward progress of empire, as it does in Herrera, but rather like a tenuous link across a barely navigable ocean between two mutually distant lands.

The few seventeenth-century Spanish maps that have come down to us tally with the way Morga imagines the world in the *Sucesos*, abandoning the geopolitical vision that culminated in Herrera's map of the Indies (see fig. 2). This is the case, for example, of the 1630 map of the North Pacific that the Portuguese cosmographer João Teixeira Albernaz prepared for Philip IV (fig. 35). The hypothetical landscape that extends across the top of the map indicates that the space above the north of the fifty-second parallel remains unexplored and suggests that the major landmasses there might be continuous, yet the Strait of Anian clearly cuts through, indicating that the mapmaker believes they are separate, or at least recognizes that they might be. In either case, however, there is no doubting that the landmasses on either side of the strait are separate continents, commensurate units of global geography. The one on the right is clearly labeled "America," not "New Spain" or "the Indies of the North," and the one on the left is clearly "Parte de Asia," not "China," "Tartary," or "the Indies of the West." Although he worked for the Hapsburg monarch Philip IV and probably prepared this map and the atlas in which it appears for use by the king and his councilors, Teixeira has apparently abandoned the tripartite geography of Velasco and Herrera, replacing it with the binary distinction between America and Asia that had become

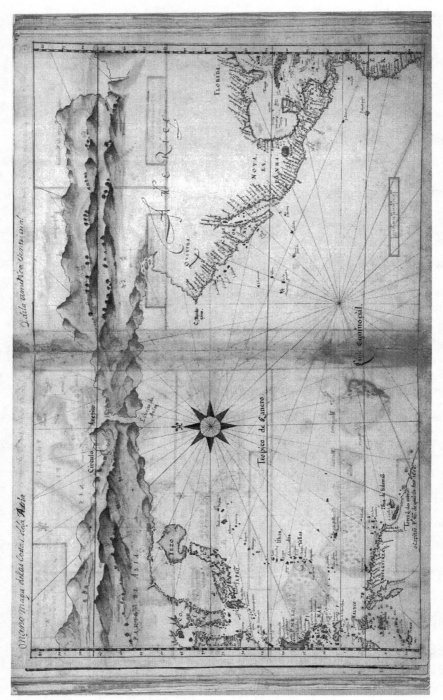

FIGURE 35. Map of the North Pacific from João Teixeira Albernaz, Jeronimo de Attayde, and Francisco de Seixas y Lovera, *Tabuas geraes de toda a navegação* (1630). Photograph: Courtesy of the Geography and Map Division, Library of Congress, Washington, DC (call no. G1015.T4 1630; control no. 78653638; image 14).

hegemonic in European geography and cartography, especially as the knowledge centers of Northern Europe began to take over European cartography (Karrow 2007).

Here as in Morga, the Spanish Pacific figures as the political, economic, and social space created by the travels of the Manila Galleons, rather than as a geographical or ethnographical space with pretensions to some sort of natural internal coherence. The map seems to acknowledge Spanish sovereignty over the islands of Luzón and the Visayas by coloring them in, while leaving all the other islands of East and Southeast Asia in outline, yet this same gesture also creates a clear contrast between the limited extent of territory actually controlled by Spain and the vast areas that lie outside its grasp. Nothing west of Manila appears on the map, making it difficult to imagine China or Cambodia as part of Pacific space. While it is certainly possible to trace the route of the galleons that tie Manila to Acapulco, the routes themselves are not depicted, as they were on López de Velasco's map, nor do any ships sail Teixeira's ocean, as they did on Ribeiro's chart (see figs. 7 and 11). While these silences probably stem from changes in the practice of cartography rather than from any deliberate attempt to mask the reality of Spain's transpacific ties, they nevertheless remain eloquent. Teixeira's map of North Pacific, briefly put, is not a map of Spain's transpacific empire, real or imagined. It is simply a map of the Pacific Basin and Rim, mostly north of the equator. The particular geopolitical imaginary examined in this book seems to have vanished even from the world-making projects of the Hapsburg crown.

Why have we bothered, then, to reconstruct a geopolitical imaginary that enjoyed only a century or so of life, and only among certain people involved in what was ultimately a failed enterprise, that of early modern transpacific empire? For one, the story of Spain's attempt to imagine the Indies as a geography that spanned the South Sea and that included what we call East and Southeast Asia as well as the New World provides an important corrective to established historical narratives regarding the so-called Age of Discovery. Those narratives have invariably emphasized the birth and development of what would eventually become the dominant vision of the world, one that soon came to oppose America in the West to Asia in the East and to center the map on an Atlantic Basin that was quickly becoming the nerve center of a new, Eurocentric global economy. By looking at the other side of the coin, so to speak, we better appreciate the contingency of that dominant vision and the halting nature of its emergences. We understand a little better that none of the moments that have been identified as the one in which "America" came into view—1507, the 1520s, the 1540s, the 1560s—was entirely decisive, that none of them managed to squash the possible alternatives. So just as maps oriented toward the south remind us that there is nothing natural or necessary to our habit of putting north at the top, the story of Spain's attempt to map the Far East as the transpacific West reminds us that no matter how natural or right a map may look, it is always the product of a particular history, reflects and supports particular interests, speaks of all sorts of opportunities foregone, and even displaces or erases alternatives that are very much still alive.

Second, I hope that this book will support ongoing efforts by scholars in numerous fields to remap today's fields of study. "America" was not invented in the sixteenth century once and for all. It has been repeatedly revised and reinvented over the course of the years, and has meant different things at different times in different discursive communities. The field of Atlantic studies has taught us to think of America as a constitutive component of a larger space that includes Europe and Africa. Spanish Pacific studies is doing something similar, but in the opposite direction and with the continent of Asia, while global history is challenging us to understand America as one part of the broadest space possible. This book supports such efforts by demonstrating that early modern participants in the Spanish imperial project did something similar to what this type of scholarship is doing today. They did not think of "America" as a closed figure, separated from Asia by a vast, empty ocean, but as an open one, connected in a variety of ways to what lay to the west. They did not think of the Pacific as a geographical and even ontological boundary between two separate parts of the world and two separate branches of the human family, but as a manageable oceanic basin that could serve to integrate the two. They found ways to generate optimistic expectations regarding the future of secular or spiritual conquest at the far end of that watery way west by appealing to the discourses of heroic masculinity, Counter-Reformation theology, early modern political theory, and millenarian historiography. They managed to do all this, moreover, even as conditions on the ground changed and developed, as knowledge of the world beyond Europe shifted and grew, and even as the conventions of geography and cartography suffered significant transformations. The connections they drew may not be the same as those that interest us today but nevertheless teach us that our efforts to map America into broader geographies, including transpacific ones, are nothing new and might even have something to tell us about how to pursue our own remappings. I should hope that my colleagues in early modern Hispanic and Latin American colonial studies, at least, will find in these pages some food for thought about the ways that the concept of "America" has shaped, or perhaps misshaped, our work as scholars and teachers.

Third, this account provides a loose model for the study of the early modern geopolitical imagination in general. It draws on the fundamental observation of Lewis and Wigen's *The Myth of the Continents*: that the West has always mapped the world, not just by dividing it into continents, but by deploying a variety of metageographies, shifting from one to the other as the occasion demands, often without explanation. It insists that if we are to understand European world-making during the century after Columbus, we must keep this sort of metageographical plurality in mind. We cannot identify a single framework, the architecture of the continents, and pretend that the transformation of that framework from its inherited tripartite form to its novel quadripartite form represents the whole story of early modern geography. From the late Middle Ages and throughout the sixteenth century, Europeans mapped the world in terms of a variety of metageographies, sometimes within the space of a single map or text. There is reason to believe, moreover, that the available frameworks multiplied over the course of the century, as the ever-expanding maritime networks of European empires became themselves a way of mapping the world. Sixteenth-century world-making was not just about continents; it was also about climates and routes.

This metageographical plurality continued to characterize the Spanish geopolitical imagination even after the particular project of mapping the Far East as the transpacific West lost the prominence it once enjoyed. Note, for example, that Teixeira's 1630 map of the world does not just divide the world into the five parts—America, Europe, Asia, Africa, and the *Terra Australis*—but also into the five climatic zones. Although the map provides no clues about what climate theory meant for the map-maker and his intended audience, climate clearly continued to matter in this discursive community. It mattered elsewhere as well, in no small measure thanks to the new lease on life given to climate theory by the work of Jean Bodin. It continued to matter, moreover, well into the eighteenth century, eventually providing the groundwork for nineteenth-century notions of "the tropics" and tropicality (Safier 2014). Indeed, the whole dispute of the New World mapped out for us by Antonello Gerbi (1973) attests to the continued authority of climate theory in the eighteenth-century Atlantic world and to the rich opportunities it afforded the geopolitical imagination.

The continued operation of metageographical plurality in the Spanish geopolitical imagination comes through loud and clear in a letter written by one Gaspar de San Agustín during 1720. San Agustín worked as a missionary in the Philippines, and he authored a lengthy history of the islands. His letter was directed to a friend in Spain, who had written the friar asking about the nature of the native inhabitants of the archipelago. San Agustín's letter has gone down in history as a notoriously racist diatribe, yet it nevertheless provides a valuable glimpse into the ways that some Spanish speakers, at least, mapped the human geography of their overseas empire and of the world as a whole in the early eighteenth century. At the beginning of the letter, the friar finds it necessary to insist that the "Indios Asiáticos" (Asiatic Indians) of the Philippines are quite different from the "Indios naturales de América" (Indians native to America), even though they are all commonly referred to as "indios" (San Agustín 1720, 2r–2v). He argues that the people of the Philippines are similar to those of other "Naciones de la India Oriental" (Nations of Oriental India), such as the Malays and the Siamese, but quite different from the politic peoples of China and Japan (3r–3v). Clearly, it was still possible in San Agustín's day to think of the Philippines as a place either in Asia or in the tropics and thereby construct different patterns of similarity or difference between Filipinos and Amerindians. America may have been invented over three hundred years prior, but it and the architecture of the continents continued to alternate with the theory of climates as the preferred framework for mapping the globe and its people.

Finally, I cannot help but reflect on the relationship between this book and the broad geopolitical shifts that have been taking place as I have been writing it, and that continue to unfold. It has been said that if the twentieth century was the American Century, then the twenty-first will be the Pacific Century, particularly as China's explosive economic growth restores that country to the position it held before the Industrial Revolution as the center of the global economy, as transpacific trade continues to expand in volume and importance, and as the Atlantic alliance that has for so long tied the destinies of Western Europe and the United States continues to fray. Some readers might find it odd to have to look to the left-hand side of Herrera's map to find East and Southeast Asia, but for others, particularly those who live in the major cities of the Pacific coast of North America, connected as they are to the Asia-Pacific region, that arrangement might not look so odd. Perhaps as the twenty-first century progresses, and the ties that bind the Americas to Asia across the Pacific continue to develop and grow, that map will not look strange at all.

Acknowledgments

My work on this book ranged across the geography of Spain's early modern overseas empire, from Madrid to Manila with stops in North and South America. Along the way, I incurred countless debts and have been looking forward to the opportunity to express my gratitude here. Jean Marc Besse, Josiah Blackmore, Sophie Eichelberger, Nicolás Wey Gómez, Dana Leibsohn, Meha Priyadarshini, and Elizabeth Wright read drafts of various chapters and provided invaluable feedback. The indefatigable Surekha Davies was exceedingly generous with her time, knowledge, and wisdom, reading and commenting on the entire manuscript twice. Rolena Adorno, James Amelang, Robert Batchelor, Adam Beaver, Zoltan Biedermann, Alison Bigelow, Jody Blanco, Cristina Branco, Renato Brandão, Timothy Brook, Chet van Duzer, Matthew Edney, Barbara Fuchs, Michael Gerli, Andrew Gray, Roland Greene, John Hebert, Ruth Hill, Elizabeth Horodowich, Carmen Hsu, Regalado José, Richard Kagan, Iris Kantor, Christina Lee, Carla Lois, Miguel Lourenço, Miguel Martínez, Oscar Mazín Gómez, Leah Middlebrook, Jorge Mojarro, Alexander Nagel, Ignacio Navarrete, Juan Pimentel, María Portuondo, Francisco Javier Pujol, Vincent Rafael, Ivonne del Valle, and Amanda Wunder have all provided advice and encouragement, posed challenging questions, and shared unpublished material or crucial bibliographical references. My thanks to them all. I am especially grateful to Timothy Brook for the time and care that he took with translating the inscriptions on the Juan Cobo map.

I have been fortunate to have had the opportunity to present aspects of this project at the meetings of the Modern Languages Association and the Renaissance Society of America, as well as at a variety of universities and research institutes in Brazil, Ecuador, the Netherlands, Portugal, Spain, and the United States. I would like to express my thanks to all who invited me to speak, who attended the events, who asked questions, and who engaged with my ideas.

I have also benefited from the sage assistance of archivists and librarians at the Archivo General de Indias, the Bancroft Library, the Biblioteca Nacional de España, the Bibliothèque Nationale de France, the John Carter Brown Library, Houghton Library, the Library of Congress, the Newberry Library, the Real Academia de la Historia, and the University of Virginia Library. My thanks to them for their guidance.

I have received financial support from the University of Virginia's College and Graduate School of Arts and Sciences, and from its Center for Global Inquiry and Innovation, as well as from the Program for Cultural Cooperation between Spain's Ministry of Culture and United States Universities, and the National Endowment for the Humanities. My thanks to these institutions for their generosity.

The majority of this book was written at Grit Coffee on Main Street in Charlottesville, Virginia. My thanks to the owners and the ever-friendly baristas for tolerating my long stays and keeping me well-caffeinated.

My greatest debt, as always, is to my friends and family. I thank my colleagues in the Department of Spanish, Italian, and Portuguese at the University of Virginia, who have always shown enthusiasm and support. Special thanks go to Allison Bigelow, Eli Carter, David Gies, Fernando Opere, Deborah Parker, Gustavo Pellón, and Alison Weber, who have been there when I most needed them as colleagues and as friends. Thanks also to my other friends at UVA and in Charlottesville, especially Max Edelson, Jim Clarke, Yuji Shinozaki, Christopher Smith, Michael Lloret, and Wes Farris. My brother and his family have always been encouraging, even when they don't really get what I do. My mother has been an unflagging source of love and support. Gracias, Mami. My dear spouse, Zoë, has put up with more than I care to admit over the course of this project's development. Thank you for believing in me, and for never failing to encourage me. Y Teresita, gracias a ti en particular, por todo. Este libro es para ti.

An earlier version of chapter 4 appeared as "(Un)Inventing America: The Transpacific Indies in Oviedo and Gómara" in *Colonial Latin American Review* 25.1 (2016), reprinted by permission of Taylor and Francis Ltd. (http://www.tandfonline.com).

Notes

Introduction

1. The modern edition of Mendoza that I use throughout abbreviates the title to *Historia del gran reino de la China* (History of the great kingdom of China). I use this abbreviated title to refer to the book henceforth.

2. We already have several accounts of "Asia in the Spanish imagination," including, for example, Swecker (1960); Rubiés (2003); and Ellis (2012).

3. For a discussion of "Iberoasia" as a framework of analysis for the two, see Valladares (2001).

Chapter 1

1. Here and throughout this book, "America" is used to refer to the entire continental landmass that dominates the western hemisphere, both north and south.

2. For the text of the *Cosmographiae introductio*, see Waldseemüller (2008). High-resolution images of Waldseemüller's 1507 and his 1516 *Carta marina* are available on the website of the Library of Congress, Washington, DC. https://www.loc.gov/collections/discovery-and-exploration/ (accessed June 20, 2018).

3. According to Lois's extensive survey of sixteenth-century maps, this arrangement can be found on all but a very few of the world maps produced during the European Renaissance (2008, 169 n. 299 and appendix.).

4. For a more nuanced account of this issue, see Harley and Woodward 1994, vol. 2, book 2:170–77.

5. I choose to engage with O'Gorman's essay, despite the decades that have passed since the publication of his text, precisely because of its canonical status. It will become apparent as this chapter progresses that, in doing so, I hope to address some broad issues in the way certain scholarly fields are configured today.

6. O'Gorman originally made his case in Spanish in 1958, and then in the revised and expanded English version that I cite here. He issued a second Spanish-language edition in 1977, revising the argument that appeared in the English version. The second Spanish edition has been reissued several times. In this book, I cite both the English and the second Spanish edition when they coincide, and note when they do not.

7. The expression disappears from the second Spanish edition.

8. Throughout this book, I use terms like "the Spicery" and "the Spice Islands" interchangeably with other terms like "the Moluccas" to refer to the five clove-producing islands known today as Tidore, Ternate, Moti, Makian, and Bacan, as well as to the surrounding region, vaguely defined. Today, these islands form part of the Indonesian province of North Maluku.

9. On the history of the name, see O. H. K. Spate (1977).

10. So much so that some historians even dismiss the term altogether. See J. R. McNeill, "Of Rats and Men: A Synoptic Environmental History of the Island Pacific," *Journal of World History* 5, no. 2 (1994): 313, although it should be noted that McNeill is interested primarily in the islands of the South Pacific. Mazumdar (1999) argues that New World crops had a significant impact upon the demographics of India and China, but he does not establish that those crops made their way to Asia exclusively or even primarily by way of the Pacific.

11. Nevertheless, there was a transpacific slave trade, which had the effect of introducing a small but significant population of Asians to colonial America. See Luengo (1996), Slack (2009), and Seijas (2014).

12. Jackson, Jonkers, and Walker (2000) provide a glimpse into the contrast. Drawing on archival materials in a variety of countries, the authors plot onto a series of world maps a host of individual attempts by European navigators to measure magnetic variation in different places and at different times between the years 1600 and 1750. The result is a rough measure of the European presence in the world's oceans during that time. While the Atlantic and Indian Oceans grow dark with data points, the Pacific remains almost empty, except for a line of points along the route of the Manila Galleons.

13. Dennis Cosgrove (2001, 95) describes it this way, as does Gillies (1994, 167).

14. Sociologist Eviatar Zerubavel (1992, 99) identifies a map from 1587 as the last to depict America as a single continuous landmass with Asia. For Acosta's argument, see Acosta (1999, Libro primero, capítulos XVI-XXII). For reproductions of these maps, see Shirley (2001).

15. The language in the Spanish version is much less strident. See O'Gorman (1986, 184–85).

16. Nagel (2013) and Horodowich (2016) examine this tradition in the Italian context.

17. Geographers Martin Lewis and Kärin Wigen define "metageography" as "the set of spatial structures through which people order their knowledge of the world: the often unconscious frameworks that organize studies of history, sociology, anthropology, economics, political science, or even natural history" (1997, ix).

18. On the galleons, see also Artes de México (1971), Benítez (1992), and Bernabéu Albert and Martínez Shaw (2013).

19. I have in mind such crucial monographs as Phelan (1959) and Chaunu (1960). The same thing could be said, however, of recent monographs in Phil-

ippine studies that should find a place on the shelf of anyone interested in early modern colonialism, like Rafael (1988) and Blanco (2009).

20. For an overview, see Armitage and Bashford (2014, 1–28). The authors cite several recent attempts at synthesis, but note that Pacific studies have remained on the margins of recent interest in Ocean studies, which has focused on the Mediterranean Sea and the Atlantic and Indian Oceans (6). On Ocean studies, see Wigen (2006) and the articles in that issue of the *American Historical Review* by Games, Horden and Purcell, and Matsuda. For recent syntheses of Pacific history in English, see Freeman (2010) and Matsuda (2012).

21. For the limitations of this thesis, see Thomas (2014).

22. Literary scholar Mercedes Maroto Camino (2005; 2008) has attempted to magnify the impact of these Spanish voyages on the European "production of the Pacific," but her work fails to add to what we already knew, that the exaggerated accounts of these voyages helped consolidate European belief in the existence of the *Terra Australis*. Her argument, I believe, is hampered from the start, when it accepts the status of the Mendaña and Queirós voyages as preludes to the eighteenth-century European encounter with the South Pacific, rather than sees them, as Richard Sheehan (2008, 353–56) does, as extensions of the strategic interests embodied by Spain's presence in the Philippines.

23. Flynn and Giráldez have consolidated their argument in Flynn and Giráldez (2010) and Giráldez (2015). They have also consolidated key twentieth-century contributions to the secondary literature in Flynn, Giráldez, and Sobredo (2001). On the far-reaching implications of Flynn and Giráldez's work, and the Sinocentric global economy it helps to map, see Frank (1998).

24. Crucial to the argument is the corrective that Flynn and Giráldez provide to Pierre Chaunu's earlier argument that the Pacific silver trade declined dramatically in volume after 1640. This corrective has spawned controversy among economic historians. See, for example, Flynn, Frost, and Latham (2002, 27–28).

25. On the fortunes of the Spanish silver dollar, see Jones (1993, 57). On the cultural impact of silver upon Ming China, see Brook (1999, esp. 210–62; 2010, 231–32).

26. For a survey of this scholarship, see Leibsohn and Priyadarshini (2016). The essays in Pierce (2009) are also quite useful.

27. O'Gorman plainly admits to the foundational role played by all this historiography, not just his own contribution, when he asserts at the outset of *The Invention of America* that "the most important problem facing the history of America is the need of giving a satisfactory explanation of the way that America as such appeared on the historical scene" (1961, 9; 1986, 15).

28. I use the terms "Amerasian continuity" and "American insularity" for clarity's sake, despite their obvious awkwardness. The notion of "Amerasia" is particularly problematic, insofar as it refers to a geographical entity, "America," whose existence goes unacknowledged by the cartography in question.

29. See the world map that accompanies Franciscus Monachus's 1523 *De orbe*

novo, or the one that appears in Gerónimo de Girava's 1556 *Dos libros de cosmographia*.

30. Instituto de Investigación Rafael Lapesa de la Real Academia Española (2013): *Corpus del Nuevo diccionario histórico (CDH)* [online]. http://web.frl.es /CNDHE (accessed June 30, 2018).

31. Instituto de Investigación Rafael Lapesa de la Real Academia Española (2013): *Corpus del Nuevo diccionario histórico (CDH)* [online]. http://web.frl.es /CNDHE (accessed June 30, 2018).

32. Contrary to popular belief, the 1494 treaty did not split the world in two "like an orange," as the Spanish saying goes. It only charted a half meridian through the Atlantic. According to Ezquerra Abadía (1975), the notion that the treaty implied the existence of an "antimeridian," an extension of the original line of demarcation into the opposite hemisphere, was concocted by Spanish officialdom in its effort to put limits on Portugal's relentless expansion into the most coveted parts of the Indies, the Spice Islands.

Chapter 2

1. My methods draw on the approach to cartography pioneered by J. Brian Harley in a series of essays from the 1990s and then anthologized in J. Brian Harley and Paul Laxton (2001), although they also draw on the work of the many scholars who have followed Harley's example, among them Christian Jacob (1992; 2006), Kivelson (2006), and Ramachandran (2015).

2. The BNF has made the Reinel map facsimile available in high resolution on its online service, Gallica. See http://gallica.bnf.fr/ark:/12148/btv1b59055673/f1 .item.r=jorge%20reinel (accessed July 30, 2018).

3. The atlas is a companion piece to Kunstmann, Spruner von Merz, and Thomas (1859). It is because of its inclusion in this atlas that the Reinel chart is sometimes known as the Kunstmann IV.

4. All latitudes and longitudes on the Reinel chart are approximate. The chart is graduated for both, depicting a series of 180 dots stretching across the equator, as well as a similar series along the line of demarcation. The marks themselves, however, are not numbered, requiring one to count them to determine coordinates. Some dots are obscured by other markings.

5. Note that the Yucatan, the Bahamas, Cuba, Hispaniola, and other Caribbean locations appear north of their true latitudes. This may be a lingering effect of Spanish attempts to falsely shift the Antilles northward, in an effort to make sure that they fell outside the territory accorded to Portugal by the 1479 Treaty of Alcáçovas, by which Portugal enjoyed rights over everything discovered in the Ocean Sea south of the Tropic of Cancer.

6. On the *Terra Australis*, see Eisler (1995) and Hiatt (2008).

7. Note the contrast with Michel de Certeau's oft-cited argument that images of ships recall the circumstances of the map's production. See Certeau (1984, 120).

8. I preserve the Portuguese spelling of Brazil in order to emphasize that the

territory in question is not the same as the modern nation-state by that name. On these maps, Brasil is the name for all of what we call South America.

9. "Chis" is the version of the word that gives us "China" that we find on a number of early Portuguese maps, according to Cortesão (1935, 71, 81, 247). On early Portuguese contact with the Chinese, which began in earnest after the conquest of Malacca in 1510, see Gruzinski (2014, 6–91).

10. For medieval maps of the world, see Woodward (1987, 286–370) and Edson (1997 and 2007).

11. The notion of multiple worlds was a heretical idea if ever there was one, but it nevertheless enjoyed a rich and complex life in medieval thought and literature. See Hiatt (2008, 96–144).

12. For more about this space, see chapter 6.

13. The manuscript containing the map is held by the Biblioteca Nazionale Centrale di Firence and has been digitized. It can be viewed by going to https://www.bncf.firenze.sbn.it/pagina.php?id=35 and browsing the inventory of manuscripts for Banco Rari 234, 54r-65v. The map has also been reproduced widely. See, for example, Wey Gómez (2008, 206-7).

14. I have modified the English translation in Anghiera (1912, 1.408) on the basis of the Latin original and the modern translations into Italian and Spanish available in Anghiera (2005, 1.446–47) and Anglería (1989, 245). For the dating of Martyr's material, see the introduction to Anglería (1989, XXXI).

15. According to Bartolomé de las Casas, who provides the only period account of the logic behind the name, Balboa came up with *Mar del sur* (South Sea) by way of a simple contrast with *Mar del norte* (North Sea), as the North Atlantic was then known (1986, 3. 355); Medina (1914, 1.128). The explanation enjoys considerable common-sense appeal when we remember that the Isthmus of Panama runs generally from east to west, which means that Spaniards in what was then called "Tierra Firme" experienced the Atlantic as an ocean lying to their north and discovered what we call the Pacific on a march toward the south. Nevertheless, we should approach Las Casas's explanation with some skepticism. The Dominican was writing during the 1530s, twenty years after Balboa's discovery, after Spain had ceded its claim to the Moluccas to Portugal in the 1529 Treaty of Zaragoza, and when the true significance of Balboa's discovery seemed to be the way it opened up the way south to the empire of the Incas. In that context, it made sense to figure the North and South Seas the way Las Casas does, as the hydrographic frame for the geographical theater of Spanish colonialism, the islands of the Caribbean, Mesoamerica, and the Andes.

16. Although that reconsideration would be far from complete for some time. See, for example, Safier (2014).

17. In one passage, the newly discovered Indies become the "imperio occidental" (western empire) and in the other, the "imperio occidental de las Indias, islas y tierra-firme del mar Océano" (western empire of the Indies, islands and mainlands of the Ocean Sea). A third passage refers to them as "las Indias de su

Magestad" (Your Magesty's Indies). See Fernández de Oviedo (1950, e.g. 207–8, 272–73).

18. This argument was at the heart of the eighteenth-century dispute over the New World. See Gerbi (1973), Marshall (1982), and Cañizares-Esguerra (2001).

19. Mack draws illuminating contrasts between these notions and other very different concepts of the sea, such as those typically held by Pacific Islanders.

20. I am paraphrasing definitions from the 1780 *Diccionario de la lengua castellana compuesto por la Real Academia Española*, available online through the *Nuevo tesoro lexicográfico de la lengua española*, http://buscon.rae.es/ntlle /SrvltGUILoginNtlle (accessed July 30, 2018).

21. The most famous Spanish instance is probably the diatribe against navigation in Góngora's *Soledad primera* (1994, verses 366–502), from the early seventeenth century. On the diatribe, see Gaylord Randel (1978), Amann (1997), and Padrón (2007).

22. Accessed July 30, 2018.

23. Accessed July 30, 2018.

24. I discuss the rest of the ships in the conclusion to this chapter.

Chapter 3

1. Nine of the extant Seville maps are of interest here. The first of these was made in 1522 by Nuño García de Toreno and depicts the disputed region around the Moluccas, while another seven depict the entire world. Two are by García de Toreno (1523, 1525); one is by Juan Vespucci (1526); and four are by Diogo Ribeiro (1525, 1527, two from 1529). All seven of these are manuscript "plane charts," a kind of nautical chart invented by the Portuguese by adding a scale of latitude to medieval portolan charts. The Seville charts also feature a scale of longitude. For an introduction to these maps, see T. Campbell (1987) and Sandman (2007).

2. Although it still fell considerably short of the true figure. The city of Guayaquil, Ecuador, on the Pacific coast of South America sits near the equator at roughly 80° W longitude, and the island of Halmahera, in Maluku, is at roughly 128° E, making for a difference between the two of about 152° of longitude.

3. O. H. K. Spate (1979, 57) concurs, identifying the Seville planispheres as evidence that the Pacific had been discovered.

4. By suggesting that "South Sea" names a relatively small, well-bounded ocean basin and "Pacific Ocean" names a vast, apparently boundless body of water, I do not mean to imply that the names were used in this way during the early modern period. This is a heuristic device that I adopt for the purposes of my argument.

5. I borrow this felicitous phrase from Schmidt (1997, 552). For more on cartography and state power, see R. Kagan and Schmidt (2007). On cartographic "deception" in the Moluccas quarrel, see Antonio Sánchez (2013, 103–19).

6. Erastothenes of Cyrene had estimated that the Earth was 252,000 stades in circumference, and the Roman geographer Strabo had passed this estimate on to posterity, but Ptolemy rejected it in favor of a smaller estimate of unknown

origin, 180,000 stades, about 70 percent of Erastothenes's more accurate number. The smaller estimate is often attributed to Posidonious of Rhodes, but see Ptolemy, Berggren, and Jones (2001, 21 n. 21).

7. On dead reckoning, see Bedini and Buisseret (1992, 221–22). For a modern discussion of this phenomenon, see Sheppard and Soule (1922, 621).

8. It is certainly not the only eyewitness account to have come down to us. For others, see Elcano et al. (2012).

9. This publication became the basis for all subsequent early modern editions, including, ironically, the Italian versions of Giambattista Ramusio (1536, 1550), and the English editions of Richard Eden (London, 1555), and Richard Hakluyt (London 1625). The most authoritative modern edition, by Antonio Canova, is based on the Ambrosiana manuscript, although it takes into account the other three extant manuscripts, all in French, available at Paris's Bibliothèque Nationale and Yale's Beinecke Library. Theodore Cachey's English translation of Pigafetta is based on this edition. I nevertheless cite from the 1969 edition prepared by the William Clements Library, which reproduces the original French text and provides a facing English translation, because I am most interested in the Pigafetta that was actually available to early modern readers, and to which Spanish writers reacted.

10. See Cachey's introduction to Pigafetta (Cachey 2007, xxiv–xxvii). Black-and-white reproductions of the maps appear throughout.

11. The Paris edition transposes the first-person narrative of the manuscript tradition into the third person. Compare Pigafetta (1999, 198–99) and Pigafetta (2007, 24–25). Subsequent parenthetical references to Pigafetta are to the 1969 edition of the Paris *princeps*.

12. On the signifying function of images of cannibalism, see Davies (2016, 65–108).

13. On the sublimity of the sea, see Mack (2013, 95–137). Although the concept of the sublime was not theorized until the eighteenth century, it has an important precursor in the early modern experiences of wonder.

14. The Ambrosiana manuscript is even more explicit about this point, suggesting as it does that if one sailed westward from the Strait of Magellan, one would encounter no land at all until one hit the strait once again. It imagines only water where others mapped the *Terra Australis*. See Pigafetta (1999, 200; 2007, 25).

15. I cite from the 1523 edition of *De Moluccis Insulis*. English translations are my own. For an English-language edition of the letter, see Transylvanus (1969).

16. It should be noted that some of the text's translators introduce "America" at junctures like this one, in the interest of helping the reader understand the geography being described, but at the cost of distorting the geography imagined by the text.

17. I take the details about Martyr's biography and the composition of the *Decades* from Ramón Alba's introduction to Anglería (1989, xxi–xxxiv).

18. Here as elsewhere, I cite from the Latin original in Anghiera (2005) and

provide a version of the English translation in Anghiera (1912) that I have modified on the basis of comparisons with the original and more recent translations to Italian and Spanish.

19. On the early modern meaning of the word "continent," see the discussion of Von Sevenborgen above.

20. Astonishingly, he makes this claim in his Seventh Decade, which was written in 1524, two years after the *Victoria* docked in Seville, in light of the survivor's reports that must have spoken of the apparent emptiness of the ocean between Chile and Guam.

21. The Ribeiro chart does not have a scale that would allow one to measure distance in leagues, but if we assume that he assigns 17.5 leagues to a degree of longitude at the equator, as the Spanish typically did, then his 135° of longitude between the coast of South America and the island of Gilolo comes out to 2362.5 leagues

22. In-text citations refer to this edition of Oviedo.

23. Oviedo's only explicit discussion of longitudes comes later in the text, in a passage that reports findings of the Loaysa expedition regarding the longitude of key points on the eastern coast of South America. The passage reports that the Ribeiro and Nuño García charts placed these points too far west (36). Note that if one were to shift South America eastward, as these findings suggest, a larger part of it would fall into Portuguese hands, but the Moluccas could also fall decisively into the Castilian demarcation, depending on how broad the South Sea was. The revelation of this detail is thus consistent with Oviedo's program of keeping the Moluccas in Spanish territory.

24. For color reproductions and commentary, see Martín Merás (1992, 83–96) as well as Cerezo Martínez (1994, 173–96) and Sandman (2007, 1107–22).

Chapter 4

1. This chapter deals with Oviedo's account of the Loaysa expedition. The other ventures were the voyages of Estevão Gómes (1524–25), Sebastian Cabot (1526–30), and Diego García de Moguer (1526–30). For details, see Morison (1971).

2. For the Santa Cruz map, which does not depict transpacific space, see his *Islario general de todas las Indias del mundo*, available online at the Biblioteca Digital Hispánica. http://bdh-rd.bne.es/viewer.vm?id=0000149359&page=1 (accessed March 11, 2019).

3. On the reluctance of Martyr and Oviedo to use the expression "New World," see Gerbi (1986, 258–61).

4. For a biography of Urdaneta, see Arteche (1968).

5. For a similar interpretation of book 20's lack of narrative coherence, see Mojarro Romero (Forthcoming). My thanks to Professor Romero for providing me the typescript of his article in advance of publication.

6. Of course, an annalistic arrangement also imposes an ideological framework on events. See White (1988, 1–25).

7. On Saavedra's failure, see Wright (1939, 480–82).

8. We have a rutter and a brief report made by one of the survivors, Vicente de Nápoles, upon his return to Spain in 1534. Both are available in Fernández de Navarrete (1825, 5. 465–86).

9. I cite from Jorge Gurría Lacroix's edition of Gómara, as it appears online at the Biblioteca Virtual Miguel de Cervantes, http://www.cervantesvirtual.com /nd/ark:/59851/bmcz8963 (accessed July 30, 2018).

10. Gómara's geographical ideas probably reflect developments in European geography and cartography. His global geography is roughly consistent, for example, with the one we find on Mercator's double cordiform map of the world from 1538 and his 1541 globe. The map is available at American Geographical Society Digital Map Collection, http://collections.lib.uwm.edu/cdm/ref/collection /agdm/id/854 (accessed July 30, 2018), and the globe on the website of the Harvard College Library, http://id.lib.harvard.edu/aleph/008913773/catalog (accessed July 30, 2018).

11. For Gómara's account of the Coronado expedition and of Spanish exploration along the Pacific coast of Mexico and California, see López de Gómara (1979).

12. Given the techniques that were used to measure distance traveled by sea, we would expect a smooth Pacific crossing to produce an underestimate of the breadth of that ocean. See the previous chapter.

13. Most histories of the Panama Canal make some mention of these early projects. The one in Bennett (1915, 96–98) is particularly thorough, but poorly documented. For an account that includes archival documentation, see Mena García (1992, 203–17).

14. The last canal proposal during the long sixteenth century was made in 1616. After that, the project would not be taken up again until the eighteenth century. See Bennett (1915, 95–96).

15. The key study of the map is Buisseret and Holzheimer (1992). See also Barnes (2007).

16. Martín Meras describes both charts, transcribes the inscriptions from the Sancho Gutiérrez planisphere, and provides an image of it (1992, 111–19). The 1550 chart of the Atlantic has been digitized by the Bibliothèque Nationale de France and can be found at http://catalogue.bnf.fr/ark:/12148/cb43591772n (accessed March 13, 2019).

Chapter 5

1. The relevant documents are available in Hidalgo Nuchera (1995) and in English in Blair and Robertson (1903, vol. 2).

2. The reference in this text to the Philippines as "Spice Islands" indicates that the early modern imagination did not clearly distinguish between the Moluccas and the Philippines as we know them today. The difference was in fact constructed by Iberian colonialism, in and through the Hispano-Portuguese rivalry for control over insular Southeast Asia. For a detailed consideration of the carto-

graphic rivalry over the Philippines during the sixteenth century, see Lourenço (2009).

3. The sole copy was lost during World War II and is known only by way of a facsimile made in 1905.

4. On science in the reign of Phillip II, see Ruiz (1999), as well as Pimentel (2000), Barrera-Osorio (2006), Pardo Tomás (2006), and Portuondo (2009).

5. For Velasco and his work, see Berthe (1998) and Portuondo (2009, 141–256). If Velasco ever wrote a word of the general history he was supposed to produce before he left his job for a more lucrative position in 1591, we know nothing of it.

6. Two editions of Velasco's *Geografía* have been published to date, López de Velasco in 1894 and 1971. The manuscript of the *Geografía* has been lost. For a discussion of Velasco's sources, see Portuondo (2009, 172).

7. Subsequent in-text citations of the *Geografía y descripción* refer to this edition. All translations are my own.

8. For the history of these debates, see Gliozzi (2000) and Romm (2001). Livingstone (2008) examines how the debate over Adamic ancestry played out in the Enlightenment. On the cartography or Arias Montano, see Shalev (2003).

9. The *Geografía y descripción* was clearly equipped with its own maps, but they have not come down to us.

10. There is no analogous section in the *Sumario*.

11. Velasco also includes a chapter explaining how the humoral complexion of Spaniards born in the Indies should come to resemble that of the native inhabitants, over the course of several generations. This section was struck from the final version of the text by a censor appointed by the Council of Indies. See López de Velasco (1894, 37–38).

12. Velasco even argues that the Spanish themselves should come to resemble the native inhabitants of the Indies, even if they did not mix with the natives, as the tropical environment came to bear on succeeding generations of settlers (37–38). The passage was censored by the Council of Indies, suggesting that the theory of climates was becoming a liability for imperial apologetics.

13. See Sepúlveda (1984, 2a:36–37) and Sarmiento de Gamboa (1988), as well as Parra (2015).

14. López de Velasco 1894, 569. I have been unable to determine why Velasco assigns northern and southern boundaries to the region that are in no way suggested by the documents governing the territorial dispute with Portugal. It seems that he defines the region by combining boundaries of three different types: a *de jure* boundary, the antimeridian; a direction, west of Mexico and Peru; and two *de facto* boundaries, the northern and southern limits of practical Spanish knowledge of the area.

15. López de Velasco 1894, 54–89. This material may have been crucial to Philip II's decision to prohibit the publication of the *Geografía y descripción*. At about the same time, the Council of Indies refused to allow Juan Escalante de Mendoza's *Itinerario de navegación de los mares y tierras occidentales* (1575) to be

published, because it did not want the text's detailed description of Spanish maritime routes to be made public. (See Escalante de Mendoza 1985, 13.)

16. The maritime chart that Velasco prepared to accompany this section of the text has not come down to us, but it probably resembled the one that accompanies one of the existing copies of the *Sumario*. (See fig. 11.)

17. On Borja and Gesio, see Portuondo (2009, 186–87).

18. The redacted manuscript is now lost, but the changes are registered in the first printed edition of the *Geografía y descripción*. On this edition, see Berthe (1998, 151–53). For the redacted passages, see López de Velasco (1894, 5, 569, 581, 582).

19. This section also appears in the *Sumario*. See Pacheco and Cárdenas (eds., López de Velasco 1864, 410–18).

20. The text makes no mention of Guam, which sometimes served as a transpacific way station (84).

21. On cartographic modes, see Edney (1993).

22. One of the copies of the *Sumario* is in the Biblioteca Nacional de España, Madrid (MS 2825), and the other in the John Carter Brown Library, Providence, RI (Cod. Spa. 7). The latter includes fourteen ink and watercolor maps, a general map of the Indies as a whole, one each of North and South America, nine depicting each of the *audiencias* or administrative districts, of the Spanish New World, one of Chile, and one of "the Islands of the West." The JCB's online map collection (JCB Map Collection, jcb.lunaimaging.com/luna/servlet/JCBMAPS~1~1, accessed July 5, 2016) includes all the maps, except for the one depicting the Islands of the West. The text of the BNE copy of the *Sumario* has been published in Pacheco and Cárdenas's edition (López de Velasco 1864, 15. 409–164).

23. Here and elsewhere, I cite the text of the *Sumario* available in Pacheco and Cárdenas.

Chapter 6

1. On China in Portuguese and Jesuit writing from 1550 to 1578, see Roque de Oliveira (2003, 509–784), Ollé (2008), and Romano (2013, 113–260). The Portuguese accounts have been anthologized by D'Intino (1989). For English translations, see Ferguson (1902) and Boxer (2010).

2. It is not unreasonable to imagine that the shift described in the previous chapter from the *Geografía y descripción* to the *Sumario* could have responded to the influence of Escalante or others like him. Escalante's *Discurso* was published after Velasco finalized the *Geografía*, but before he is believed to have composed the *Sumario*.

3. For this tropology, see Manel Ollé (2000, 30).

4. Iberian Sinophobia actually begins with the earliest Portuguese accounts, the letters written in the 1520s or 1530s by the Portuguese captives Vasco Calvo and Chrístovão Vieira and the letter written in 1562 by the Portuguese captive Amaro Pereira. All three are anthologized in D'Intino (1989). The most prominent pieces of Castilian Sinophobia include the writings of Francisco Sande and

Alonso Sánchez available on the website "La China en España: Elaboración de un corpus digitalizado de documentos españoles sobre China de 1555 a 1900," https://www.upf.edu/asia/projectes/che/principal.htm (accessed January 26, 2018).

5. On early modern European Sinophobia and the yellowing of the Chinese, see Mungello (2009, 123–64) and Keevak (2011). For the "yellowing" of the Japanese, see Kowner (2014). On the history of "oriental despotism" see Valensi (1993) and Rubiés (2005).

6. I develop Ollé's argument in Padrón (2014).

7. Manel Ollé (2002, 72 n. 2) provides a brief sketch of Sande's life.

8. The following draws upon Sande (1576).

9. Pereira's letter first appeared in *Nuovi Avisi Delle Indie di Portogallo venuti nuovamente dalli Reverendi Padri della Compagnia de Gesú*, Part IV (Venice, 1563), and in English translation in 1577 and 1625.

10. There is an ample bibliography on torture and capital punishment in the Chinese judicial and penal systems, as well as on the uses to which Europeans put accounts and imagery of Chinese practices. See, for example, Mühlhahn (2009, 14–57), Brook et al. (2008, 159–71), and Schmidt (2015, 294–96).

11. For Aristotle's definition of the tyrant, Aristotle (2013, 5.10.9). On the tyrant's use of fear, Aquinas (2002, 14–15).

12. Subsequent in-text citations to the *Discurso de la navegación* refer to this edition.

13. See Lara Vilà's introduction to Escalante (2008, 38).

14. Subsequent in-text citations refer to this edition of the *Historia de la gran China*.

15. On Renaissance humanism, utopian discourse, and Mendoza's China, see Hsu (2010).

16. For recent discussions of Castilian-Portuguese relations during the Union of the Crowns, see Martínez Shaw and Martínez Torres (2014). On the frontier between the empires in Iberia and America, see Herzog (2015).

17. Boxer (1978, 77–84), provides a general introduction to these arrangements and their history. Ross (2003, 38–39) reminds us that the relationship between the Jesuits and royal authority could be and often was quite contentious.

18. For Mendoza's biography, see Santiago Vela (1913, 201–6).

19. On Mendoza's reliance on Escalante, see Roque de Oliveira (2003, 863) and Vilà (2013, 78).

20. On ethnography and the authority derived from narratives of "being there," see Geertz (1988).

21. Compare the manuscript account of the same voyage by the soldier Miguel de Loarca, who uses this occasion to comment upon how the mandarins inspire the populace to adore them. See Luarca [Loarca] (2002).

22. For Cruz's account, see D'Intino (1989, 246).

23. González de Mendoza and Loyola (1586). There are also at least two modern editions, Loyola (2002) and Loyola (2011).

24. My thanks to Rui Loureiro, Zoltán Biederman, and Liam Brockey for their help in identifying this obscure Lisbon publisher.

25. On the ease of "damas," see Covarrubias Orozco (1611, 594).

26. The sole surviving copy of Cobo's *Shilu* is held by the Biblioteca Nacional in Madrid and can be accessed on their Biblioteca Digital Hispánica at http://catalogo.bne.es/uhtbin/cgisirsi/?ps=ubDmOyLjiX/BNMADRID/227730373/9 (accessed July 31, 2018). I have relied on the text's only modern edition, which includes translations to Spanish and English. See Cobo (1986).

27. See Costa (1967, 37–57), Vega y de Luque (1979, 1980, 1981, 1982), Ollé (1998; 2000, 105–40; 2002, 89–225; and 2008, 54–56), Romano (2013, 103–11), and Gruzinski (2014, 218–37).

Chapter 7

1. Boxer (1974) remains fundamental to our understanding of Christianity in sixteenth-century Japan. See also Elison et al. (1973), Cabezas et al. (1995), Ross (1994), and Sola (2012).

2. Small numbers of Capuchins had been in Japan since at least 1585. For relations between Manila and Japan, see Sola (2012) and Tremml-Werner (2015), as well as Correia (2001).

3. On the controversy, see Brockey (2016).

4. The Jesuits got into print first with Fróis (1599), followed shortly afterward by the Franciscans with Santa María (1601).

5. Its single volume of 728 pages is divided into six books devoted to the following topics: 1) the discovery and conquest of the Philippines and the beginnings of the Franciscan mission there; 2) Franciscan efforts to establish missions in China, Tartary, Cochinchina, Siam, and Cambodia; 3) biographies of the friars of the province; 4) the history of the Franciscan mission in Japan and especially Kyoto; 5) the story of the Nagasaki martyrdom; and 6) biographies of the martyrs. I cite the only modern edition of the text, Ribadeneira (1970), which includes a faulty English translation. The translations I provide are my own.

6. See also Lois (2008, 140).

7. On Valignano, see Üçerler (2003) and Moran (2012). On Frois, see Fróis (2014).

8. For more details, see Wroth (1944, 344).

9. On the *gyogi* map, see Nakamura (1939) and Kish (1949).

10. Ribadeneira's approach to idolatry as a false religion ultimately motivated by the direct action of the devil in the world is typical of Counter-Reformation ethnography, but it is not the only way that the Christian tradition approached idolatry. For more, see Johnson (2006), Rubiés (2006), and the other articles that appear alongside them in the special issue of the *Journal of the History of Ideas* devoted to idolatry.

11. For a more tolerant view of Siamese Buddhism, which makes no mention of the devil or of "idols," see Barros (1563, 38–40).

12. The Indian origins of Buddhism were clear to the Jesuits (Lach 1965, 1.2: 831).

13. On medieval Franciscan missions to Asia, see Phillips (1988, 83–101).

14. These were not the first Franciscans to arrive in New Spain. For details, see Rubial García (1996, 99).

15. For a comparison of the arguments, see Rubial García (1996, 130). For the roots of the millenarianism typical of the Franciscans in New Spain, see Baudot (1995, 71–91) and (1990, 13–36). Frost (2002, 131–58) provides a useful corrective to Baudot's emphasis on the novelty of the New World in Franciscan thought.

16. On the life and works of Motolinía, including a discussion of the complex textual history of his writings, see Baudot (1995, 246–398).

17. Elsewhere, Mendieta has us delve into Fray Martín's world by describing the surreal dream vision that propelled the founder to seek these people out.

18. It is also possible that Mendieta is referring to East Asia in general, following the common usage of the name "China" in New Spain. Domingo Chimalpahin, for example, makes several references to Manila as a city in China (Chimalpahin Cuauhtlehuanitzin and Tena 2001, 111, 199, 217). He never uses the name "Filipinas."

19. When Baudot claims that China is the true destination in Motolinía's text, he is reading Motolinía through the filter provided by Mendieta and thus obscuring the way in which Franciscan historians repeatedly rewrote this episode, and with it, the entire trajectory of history's march toward the millennium, in order to adjust to the realities and opportunities of their day.

20. His discretion at this juncture can be taken as a sign of the times. By the late sixteenth century millenarianism had become suspect as a legitimate strain of Catholic thought. There is reason to believe, argue Baudot and Phelan, that the crown associated it with separatist tendencies, and that these suspicions led to the confiscation of Motolinía's writings and the decision to prohibit publication of Mendieta's work. See Baudot (1995, 518–19) and Phelan (1970, 112). The Jesuits scoffed at it, and José de Acosta even published a treatise on the end times rejecting the idea that human beings could actually predict the timeframe of the Second Coming.

Chapter 8

1. For Herrera and his work, including editions and translations, see Cuesta Domingo (2015).

2. All subsequent in-text citations refer to this edition of Herrera's *Historia general*.

3. For reproductions and discussion of the other maps, see Suárez (1999, 164–97).

4. Elsewhere, he remarks that the Indies of the South are often called "America" after Amerigo Vespucci, but that the appellation is unjustified, presumably because it honors Vespucci over the real discoverer, Columbus (1.175).

5. Although the extant manuscripts of the friar's text do not include the

cosmographical digression we are discussing, the tone and style of the passage scream Las Casas, suggesting that Herrera had access to a copy of the manuscript different from the ones that have come down to us today. My thanks to Rolena Adorno for suggesting this possibility.

6. On Herrera's authorship, see Kagan (2010, 258). Christina Findlay (2014) provides the fullest discussion and analysis of these images available to date.

7. "Storyline" should be interpreted here in both a temporal and spatial sense. The term, as I use it here, refers at once to the series of events that constitute a particular narrative, as well as to the series of places through which that narrative moves, the narrative itinerary, or what narratology calls the "storyspace."

8. To put it in the technical language of narratology, the narrator draws the reader's attention to the fact that the different "storyspaces" ("the space relevant to the plot, as mapped by the actions and thoughts of the characters") all unfold in the same "storyworld" (the "coherent, unified, ontologically full, and materially exisiting geographical entity" in which everything unfolds). The storyspace, in each case, is the itinerary followed by each expedition. The storyworld is the larger geography in which both itineraries unfold, specifically, the "Islands and Mainland of the Ocean Sea."

9. All subsequent in-text citations refer to this edition of the *Conquista de las Malucas*.

10. As Sheehan notes, the takeover was not complete. Portugal continued to enjoy a monopoly on the spice trade from Ternate and Tidore, and the Jesuit missionary enterprise in the islands continued to be directed from the Jesuit Province of Cochin, in Portuguese territory. Spanish Jesuits did not take over until after Portugal regained its independence (2008, 335).

11. On the allegory of the four parts of the world as women, see Corbeiller (1961). On the imperial politics of the image of America, see Certeau (1988). For the importance of gender in such images, see Montrose (1993), as well as the essays in Wiesner-Hanks (2015). For the feminization of territory, see Kolodny (1975).

12. On the Black Legend of Portugal in Asia more generally, see Winius (1985).

13. For Sarmiento's account of the expedition, see Sarmiento de Gamboa (2000).

14. What Argensola presents as the first Dutch expedition was actually the second, but since the first did not reach Tidore and Ternate, he omits it from his account. For more on early Dutch voyages to Southeast Asia, see Masselman (1963, 86–118).

15. Uppercase in original. Seed (1995, 165–67) compares these actions to similar Dutch gestures elsewhere.

16. John Lloyd Stephens provides an example of how Argensola's text was "orientalized" in the eighteenth century, by eliminating the word "indios" from his translation of this passage. See Leonardo y Argensola (1708, 1).

17. Reyes and Clarence-Smith (2012, 134) use the word "pogrom." Historian Ray Huang calls it "the Spanish atrocity." See Fairbank and Twitchett (1978, 7.

562). Kamen refers to the event as a "massacre" of the Chinese by the Spanish (2003, 208). The number seems exorbitant, but it has become canonical. For discussion of the sources, see Gil (2011, 480 n. 114).

18. For a bibliography of print and manuscript soruces, see Gil (2011, 468 n. 67).

19. For a clear example of Benavides's Sinophobic outlook, see his July 5 letter to the king, reproduced in Colín (1900, 2.413–14).

20. Morga's *Sucesos de las Islas Filipinas* enjoyed only one edition, and few copies of it remain. The first modern edition, Morga (1868), is actually an English translation. There is also an edition prepared by José Rizal, the hero of Philippine Independence, Morga (1890). The Hakluyt Society reissued this edition in 1971, with additional notes by J. S. Cummins (Morga 1971). The noted Philippinist Wenceslao Retana used Rizal's edition as the basis for his own, Morga (1909). There is also a serviceable English edition in Blair and Robertson (1903, vols. 15 and 16). This translation was also published as a separate title. I cite from Morga (2007), which is based on Retana's edition.

21. For Acuña's assessment, see Gil (2011, 473).

22. For more on this figure, see Crossley (2011).

Works Cited

Primary Sources

Acosta, José de. 1999. *Historia natural y moral de las Indias*. Edited by P. Francisco Mateos. Biblioteca Virtual Miguel de Cervantes. http://www.cervantesvirtual.com/FichaObra.html?Ref=600. Accessed July 31, 2018. Web.

Anghiera, Pietro Martire d'. 1534. *Libro primo della historia del' Indie Occidentali*. Venice: Stefano Nicolini da Sabbio. Print.

——. 1912. *The Eight Decades of Peter Martyr D'Anghera*. Translated by Francis Augustus MacNutt. Project Gutenberg Australia. Web.

——. 2005. *De orbe novo decades. I–VIII*. Edited by Rosanna Mazzacane and Elisa Magioncalda. Genova: Università di Genova, Facoltà di lettere, Dipartimento di archeolologia, filologia classica e loro tradizioni. Print.

Anglería, Pedro Mártir de. 1989. *Décadas del Nuevo Mundo*. Translated by Joaquín Torres Asencio. Edited by Ramón Alba. Madrid: Ediciones Polifemo. Print.

Anonymous. 1983. "Libro segundo que trata del fundamento y principio del armada que llevo Ruy Lopez de Villalobos . . . ." In *El viaje de Don Ruy López de Villalobos a las Islas del Poniente*, edited by Consuelo Varela, 35–115. Milan: Cisalpino-Goliardica. Print.

Antonio, Juan Francisco de San. 1738. *Chronicas de la apostolica provincia de S. Gregorio de Religiosos Descalzos de N.S.P.S. Francisco en las Islas Philipinas, China, Japon, &c: Parte primera, en que se incluye la descripcion de estas islas . . .* Manila. Print.

Aquinas, Saint Thomas. 2002. *Aquinas: Political Writings*. Cambridge: Cambridge University Press. Print.

Argensola, Bartolomé Leonardo de. 1708. *The Discovery and Conquest of the Molucco and Philippine Islands*, translated by John Stevens. London. Print.

——. 1996. *Alteraciones populares de Zaragoza, año 1591*. Edited by Gregorio Colas Latorre. Zaragoza: Institución "Fernando el Católico." Print.

——. 2009. *Conquista de las islas Malucas*. Madrid: Miraguano. Print.

Aristotle. 2013. *Aristotle's Politics*. Edited by Carnes Lord. 2nd ed. Chicago: University of Chicago Press. Print.

Audiencia de Nueva España. 1564. "Documento 17: Carta de La Audiencia de

México Al Rey, México, 12 de Septiembre de 1564." In Hidalgo Nuchera, *Los primeros de Filipinas*, 132–33. Print.

Balbuena, Bernardo de. 1997. *La grandeza mexicana y compendio apologético en alabanza de la poesía*. Edited by Luis Adolfo Domínguez. Mexico City: Porrúa. Print.

Barros, João de. 1552. *Asia de Joam de Barros: Dos fectos que os Portugueses fizeram no descobrimento & conquista dos mares & terras do Oriente*. Vol. 1. 2 vols. Lisbon: Germâo Galharde. Print.

———. 1563. *Terceira decada da Asia de Joam de Barros: Dos feytos que os Portugueses fizeram no descobrimento & conquista dos mares & terras do Oriente*. Lisbon: Joam de Barriera. Print.

Blair, Emma Helen, and James Alexander Robertson. 1903. *The Philippine Islands, 1493–1898: Explorations by Early Navigators, Descriptions of the Islands and Their Peoples, Their History and Records of the Catholic Missions . . .* 55 vols. Cleveland: Arthur H. Clark. Print.

Bry, Theodor de, Antonio de Herrera y Tordesillas, and José de Acosta. 1624. *America sive descriptio Indiae Occidentalis*. Frankfurt: Sumptibus haeredum I.T. de Bry. Print.

Carvalho, Joaquim Barradas de. 1991. *Esmeraldo de situ orbis de Duarte Pacheco Pereira: Edition critique et commentée*. Lisbon: Fundação Calouste Gulbenkian Serviço de Educação. Print.

Casas, Bartolomé de las. 1967. *Apologética historia sumaria*. Edited by Edmundo O'Gorman. Mexico City: UNAM Instituto de Investigaciones Históricas. Print.

———. 1986. *Historia de las Indias*. Edited by André Saint-Lu. 3 vols. Caracas, Venezuela: Biblioteca Ayacucho. Print.

Castanheda, Fernão Lopes de. 1552. *Ho primeiro livro da Historia do descobrimento e conquista da India pelos Portugueses . . .* Coimbra: J. de Barreyra and J. Alvarez. Print.

Chimalpahin Cuauhtlehuanitzin, Domingo Francisco de San Antón Muñón. 2001. *Diario*. Translated and edited by Rafael Tena. *Cien de México*. México, D.F.: Consejo Nacional para la Cultura y las Artes. Print.

Cobo, Juan. 1986. *Bian Zhengjiao Zhenchuan Shilu = Apologia de la verdadera religion*. Edited by Fidel Villaroel. Manila: University of Santo Tomás Press. Print.

Colón, Cristobal. 1992. *Textos y documentos completos*. Edited by Consuelo Varela. Madrid: Alianza. Print.

Consejo de Indias. 1570. "Real cédula a Martín Enríquez [de Almansa Ulloa], Virrey de Nueva España, para que informe sobre la merced de 2.000 pesos a Pedro Salcedo, que solicita su hijo, Felipe de Salcedo Legazpi, . . ." Madrid: Archivo General de Indias. ES.41091.AGI/23.10.1105//MEXICO,1090,L.6,F.156R-156V. At http://pares.mcu.es/. Accessed June 6, 2019. Web.

Cortés, Hernán. 1993. *Cartas de relación*. Edited by Angel Delgado Gómez. Madrid: Castalia. Print.

Covarrubias Orozco, Sebastián de. 1611. *Tesoro de la lengua castellana o española, compuesto por el licenciado Don Sebastian de Cobarruvias Orozco* . . . Madrid: L. Sanchez. At *Nuevo Tesoro lexicográfico de la lengua española*. http://buscon .rae.es/ntlle/SrvltGUILoginNtlle. Accessed July 31, 2018. Web.

D'Intino, Raffaella, ed. 1989. *Enformação das cousas da China textos do século XVI*. Lisbon: Imprensa Nacional-Casa da Moeda. Print.

Elcano, Juan Sebastián de, Maximiliano Transilvano, Francisco Albo, Duarte Barbosa, Ginés de Mafra, Antonio Pigafetta, and Ginovés anónimo. 2012. *La primera vuelta al mundo*. Madrid: Miraguano. Print.

Escalante, Bernardino de. 1992. *Diálogos del arte militar*. Ed. Universidad de Cantabria. Print.

——. 1995. *Discursos de Bernardino de Escalante al Rey y sus ministros (1585– 1605)*. Edited by José Luis Casado Soto. Santander: Editorial de la Universidad de Cantabria. Print.

——. 2008. *Navegación a Oriente y noticia del reino de China*. Edited by Lara Vilà. Córdoba, Almuzara: Fundación Biblioteca de Literatura Universal. Print.

Escalante Alvarado, García de. *Viaje a las islas de Poniente*. Santander: Ed. Universidad de Cantabria, 2015. Ebook. http://www.digitaliapublishingcompany .com. Accessed June 5, 2019. Web.

Escalante de Mendoza, Juan de. 1985. *Itinerario de navegación de los mares y tierras occidentales 1575*. Madrid: Museo Naval. Print.

Ferguson, Donald. 1902. *Letters from Portuguese Captives in Canton, Written in 1534 & 1536: With an Introduction on Portuguese Intercourse with China in the First Half of the Sixteenth Century*. Bombay: Education Society Steam Press. Print.

Fernández de Enciso, Martín. 1948. *Suma de geografía*. Madrid. Print.

Fernández de Lizardi, José Joaquín. 1997. *El periquillo sarniento*. Edited by Carmen Ruiz Barrionuevo. Madrid: Cátedra. Print.

Fernández de Navarrete, Martín. 1825. *Coleccion de los viages y descubrimientos que hicieron por mar los españoles desde fines del siglo XV* . . . Madrid: Imprenta real. Print.

Fernández de Oviedo y Valdés, Gonzalo. 1852. *Historia general y natural de las Indias, islas y tierra-firme del Mar Océano*. Edited by José Amador de los Rios. Madrid: Real Academia de la Historia. Print.

——. 1950. *Sumario de la natural historia de las Indias*. Edited by José Miranda. Biblioteca Americana: Cronistas de Indias. Mexico: Fondo de Cultura Económica. Print.

Fernández de Quirós, Pedro. 1986. *Descubrimiento de las regiones austriales*. 1a ed. Madrid: Historia 16.

——. 1991. *Memoriales de las Indias australes*. 1. ed. Madrid: Historia 16.

Fróis, Luís. 1599. *Relatione della gloriosa morte di xxvi. Posti in croce per commandamento del re di Giappone, alli 5 di febraio 1597* . . . Rome: Luigi Zannetti. Print.

——. 2014. *The First European Description of Japan, 1585: A Critical English-Language Edition of Striking Contrasts in the Customs of Europe and Japan by*

Luis Frois, S.J. Edited by Richard K Danford, Robin Gill, and Daniel T Reff. London: Routledge. Print.

Girava, Gerónimo. 1556. *Dos libros de cosmographia compuestos nuevamente por Hieronymo Girava tarragones.* Milan, 1556. Print.

Góngora y Argote, Luis de. 1994. *Soledades.* Edited by Robert Jammes. Madrid: Castalia. Print.

González de Mendoza, Juan. 1990. *Historia del gran reino de la China.* Madrid: Miraguano; Polifemo. Print.

González de Mendoza, Juan, and Martín Ignacio de Loyola. 1586. *Itinerario y compendio de las cosas notables que ay desde España, hasta el reyno dela China, y dela china à España, boluiendo por la India Oriental, . . .* Lisbon: San Phelippe el Real. http://archive.org/details/itinerarioycompeoogonz. Accessed July 31, 2018. Web.

Guzmán, Luis. 1601. *Historia de las missiones que han hecho los religiosos de la Compañia de Iesus para predicar el sancto euangelio en la India Oriental y en los reynos de la China y Iapon.* 2 vols. Alcalá de Henares. Print.

Herrera y Tordesillas, Antonio de. 1991. *Historia general de los hechos de los Castellanos en las islas y tierrafirme del Mar Océano, o, "Décadas."* Edited by Mariano Cuesta Domingo. 4 vols. Madrid: Universidad Complutense de Madrid. Print.

Hidalgo Nuchera, Patricio, ed. 1995. *Los primeros de Filipinas: Crónicas de la conquista del Archipiélago de San Lázaro.* Madrid: Miraguano; Polifemo. Print.

López de Gómara, Francisco. 1912. *Annals of the Emperor Charles V.* Translated and edited by Roger Bigelow Merriman. Oxford: Clarendon Press. Print.

———. 1999. *Historia general de las Indias y vida de Hernán Cortés.* Edited by Jorge Gurría Lacroix. Alicante: Biblioteca Virtual Miguel de Cervantes. http://www .cervantesvirtual.com/nd/ark:/59851/bmcz8963. Accessed July 31, 2018. Web.

López de Velasco, Juan. 1864. "Demarcación y división de las Indias." In *Colección de documentos inéditos relativos al descubrimiento, conquista y colonización de las posesiones españolas en América y Oceania: sacados, en su mayor parte, del Real Archivo de las Indias.* Edited by Joaquín Pacheco and Francisco de Cárdenas, 409–572. Madrid: Imprenta Quirós. Print.

———. 1894. *Geografía y descripción universal de las Indias.* Edited by Justo Zaragoza. Madrid: Real Academia de la Historia. Print.

———. 1971. *Geografía y descripción universal de las Indias.* Edited by Marcos Jiménez de la Espada. *Biblioteca de Autores Españoles* 248. Madrid: Atlas. Print.

Loyola, Martín Ignacio. 2002. *Viaje alrededor del mundo.* Edited by José Ignacio Tellechea Idígoras. Madrid: Dastin. Print.

———. 2011. *Viaje alrededor del Mundo.* Linkgua digital. Web.

Luarca [Loarca], Miguel de. 2002. *Verdadera relación de la grandeza del Reino de China (1575).* Edited by Santiago García Castañon. Asturias: Eco de Luarca. Print.

Magellan, Ferdinand, and Ruy Falero. 1954. "Memorial presentado al Rey (al parecer por Magallanes y Falero) sobre el descubrimiento de las Islas del Maluco, que habían propuesto, y las mercedes que pedían se les concediesen (Archivo de Indias. Legajo 10, Papeles Del Maluco, 1519, 1547)." In *Obras*. Edición y estudio preliminar de Carlos Seco Serrano. Edited by Martín Fernández de Navarrete, 2 (76):472–74. *Biblioteca de autores españoles desde la formación del lenguaje hasta nuestros días* (Continuacion) 75–77. Madrid: Ediciones Atlas. Print.

Mendieta, Gerónimo de. 1870. *Historia eclesiástica indiana*. Mexico City. Google Books. Web.

Monachus, Franciscus. 1526–30. *De Orbis situ ac descriptione*. Antwerp. Print.

Morga, Antonio de. 1890. *Sucesos de las islas Filipinas*. Edited by José Rizal. Paris: Garnier hermanos. Print.

———. 1909. *Sucesos de las Islas Filipinas*. Edited by Wenceslao E. Retana. Madrid: Librería General de Victoriano Suárez. Print.

———. 2007. *Sucesos de las Islas Filipinas*. Edited by Francisca Perujo. Mexico City: Fondo de Cultura Económica.

Motolinía, Toribio. 1985. *Historia de los indios de la Nueva España*. Edited by Georges Baudot. Madrid: Castalia. Print.

Navarrete, Martín Fernández de. 1971. "Asientos y capitulaciones hechas por S.M. con el adelantado don Pedro de Alvarado, sobre descubrimiento, conquista y población de las islas que estuviesen en la mar del sur hacia poniente, en los años 1538 y 1539 y con el virrey de Nueva España don Antonio de Mendoza en el de 1541." In *Colección de documentos y manuscriptos compilados por Fernández de Navarrete. Prólogo Del Almirante Julio Guillén Tato*, 15: 465–507. Nendeln, Liechtenstein: Kraus-Thomson. Print.

Ortelius, Abraham. 1570. *Theatrum orbis terrarum*. Antwerp. Print.

Pacheco Pereira, Duarte. 1988. *Esmeraldo de situ orbis*. 3rd ed. Edited by Damião Peres. Lisbon: Academia Portuguesa da História. Print.

Pereira, Galiote, and Gaspar da Cruz. 1989. *Primeiros escritos portugueses sobre a China*. Edited by Rui Loureiro and Maria da Graça. Lisbon: Biblioteca de expansão portuguesa. Print.

Pigafetta, Antonio. 1969. *The Voyage of Magellan: The Journal of Antonio Pigafetta. A Translation by Paula Spurlin Paige from the Edition in the William L. Clements Library, University of Michigan, Ann Arbor*. Translated by Paula Spurlin Paige. Englewood Cliffs, NJ: Prentice-Hall. Print.

———. 1999. *Relazione del primo viaggio attorno al mondo*. Edited by Andrea Canova. Padua: Editrice Antenore. Print.

———. 2007. *The First Voyage around the World, 1519–1522: An Account of Magellan's Expedition*. Edited by Theodore J. Cachey. Toronto: University of Toronto Press. Print.

Ptolemy, J. Lennart Berggren, and Alexander Jones. 2001. *Ptolemy's Geography: An Annotated Translation of the Theoretical Chapters*. Princeton, NJ: Princeton University Press. Print.

Rada, Martín de. 1569. "Copia de carta del P. Martín de Rada al Virrey de México, dándole importantes noticias sobre Filipinas." Cebu, Philippines. Archivo

General de Indias. *La China en España. Elaboración de un corpus digitalizado de documentos españoles sobre China de 1555 a 1900.* https://www.upf.edu/asia/projectes/che/s16/rada1569.htm. Accessed July 31, 2018. Web.

———. 1575. "Relaçion verdadera delascosas del Reyno de TAIBIN por otro nombre China y del viaje que ael hizo el muy Reverendo padre fray martin de Rada provinçial . . ." Fonds Espagnol. Bibliothèque Nationale, Paris. *La China en España. Elaboración de un corpus digitalizado de documentos españoles sobre China de 1555 a 1900.* http://www.upf.edu/asia/projectes/che/s16/radapar.pdf. Accessed July 31, 2018. Web.

Real Academia Española. 1726–39. *Diccionario de Autoridades.* At *Nuevo Tesoro lexicográfico de la lengua española.* http://buscon.rae.es/ntlle/SrvltGUILoginNtlle. Accessed July 31, 2018. Web.

Ribadeneira, Marcelo de. 1970. *Historia del archipiélago y otros reynos. History of the Philippines and Other Kingdoms.* 2 vols. Manila: Historical Conservation Society. Print.

Rodríguez Maldonado, Miguel. 1606. *Relacion verdadera del levantamiento de los Sangleyes en las Filipinas, y el milagroso castigo de su rebelion: con otros sucessos de aquellas Islas.* Seville: Clemente Hidalgo. http://archive.org/details/A109085086. Accessed July 31, 2018. Web.

Rubios, Juan López de Palacios. 1954. *De las islas del mar Océano.* Fondo de Cultura Económica. Print.

Ruiz Vega (Coleccionista), Luis. 1587. "Informaciones sobre el Japón." Archivo Histórico Nacional. ES.28079.AHN/5.1.15//DIVERSOS-COLECCIONES,26,N.9. At http://pares.mcu.es/. Accessed June 6, 2019. Web.

San Agustín, Gaspar de, OSA. 1720. "Carta que Fr. Gaspar de San Agustín, Religioso de la orden de Agustinos escribió a un amigo suyo que desde España le preguntó el natural y geno de los Yndios naturales de Filipinas." Manila. Biblioteca Nacional de España. MSS/7861. Print. (Also available in Blair and Robertson 1903, 40. 183–295).

Sanchez, Alonso. 1583. "Relación breve de la jornada del Padre Alonso Sánchez de la Compañía de Jesús. Por orden y parecer de Don Gonzalo Ronquillo de Peñalosa Gobernador delas Philippinas y del Sr. Obispo y officiales de su Magestad desde la Isla de Luzón de la ciudad de Manila para la China." Manila. Archivo General de Indias. *La China en España. Elaboración de un corpus digitalizado de documentos españoles sobre China de 1555 a 1900.* http://www.upf.edu/asia/projectes/che/s16/sanchez2.htm. Accessed July 31, 2018. Web.

Sánchez, Alonso. 1588. "Relación de las cosas particulares de la China, la qual escribio el P. Sanchez de la Compañía de Jesús que se la pidieron para leer a su Magestad el rey don Felipe II estando indispuesto." Archivo General de Indias. *La China en España. Elaboración de un corpus digitalizado de documentos españoles sobre China de 1555 a 1900.* http://www.upf.edu/asia/projectes/che/s16/sanchez.htm. Accessed July 31, 2018. Web.

Sande, Francisco de. 1576. "Carta a Felipe II del Gobernador de Filipinas, doctor Sande. Da cuenta de su llegada y accidentes de su viaje; de la falta que hay allí de todo, y habla de Religiosos, minas, de la China, Mindanao, Borneo, etc.," June 7, 1576. Archivo General de Indias. *La China en España. Elaboración de un corpus digitalizado de documentos españoles sobre China de 1555 a 1900.* http://www.upf.edu/asia/projectes/che/s16/sande1576.htm. Accessed July 31, 2018. Web.

———. 1580. "Carta de Francisco de Sande sobre conquista de China." Letter. Manila. Archivo General de Indias, Filipinas,6,R.3,N.38. Accessed July 31, 2018. Web.

Santa Cruz, Alonso de. 2003. *Islario de Santa Cruz.* Edited by Mariano Cuesta Domingo. Madrid: Real Sociedad Geográfica. Print.

Santa María, Juan de. 1601. *Relacion del martirio que seys padres descalços Franciscos, tres hermanos de la Compañia de Iesus, y diecisiete iapones christianos padecieron en Iapon* . . . Madrid: Herederos de Iuan Iñiguez de Lequerica; En casa del Licenciado Varez de Castro. Biblioteca Digital Hispánica. http://bdh-rd.bne.es/viewer.vm?id=0000120382&page=1. Accessed July 31, 2018. Web.

Santiago Vela, Gregorio de. 1913. *Ensayo de una biblioteca ibero-americana de la Orden de San Agustin.* Madrid: Impr. del Asilo de Huérfanos del S.C. de Jesús. http://archive.org/details/ensayodeunabiblio3santuoft. Accessed July 31, 2018. Web.

Sanz, Carlos, ed. 1958. *Primer documento impreso de la historia de las Islas Filipinaz; Relata la expedición de Legazpi, que llegó a Cebú en 1565, estampado en Barcelona el año de 1566.* Madrid: Graficas Yagues. Print.

Sarmiento de Gamboa, Pedro. 1988. *Historia de los Incas.* Madrid: Miraguano Ediciones: Ediciones Polifemo. Print.

———. 2000. *Viajes al Estrecho de Magallanes.* Edited by Juan Bautista. Las Rozas, Madrid: Dastin. Print.

———. 2015. *Sumaria relación.* Edited by Joaquín Zuleta Carrandi. Bibioteca Indiana 40. Pamplona, Madrid, and Frankfurt: Iberoamericana Vervuert. Print.

Sepúlveda, Juan Ginés de. 1984. *Démócrates Segundo, o, de las justas causas de la guerra contra los indios.* Vol. 2a. Madrid: Consejo Superior de Investigaciones Cientâificas, Instituto Francisco de Vitoria. Print.

Tordesillas, Agustín de. 1578. "Relación de el viaje que hezimos en China nuestro hermano Fray Pedro de Alpharo . . ." Real Academia de la Historia. *La China en España. Elaboración de un corpus digitalizado de documentos españoles sobre China de 1555 a 1900.* https://www.upf.edu/asia/projectes/che/s16/tordes.htm. Accessed July 31, 2018. Web.

Torquemada, Juan de. 1971. *Monarquía Indiana* - UNAM IIH. 1971. http://www.historicas.unam.mx/publicaciones/publicadigital/monarquia/. Accessed July 31, 2018. Web.

Transylvanus, Maximilian. 1523. *De Moluccis Insulis itemq[ue] aliis pluribus mira[n]dis, quae novissima Castellanorum navigatio Serenis Imperatoris Caroli V auspicio suscepta nuper invenit: Maximiliani Transylvani ad Reverendiss.*

Cardinalem Saltzburgensem epistola lectu perquam iucunda. Cologne. http://
archive.org/details/demoluccisinsulioomaxi. Accessed July 31, 2018. Web.

———. 1969. "De Moluccis Insulis." In *First Voyage around the World*, edited by
Carlos Quirino. 110–30. Manila: Filipiniana Book Guild. Print.

Urdaneta, Andrés de. 1560. "Documento 5: Parecer de Urdaneta que acompaña
la carta anterior [de Urdaneta al Rey]." In Hidalgo Nuchera, *Los primeros de
Filipinas*, 85–86.

———. 1561. "Documento 8: Derrotero muy especial, hecho por Fray Andrés de
Urdaneta, de la navegación que había de hacer desde el puerto de Acapulco
para las islas de pontiente, 1561." In Hidalgo Nuchera, *Los primeros de Filipinas*,
90–97. Print.

———. 1954. "Relación escrita y presentada al Emperador por Andrés de Urdaneta
de los sucesos de la armada del Comendador Loaisa desde 24 de julio de 1525
hasta el año 1535 (AGI Legajo 10, Papeles Del Maluco Desde 1519 a 1547)." In
Obras. Edición y Estudio Preliminar de Carlos Seco Serrano, edited by Martín
Fernández de Navarrete, 3 (77):226–50. *Biblioteca de autores españoles desde
la formación del lenguaje hasta nuestros días* (Continuacion) 75–77. Madrid:
Ediciones Atlas. Print.

Varela, Consuelo. 1983. *El viaje de Don Ruy López de Villalobos a las Islas del Poniente*. Milan: Cisalpino-Goliardica. Print.

Vega Carpio, Lope Felix de. 2006. *Los mártires de Japón*. Edited by Christina Hyo
Jung Lee. Newark, DE: Juan de la Cuesta. Print.

———. 1965. *Triunfo de la fee en los Reynos del Japón*. Edited by J. S. Cummins.
Colección Támesis Serie B Textos, 1. London: Tamesis Books. Print.

Vera, Santiago de. 1859. "Memorial de la junta de Manila al Consejo de Indias." In
*Documentos, datos, y relaciones para la historia de Filipinas hasta hora inéditos
fielmente copiados de los originales existentes en los archivos y bibliotecas por
Ventura del Arco, auditor de Marina oficial d la Contaduría General del Ejército*,
edited by Ventura del Arco. Vol. 1. 1586–1630. Chicago: Newberry Library.
Print.

Vitoria, Francisco de. 1991. *Francisco de Vitoria: Political Writings*. Edited by
Anthony Pagden and Jeremy Lawrance. Cambridge: Cambridge University
Press. Print.

Waldseemüller, Martin. 2008. *The Naming of America: Martin Waldseemüller's
1507 World Map and the Cosmographiae Introductio*. Edited by John W. Hessler.
London: GILES, in association with the Library of Congress. Print.

Secondary Sources

Amann, Elizabeth M. 1997. "Orientalism and Transvestism: Góngora's 'Discurso
contra las navegaciones' (Soledad Primera)." *Caliope: Journal of the Society for
Renaissance & Baroque Hispanic Poetry* 3 (1): 18–34.

Andaya, Barbara Watson. 2015. *A History of Early Modern Southeast Asia*. Cambridge and New York: Cambridge University Press. Print.

Andaya, Leonard Y. 1993. *The World of Maluku: Eastern Indonesia in the Early Modern Period*. Honolulu: University of Hawaii Press. Print.

Armitage, David, and Alison Bashford, eds. 2014. *Pacific Histories: Ocean, Land, People*. New York: Palgrave Macmillan. Print.

Arteche, José. 1968. *Urdaneta, el dominador de los espacios del Oceano Pacífico*. 2nd ed. San Sebastian: Sociedad Guipuzcoana de ediciones y publicaciones. Print.

Artes de México. 1971. *El galeón de Manila*. México: Artes de México. Print.

Ballesteros Gaibrois, Manuel. 1981. *Gonzalo Fernández de Oviedo*. Madrid: Fundación Universitaria Española. Print.

Barnes, Jerome R. 2007. "Giovanni Battista Ramusio and the History of Discoveries: An Analysis of Ramusio's Commentary, Cartography, and Imagery in *Delle navigationi et viaggi*." PhD Dissertation, University of Texas at Arlington. Print.

Barrera-Osorio, Antonio. 2006. *Experiencing Nature: The Spanish American Empire and the Early Scientific Revolution*. Austin: University of Texas Press. Print.

Basch, Martín Almagro, Luis Suárez Fernández, Vicente Palacio Atard, and Francisco Morales Padrón. 1962. *Manual de historia universal: Historia de América, por Francisco Morales Padrón*. Madrid: Espasa Calpe. Print.

Baudot, Georges. 1990. *La pugna franciscana por México*. México, D.F.: Alianza Editorial Mexicana: Consejo Nacional para la Cultura y las Artes. Print.

———. 1995. *Utopia and History in Mexico: The First Chroniclers of Mexican Civilization (1520–1569)*. Niwot: University Press of Colorado. Print.

Baxandall, Michael. 1980. *The Limewood Sculptors of Renaissance Germany*. New Haven, CT: Yale University Press. Print.

Bedini, Silvio A., and David Buisseret, eds. 1992. *The Christopher Columbus Encyclopedia*. New York: Simon & Schuster. Print.

Benítez, Fernando. 1992. *El galeón del Pacífico: Acapulco-Manila, 1565–1815*. Mexico City: Gobierno Constitucional del Estado de Guerrero. Print.

Bennett, Ira E. 1915. *History of the Panama Canal*. Washington, DC: Washington Historical Publishing Co. Print.

Bernabéu Albert, Salvador, and Carlos Martínez Shaw. 2013. *Un océano de seda y plata: el universo económico del Galeón de Manila*. Seville: Consejo Superior de Investigaciones Científicas. Print.

Berthe, Jean-Pierre. 1998. "Juan López de Velasco (ca. 1530–1598), Cronista y osmógrafo Mayor del Consejo de Indias: Su personalidad y su obra geográfica." *Relaciones: Estudios de historia y sociedad* XIX (75): 143–72. Print.

Besse, Jean-Marc. 2003. *Les grandeurs de la terre: Aspects du savoir géographique à la Renaissance*. Collection Sociétés, Espaces, Temps. Lyon: Ens. Print.

Biedermann, Zoltán. 2016. "The Temporal Politics of Spiritual Conquest: History, Geography and Franciscan Orientalism in the *Conquista espiritual do Oriente* of Friar Paulo Da Trindade." *Culture & History Digital Journal* 5 (2): e014.

http://cultureandhistory.revistas.csic.es/index.php/cultureandhistory/article
/view/101. Accessed July 31, 2018. Web.

Blackmore, Josiah. 2002. *Manifest Perdition: Shipwreck Narrative and the Disruption of Empire*. Minneapolis: University of Minnesota Press. Print.

Blanco, John D. *Frontier Constitutions: Christianity and Colonial Empire in the Nineteenth-Century Philippines*. 2009. Berkeley: University of California Press. Print.

Borao, José Eugenio. 1998. "The Massacre of 1603: Chinese Perception of the Spaniards in the Philippines." *Itinerario* 23 (1): 22–39. Print.

Boruchoff, David A. 2012. "The Three Greatest Inventions of Modern Times: An Idea and Its Public." In *Entangled Knowledge: Scientific Discourses and Cultural Difference*, 133–36. Münster and New York: Waxmann. Print.

Bourne, Edward Gaylord. 1904. *Spain in America, 1450–1580*. New York: Harper & Brothers. Print.

———. 1903. "Historical Introduction." In Blair and Robertson, *The Philippine Islands, 1493–1803*. Vol. 1. Cleveland: Arthur H. Clark. Print.

Boxer, Charles R. 1974. *The Christian Century in Japan: 1549–1650*. Reprint. Berkeley: University of California Press. Print.

———. 1978. *The Church Militant and Iberian Expansion, 1440–1770*. Baltimore: Johns Hopkins University Press. Print.

———. 2010. *South China in the Sixteenth Century, Being the Narratives of Galeote Pereira, Fr. Gaspar Da Cruz, O.P. [and] Fr. Martín De Rada, O.E.S.A. (1550–1575)*. London: Hakluyt Society. Print.

Bradley, Peter T. 2009. *Spain and the Defence of Peru, 1579–1700*. Lexington, KY: Lulu.com. Print.

Broc, Numa. 1986. *La geographie de la Renaissance, 1420–1620*. Paris: Les Éditions du C.T.H.S.

Brockey, Liam. 2007. *Journey to the East: The Jesuit Mission to China, 1579–1724*. Cambridge, MA: Belknap Press of Harvard University Press. Print.

———. 2016. "Conquests of Memory: Franciscan Chronicles of the East Asian Church in the Early Modern Period." *Culture & History Digital Journal* 5 (2): 015. Print.

Brook, Timothy. 1999. *The Confusions of Pleasure: Commerce and Culture in Ming China*. Berkeley: University of California Press. Print.

———. 2010. *The Troubled Empire: China in the Yuan and Ming Dynasties*. Cambridge, MA: Harvard University Press. Print.

Brook, Timothy, Jérôme Bourgon, and Gregory Blue. 2008. *Death by a Thousand Cuts*. Cambridge, MA: Harvard University Press.

Brotton, Jerry. 2004. *Trading Territories: Mapping the Early Modern World*. London: Reaktion Books. Print.

Buisseret, David, and Arthur Holzheimer. 1992. *The "Ramusio" Map of 1534: A Facsimile Edition*. Chicago: Newberry Library. Print.

Cabezas, Antonio, Universidad de Valladolid, Instituto de Estudios Japoneses, and Kokusai Kōryū Kikin. 1995. *El siglo ibérico del Japón: la presencia hispano-*

portuguesa en Japón (1543–1643). Valladolid: Instituto de Estudios Japoneses, Universidad de Valladolid: Secretariado de Publicaciones, Universidad de Valladolid. Print.

Cachey, Theodore J. 2007. "Introduction." In Pigafetta, *The First Voyage around the World, 1519–1522: An Account of Magellan's Expedition*, ix–xxxvi. Toronto: University of Toronto Press. Print.

Campbell, Mary B. 2004. *Wonder and Science: Imagining Worlds in Early Modern Europe*. Ithaca, NY: Cornell University Press. Print.

Campbell, Tony. 1987. "Portolan Charts from the Late Thirteenth Century to 1500." In *The History of Cartography*, 371–446. Chicago: University of Chicago Press. Print.

Cañizares-Esguerra, Jorge. 2001. *How to Write the History of the New World: Histories, Epistemologies, and Identities in the Eighteenth-Century Atlantic World*. Stanford, CA: Stanford University Press. Print.

Carey, Daniel, ed. 2004. *Asian Travel in the Renaissance*. Oxford: Society for Renaissance Studies and Blackwell. (Reprint articles that originally appeared in *Renaissance Studies* 17:3). Print.

Carroll, Lewis. 1939. "The Hunting of the Snark." In *The Complete Works of Lewis Carroll with an Introduction by Alexander Woollcott and the Illustrations by John Tenniel*, 757. London and New York: Random House / Nonesuch Press. Print.

Cerezo Martínez, Ricardo. 1994. *La cartografía nautica española en los siglos XIV, XV, y XVI*. Madrid: CSIC. Print.

Certeau, Michel de. 1984. *The Practice of Everyday Life*. Berkeley: University of California Press. Print.

———. 1988. *The Writing of History*. New York: Columbia University Press. Print.

Chaplin, Joyce E. 2012. *Round about the Earth: Circumnavigation from Magellan to Orbit*. 1st Simon & Schuster hardcover ed. New York: Simon & Schuster. Print.

Chaunu, Pierre. 1960. *Les Philippines et Le Pacifique Des Iberiques: Introduction Methodologique et Indices d'activite*. Paris: S.E.V.P.E.N. Print.

Christina Findlay, Connett. 2014. "Cartography as a Strategy of Empire in *Historia general de los hechos de los Castellanos en las islas i tierra firme del Mar Oceano* by Antonio de Herrera y Tordesillas." PhD Dissertation, Universitat de València, Valencia, Spain. Print.

Colín, Francisco. 1900. *Labor evangelica, ministerios apostólicos de los obreros de la Compañia de Jesús, fundación, y progressos de su provincia en las Islas Filipinas*. Edited by Pablo Pastells. 3 vols. Barcelona: Impr. y litografía de Henrich y compañía. Print.

Cooper, Michael. 2005. *The Japanese Mission to Europe, 1582–1590: The Journey of Four Samurai Boys through Portugal, Spain and Italy*. Folkestone, Kent, UK: Global Oriental. Print.

Corbeiller, Clare Le. 1961. "Miss America and Her Sisters: Personifications of the Four Parts of the World." *Metropolitan Museum of Art Bulletin*, New Series, 19 (8): 209–23. Print.

Correia, Pedro Lage Reis. 2001. "Alessandro Valignano's Attitude towards Jesuit and Franciscan Concepts of Evangelization in Japan (1587–1597)" *Bulletin of Portuguese / Japanese Studies*, no. 2 (June): 79–108. Print.

Cortesão, Armando. 1935. *Cartografia e cartógrafos portugueses dos séculos XV e XVI*. 2 vols. Lisbon: Edição da "Seara Nova." Print.

Cosgrove, Denis. 2001. *Apollo's Eye: A Cartographic Genealogy of the Earth in the Western Imagination*. Baltimore and London: Johns Hopkins University Press. Print.

Costa, Horacio de la. 1967. *The Jesuits in the Philippines, 1581–1768*. Cambridge, MA: Harvard University Press. Print.

Crossley, John N. 2011. *Hernando de Los Ríos Coronel and the Spanish Philippines in the Golden Age*. Farnham, Surrey, England; Burlington, VT: Ashgate. Print.

Cuesta Domingo, Mariano. 2015. *Antonio de Herrera y Tordesillas: Estudio crítico*. Madrid: Fundación Ignacio Larramendi. Print.

Currier, Charles Warren. 1898. *History of Religious Orders . . . Together with a Brief History of the Catholic Church in Relation to Religious Orders*. New York: Murphy & McCarthy. Print.

Daston, Lorraine, and Katharine Park. 2001. *Wonders and the Order of Nature, 1150–1750*. New York: Zone Books. Print.

Davies, Surekha. 2003. "The Navigational Iconography of Diogo Ribeiro's 1529 Vatican Planisphere." *Imago Mundi: The International Journal for the History of Cartography* 55 (1): 103–12. Print.

———. 2016. *Renaissance Ethnography and the Invention of the Human: New Worlds, Maps and Monsters*. Cambridge: Cambridge University Press. Print.

Disney, A. R. 2009. *A History of Portugal and the Portuguese Empire*. 2 vols. New York: Cambridge University Press. Print.

Edney, Matthew H. 1993. "Cartography without 'Progress': Reinterpreting the Nature and Historical Development of Mapmaking." *Cartographica* 30 (2/3): 54–68. Print.

Edson, Evelyn. 1997. *Mapping Time and Space: How Medieval Mapmakers Viewed Their World*. The British Library Studies in Map History. London: British Library. Print.

———. 2007. *The World Map, 1300–1492: The Persistence of Tradition and Transformation*. Baltimore: Johns Hopkins University Press. Print.

Eire, Carlos M. N. 1986. *War against the Idols: The Reformation of Worship from Erasmus to Calvin*. Cambridge and New York: Cambridge University Press. Print.

Eisler, William. 1995. *The Furthest Shore: Images of Terra Australis from the Middle Ages to Captain Cook*. Cambridge: Cambridge University Press. Print.

Elison, George. 1973. *Deus Destroyed; The Image of Christianity in Early Modern Japan*. Cambridge, MA: Harvard University Press. Print.

Ellis, Robert Richmond. 2012. *They Need Nothing: Hispanic-Asian Encounters of the Colonial Period*. Toronto: University of Toronto Press. Print.

Estrada de Gerlero, Elena Isabel. 2000. "Los protomártires del Japón en la hagiografía novohispana." In *Los pinceles de la historia. De la patria criolla a la nación mexicana. 1750–1850*, 72–89. Mexico City: Consejo Nacional para la Cultura y las Artes. Print.

Ezquerra Abadía, Ramón. 1975. "La idea del antimeridiano." In *A viagem de Fernão de Magalhães e a questão das Molucas: Actas do II colóquio luso-espanhol de história ultramarina*, edited by A. Teixeira da (Avelino Teixeira) Mota. Lisboa: Junta de Investigações Científicas do Ultramar. Print.

Fairbank, John King, and Denis Twitchett. 1978. *The Cambridge History of China: The Ming Dynasty, 1368–1644*. 2 vols. Cambridge: Cambridge University Press. Print.

Fernández-Armesto, Felipe. 1987. *Before Columbus: Exploration and Colonization from the Mediterranean to the Atlantic, 1229–1492*. Philadelphia: University of Pennsylvania Press. Print.

———. 2007a. "Maps and Exploration in the Sixteenth and Seventeenth Centuries." In *Cartography in the European Renaissance*, edited by David Woodward, 738–70. Print.

———. 2007b. *Pathfinders: A Global History of Exploration*. Oxford: Oxford University Press. Print.

Flint, Richard, ed. 2005. *Documents of the Coronado Expedition*. 1st ed. Dallas: Southern Medthodist University Press. Print.

Flint, Richard and Shirley Cushing Flint. 2008. *No Settlement, No Conquest: A History of the Coronado Entrada*. Albuquerque: University of New Mexico Press. Print.

Flynn, Dennis O., Lionel Frost, and A. J. H. Latham. 2002. *Pacific Centuries: Pacific and Pacific Rim Economic History since the Sixteenth Century*. London: Routledge. Print.

Flynn, Dennis O., and Arturo Giráldez. 1996. "China and the Spanish Empire." *Revista de Historia Económica (Second Series)* 14 (2): 309–38. Print.

———. 2010. *China and the Birth of Globalization in the Sixteenth Century*. London: Ashgate Variorum.

Flynn, Dennis O, Arturo Giráldez, and James Sobredo, eds. 2001. *European Entry into the Pacific: Spain and the Acapulco-Manila Galleons. The Pacific World*, 4. Aldershot: Ashgate. Print.

Frank, Andre Gunder. 1998. *Reorient: Global Economy in the Asian Age*. Berkeley: University of California Press. Print.

Freedberg, David. 1991. *The Power of Images: Studies in the History and Theory of Response*. Chicago: University of Chicago Press. Print.

Freeman, Donald B. 2010. *The Pacific*. London; New York: Routledge. Print.

Frost, Elsa Cecilia. 2002. *La historia de Dios en las Indias: Visión franciscana del Nuevo Mundo*. México, D.F.: Tusquets Editores. Print.

Fuchs, Barbara. 2009. "*La grandeza mexicana* de Balbuena y el imaginario de una 'metropolis colonial.'" *Revista Iberoamericana* 75 (228): 675–95. Print.

García-Abásolo, Antonio F. 2012. *Murallas de piedra y cañones de seda: chinos en el Imperio español (siglos XVI–XVIII)*. Córdoba: Servicio de Publicaciones, Universidad de Córdoba. Print.

Gaylord Randel, Mary. 1978. "Metaphor and Fable in Góngora's Soledad Primera." *Revista Hispánica Moderna* 40: 97–112. Print.

Geertz, Clifford. 1988. *Works and Lives: The Anthropologist as Author.* Stanford, CA: Stanford University Press. Print.

Gerbi, Antonello. 1973. *The Dispute of the New World: The History of a Polemic, 1750–1900.* Translated by Jeremy Moyle. Revised and enlarged. Pittsburgh, PA: University of Pittsburgh Press. Print.

———. 1986. *Nature in the New World: From Christopher Columbus to Gonzalo Fernández de Oviedo.* Translated by Jeremy Moyle. Pittsburgh, PA: University of Pittsburgh Press. Print.

Gil, Juan. 1987. *El libro de Marco Polo anotado por Cristobal Colón; El libro de Marco Polo versión de Rodrigo de Santaella.* Madrid: Alianza. Print.

———. 1991. *Hidalgos y samurais: España y Japón en los siglos XVI y XVII.* Madrid: Alianza Editorial. Print.

———. 1995. "Los pasajes cosmográficos de los Comentarios a los Psalmos de Jaime Pérez de Valencia." *Mar Oceana: Revista del humanismo español y latino-americano* 2: 259–82. Print.

———. 2011. *Los chinos en Manila: siglos XVI y XVII.* Macau: Centro Científico e Cultural de Macau. Print.

Gillies, John. 1994. *Shakespeare and the Geography of Difference.* New York: Cambridge: Cambridge University Press. Print.

Gillis, John. 2009. *Islands of the Mind: How the Human Imagination Created the Atlantic World.* New York: Palgrave Macmillan. Print.

Giráldez, Arturo. 2015. *The Age of Trade: The Manila Galleons and the Dawn of the Global Economy.* Lanham, MD: Rowman & Littlefield. Print.

Gliozzi, Giuliano. 2000. *Adam et le nouveau monde: la naissance de l'anthropologie comme idéologie coloniale; des généalogies bibliques aux théories raciales; 1500–1700.* Translated by Arlette Estève and Pascal Gabellone. Nîmes: Théétète Éd. Print.

Goodman, David. 1988. *Power and Penury: Government, Technology, and Science in Philip II's Spain.* Cambridge; New York: Cambridge University Press. Print.

Green, Otis H. 1952. "Bartolomé Leonardo de Argensola y el reino de Aragón." *Archivo de Filología Aragonesa* 4:7–112. Print.

Gruzinski, Serge. 2004. *Les quatre parties du monde: Histoire d'une mondialisation.* Paris: Seuil. Print.

———. 2014. *The Eagle and the Dragon: Globalization and European Dreams of Conquest in China and America in the Sixteenth Century.* Translated by Jean Birrell. Cambridge: Polity Press. Print.

Harley, J. Brian, and Paul Laxton. 2001. *The New Nature of Maps: Essays in the History of Cartography.* Baltimore: Johns Hopkins University Press. Print.

Harley, J. Brian, and David Woodward, eds. 1987. *Cartography in Prehistoric, Ancient, and Medieval Europe and the Mediterranean. The History of Cartography,* vol. 1. Chicago: University of Chicago Press.

———. 1994. *Cartography in the Traditional East and Southeast Asian Societies.* Vol. 2, book 2. *The History of Cartography.* Chicago: University of Chicago Press. Print.

Headley, John M. 1995. "Spain's Asian Presence, 1565–1590: Structures and Aspirations." *Hispanic American Historical Review* 75 (4): 623–46. Print.

Helms, Mary W. 1988. *Ulysses' Sail: An Ethnographic Odyssey of Power, Knowledge, and Geographical Distance.* 1st ed. Princeton, NJ: Princeton University Press. Print.

Herzog, Tamar. 2015. *Frontiers of Possession: Spain and Portugal in Europe and the Americas.* Cambridge and London: Harvard University Press. Print.

Hessler, John W., and Chet A. Van Duzer. 2012. *Seeing the World Anew: The Radical Vision of Martin Waldseemüller's 1507 and 1516 World Maps.* Delray Beach, FL: Levenger Press and the Library of Congress. Print.

Hiatt, Alfred. 2008. *Terra Incognita: Mapping the Antipodes before 1600.* Chicago: University of Chicago Press. Print.

Horodowich, Elizabeth. 2016. *The Venetian Discovery of America: Geographic Imagination in the Age of Encounters.* Cambridge: Cambridge University Press. Print.

Horsley, Margaret Wyant. 1950. "Sangley: The Formation of Anti-Chinese Feeling in the Philippines, a Cultural Study of the Stereotypes of Prejudice." PhD Dissertation, Columbia University, New York. Print.

Hsu, Carmen. 2010. "La imagen humanística del gran reino Chino de Juan González de Mendoza." *Bulletin of Hispanic Studies* 87 (2): 187–202. Print.

Jackson, Andrew, Art R. T. Jonkers, and Matthew R. Walker. 2000. "Four Centuries of Geomagnetic Secular Variation from Historical Records." *Philosophical Transactions: Mathematical, Physical and Engineering Sciences* 358 (1768): 957–90. Print.

Jacob, Christian. 1992. *L'empire des cartes: Approche théorique de la cartographie à travers l'histoire.* Paris: Albin Michel. Print.

———. 2006. *The Sovereign Map: Theoretical Approaches in Cartography throughout History.* Translated by Edward H. Dahl. Chicago: University of Chicago Press. Print.

Johnson, Carina L. 2006. "Idolatrous Cultures and the Practice of Religion." *Journal of the History of Ideas* 67 (4): 597–621. Print.

———. 2011. *Cultural Hierarchy in Sixteenth-Century Europe: The Ottomans and Mexicans.* Cambridge: Cambridge University Press. Print.

Jones, Eric Lionel. 1993. *Coming Full Circle: An Economic History of the Pacific Rim.* Boulder, CO: Westview Press. Print.

Kagan, Richard L. 2010. *Los cronistas y la corona: la política de la historia en España en las edades media y moderna.* Translated by Pablo Sánchez León. Madrid: Centro de Estudios Europa Hispánica y Marcial Pons Historia. Print.

Kagan, Richard, and Benjamin Schmidt. 2007. "Maps and the Early Modern State: Official Cartography." In *Cartography in the European Renaissance,* edited by David Woodward, 1: 661–79. Print.

Kamen, Henry. 2003. *Empire: How Spain Became a World Power, 1492–1763.* 1st American ed. New York: Harper Collins. Print.

Karrow, Robert. 2007. "Centers of Map Publishing in Europe, 1472–1600." In *Cartography in the European Renaissance*, edited by David Woodward, 3:611–21. Print.

Keevak, Michael. 2011. *Becoming Yellow: A Short History of Racial Thinking.* Princeton, NJ: Princeton University Press. Print.

Kelly, Celsus. 1965. *Calendar of Documents: Spanish Voyages in the South Pacific from Alvaro de Mendaña to Alejandro Malaspina, 1567–1794, and the Franciscan Missionary Plans for the Peoples of the Austral Lands, 1617–1634.* Madrid: Franciscan Historical Studies (Australia) in association with Archive Ibero-Americano (Madrid). http://catalog.hathitrust.org/Record/008463215. Accessed July 31, 2018. Web.

Kelsey, Harry. 1984. "Mapping the California Coast the Voyages of Discovery 1533–1543." *Arizona and the West* 26 (4): 307–24. Print.

———. 1998. *Juan Rodríguez Cabrillo.* San Marino, CA: Huntington Library. Print.

Kimmel, Seth R. 2010. "Interpreting Inaccuracy: The Fiction of Longitude in Early Modern Spain." *Journal of Medieval and Early Modern Studies* 40 (2): 299–323. Print.

Kish, George. 1949. "Some Aspects of the Missionary Cartography of Japan during the Sixteenth Century." *Imago Mundi* 6:39–47. Print.

Kivelson, Valerie. 2006. *Cartographies of Tsardom: The Land and Its Meanings in Seventeenth-Century Russia.* Ithaca, NY: Cornell University Press. Print.

Kolodny, Annette. 1975. *The Lay of the Land: Metaphor as Experience and History in American Life and Letters.* Chapel Hill: University of North Carolina Press. Print.

Kowner, Rotem. 2014. *From White to Yellow: The Japanese in European Racial Thought, 1300–1735.* Montreal: McGill-Queen's University Press. Print.

Kunstmann, Friedrich, Karl Spruner von Merz, and Georg Martin Thomas. 1859. *Die Entdeckung Amerikas.* Monumenta saecularia. München: In commission bei A. Asher & Cie in Berlin. Print.

Lach, Donald F. 1965. *Asia in the Making of Europe.* 3 vols. Chicago: University of Chicago Press. Print.

Lapsanski, Duane V. 1977. *Evangelical Perfection: An Historical Examination of the Concept in the Early Franciscan Sources.* St. Bonaventure, NY: Franciscan Institute, St. Bonaventure University. Print.

Lee, Christina Hyo Jung. 2006. "Introduction." In *Los mártires de Japón*, by Lope de Vega, edited by Christina Hyo Jung Lee, 9–44. Print.

———. 2008. "The Perception of the Japanese in Early Modern Spain: Not Quite 'The Best People Yet Discovered.'" *eHumanista* 11. http://www.ehumanista.ucsb.edu/sites/secure.lsit.ucsb.edu.span.d7_eh/files/sitefiles/ehumanista/volume11/15%20Lee.pdf. Accessed July 31, 2018. Web.

Lefebvre, Henri. 1991. *The Production of Space.* Translated by Donald Nicholson-Smith. Cambridge: Blackwell. Print.

Lehmann, Martin. 2013. "Amerigo Vespucci and His Alleged Awareness of America as a Separate Land Mass." *Imago Mundi* 65 (1): 15–24. Print.

Leibsohn, Dana, and Meha Priyadarshini. 2016. "Transpacific: Beyond Silk and Silver." *Colonial Latin American Review* 25 (1): 1–15. Print.

León-Portilla, Miguel. 2005. *Hernán Cortés y la Mar del Sur.* Madrid: Algaba. Print.

Lestringant, Frank. 1991. "The Crisis of Cosmography at the End of the Renaissance." In *Humanism in Crisis: The Decline of the French Renaissance*, 153–79. Ann Arbor: University of Michigan Press. Print.

Lewis, Martin W. 1999. "Dividing the Ocean Sea." *Geographical Review* 89 (2): 188–214. Print.

Lewis, Martin W., and Karen E. Wigen. 1997. *The Myth of the Continents: A Critique of Metageography.* Berkeley: University of California Press. Print.

Livingstone, David N. 2008. *Adam's Ancestors: Race, Religion, and the Politics of Human Origins.* Baltimore: Johns Hopkins University Press. Print.

Lois, Carla. 2008. "Plus Ultra Equinoctaliem." PhD Dissertation, University of Buenos Aires, Argentina. Print.

Lourenço, Miguel. 2009. "De São Lázaro às Filipinas: imagens de um arquipélago na cartografia náutica ibérica do século XVI." In *Historias de la cartografía de Iberoamérica: nuevos caminos, viejos problemas*, edited by Héctor Mendoza Vargas and Carla Lois, 387–422. México, D.F.: Instituto de Geografía, UNAM: INEGI. Print.

Luengo, José María. 1996. *A History of the Manila-Acapulco Slave Trade (1565–1815).* Tubigon, Bohol, Philippines: Mater Dei Publications. Print.

Mack, John. 2013. *The Sea: A Cultural History.* London: Reaktion Books. Print.

Maroto Camino, Mercedes. 2005. *Producing the Pacific: Maps and Narratives of Spanish Exploration (1567–1606).* Amsterdam and New York: Rodopi. Print.

———. 2008. *Exploring the Explorers: Spaniards in Oceania, 1519–1794.* Manchester; New York: Manchester University Press/Palgrave Macmillan. Print.

Marshall, P. J. 1982. *The Great Map of Mankind: Perceptions of New Worlds in the Age of Enlightenment.* Cambridge, MA: Harvard University Press. Print.

Martínez Shaw, Carlos, and José Antonio Martínez Torres. 2014. *España y Portugal en el mundo, 1581–1668.* Madrid: Polifemo. Print.

Martín Merás, Luisa. 1992. *Cartografía marítima hispánica: La imagen de América.* Madrid: Lunwerg for Ministerio de Obras Públicas, Transportes y Medio Ambiente. Print.

Masselman, George. 1963. *The Cradle of Colonialism.* New Haven, CT: Yale University Press. Print.

Matsuda, Matt K. 2012. *Pacific Worlds: A History of Seas, Peoples, and Cultures.* Cambridge; New York: Cambridge University Press. Print.

Mazumdar, Sucheta. 1999. "The Impact of New World Crops on the Diet and Economy of China and India, 1600–1900." In *Food in Global History*, edited by Raymond Grew, 58–78. Boulder, CO: Westview Press. Print.

McNeill, J. R. 1994. "Of Rats and Men: A Synoptic Environmental History of the Island Pacific." *Journal of World History* 5 (2): 299–349. Print.

Medina, José. 1914. *El descubrimiento del Océano Pacífico: Vasco Núñez de Balboa, Hernando de Magallanes y sus compañeros.* 2 vols. Santiago de Chile: Imprenta Universitaria. Print.

Mena García, María del Carmen. 1992. *La ciudad en un cruce de caminos: Panamá y sus orígenes urbanos.* Seville: Consejo Superior de Investigaciones Científicas, Escuela de Estudios Hispano-Americanos. Print.

Mercene, Floro L. 2007. *Manila Men in the New World: Filipino Migration to Mexico and the Americas from the Sixteenth Century.* Diliman, Quezon City: University of the Philippines Press. Print.

Mignolo, Walter. 1982. "Cartas, crónicas y relaciones del descubrimiento y la conquista." In *Historia de la literatura hispanoamericana*, Vol. 1: *Época colonial*, edited by Manuel Alvar and Luis Íñigo Madrigal, 57–116. Madrid: Cátedra. Print.

———. 2003. *The Darker Side of the Renaissance: Literacy, Territoriality, and Colonization.* 2nd ed. Ann Arbor: University of Michigan Press. Print.

Mojarro Romero, Jorge. Forthcoming. "Filipinas en la temprana historiografía indiana." *Revista de Indias.* Print.

Montero y Vidal, José. 1887. *Historia general de Filipinas desde el descubrimiento de dichas islas hasta nuestros días.* Madrid: Imprenta y fundición de M. Tello. Print.

Montrose, Louis. 1993. "The Work of Gender in the Discourse of Discovery." In *New World Encounters*, 177–217. Berkeley: University of California. Print.

Moorehead, Alan. 2000. *The Fatal Impact: An Account of the Invasion of the South Pacific, 1767–1840.* London; Toronto: Penguin Books. Print.

Morales, Alfredo J. 2011. "Desde Manila. El 'Aspecto Symbólico del Mundo Hispánico' de Vicente de Memije y Laureano Atlas." In *Arte en los confines del imperio: Visiones hispánicas de otros mundos*, 353–73. Castelló: Universitat Jaume I. Print.

Morales Padrón, Francisco. 1963. *Historia del descubrimiento y conquista de América.* Colección Mundo Científico, Serie Historia. Madrid: Editorial Nacional. Print.

Moran, J. F. 2012. *The Japanese and the Jesuits: Alessandro Valignano in Sixteenth Century Japan.* London and New York: Routledge. Print.

Morison, Samuel Eliot. 1971. *The European Discovery of America.* 2 vols. New York: Oxford University Press. Print.

Mühlhahn, Klaus. 2009. *Criminal Justice in China: A History.* Cambridge, MA: Harvard University Press.

Mungello, D. E. 2009. *The Great Encounter of China and the West, 1500–1800.* 3rd ed. Lanham, MD: Rowman & Littlefield. Print.

Nagel, Alexander. 2013. *The Seventeenth Gerson Lecture: Some Discoveries of 1492: Eastern Antiquities and Renaissance Europe.* Groningen: University of Groningen. Print.

Nocentelli, Carmen. 2010. "Spice Race: The Island Princess and the Politics of Transnational Appropriation." *PMLA* 125 (3): 572–895. Print.

———. 2013. *Empires of Love: Europe, Asia, and the Making of Early Modern Identity*. Philadelphia: University of Pennsylvania Press. Print.

Nowell, Charles E. 1947. "The Discovery of the Pacific: A Suggested Change of Approach." *Pacific Historical Review* 16 (1): 1–10. Print.

Nunn, George E. 1929. *Origin of the Strait of Anian Concept*. Tall Tree Library 3. Philadelphia: Priv. print. Print.

O'Gorman, Edmundo. 1961. *The Invention of America: An Inquiry into the Historical Nature of the New World and the Meaning of Its History*. Bloomington: Indiana University Press. Print.

———. 1972. *Cuatro historiadores de Indias, Siglo XVI*. Vol. 1. SepSetentas, 51. Mexico City: Secretaría de Educación Pública. Print.

———. 1986. *La Invención de América*. Mexico City: Fondo de Cultura Económica. Print.

Ollé, Manel. 1998. "Estrategias filipinas respecto a China: Alonso Sánchez y Domingo Salazar en la empresa de China (1581–1593)." PhD Dissertation, Universitat Pompeu Fabra, Barcelona. Print.

———. 2000. *La invención de China: Percepciones y estrategias filipinas respecto a China durante el siglo XVI*. Wiesbaden: Otto Harrassowitz Verlag. Print.

———. 2002. *La empresa de China: De la Armada Invencible al Galeón de Manila*. 1st ed. Barcelona: Acantilado. Print.

———. 2008. "The Jesuit Portrayals of China between 1583–1590." *Bulletin of Portuguese-Japanese Studies* 16:45–57. Print.

Owens, Sarah E. 2017. *Nuns Navigating the Spanish Empire*. Albuquerque: University of New Mexico Press. Print.

Padrón, Ricardo. 2004. *The Spacious Word: Cartography, Literature, and Empire in Early Modern Spain*. Chicago: University of Chicago Press. Print.

———. 2006. "The Blood of Martyrs Is the Seed of the Monarchy: Empire, Utopia, and the Faith in Lope's Triunfo de La Fee En Los Reynos de Japón." *Journal of Medieval and Early Modern Studies* 36 (3): 517–37. Print.

———. 2007. "Against Apollo: Góngora's *Soledad Primera* and the Mapping of Empire." *MLQ* 68 (1): 363–93

———. 2011. "From Abstraction to Allegory: The Imperial Cartography of Vicente de Memije." In *Early American Cartographies*, edited by Martin Brückner, 35–66. Chapel Hill: University of North Carolina Press. Print.

———. 2014. "Producing China: Sinophobia vs. Sinophilia in the Sixteenth-Century Iberian World." *Review of Culture (Instituto Cultural do Governo da R.A.E de Macau)* 46: 94–107. Print.

Pagden, Anthony. 1986. *The Fall of Natural Man: The American Indian and the Origins of Comparative Ethnology*. Cambridge Iberian and Latin American Studies. Cambridge and New York: Cambridge University Press. Print.

———. 1990. *Spanish Imperialism and the Polticial Imagination: Studies in European and Spanish-American Social and Political Theory, 1513–1830*. New Haven, CT: Yale University Press. Print.

Pardo Tomás, José. 2006. *Un lugar para la ciencia: Escenarios de práctica científica*

en la sociedad hispana del siglo XVI. La Orotava, Tenerife: Fundación Canaria
Orotava de Historia de la Ciencia. Print.

Parra, Richard. 2015. *La tiranía del Inca: el Inca Garcilaso y la escritura política en
el Perú colonial (1568–1617)*. Petroperú, Ediciones Copé. Print.

Parry, J. H. 1974. *The Discovery of the Sea*. New York: Dial Press. Print.

——. 1990. *The Spanish Seaborne Empire*. Berkeley: University of California
Press. Print.

Pérez-Mallaína Bueno, Pablo Emilio. 1998. *Spain's Men of the Sea: Daily Life on
the Indies Fleets in the Sixteenth Century*. Translated by Carla Rahn Phillips.
Baltimore: Johns Hopkins University Press. Print.

Phelan, John Leddy. 1959. *The Hispanization of the Philippines: Spanish Aims
and Filipino Responses, 1565–1700*. Madison: University of Wisconsin Press.
Print.

——. 1970. *The Millennial Kingdom of the Franciscans in the New World*. Berkeley: University of California Press. Print.

Phillips, J. R. S. 1988. *The Medieval Expansion of Europe*. New York: Oxford University Press. Print.

Pierce, Donna, and Frederick and Jan Mayer Center for Pre-Columbian and
Spanish Colonial Art, Denver Art Museum. 2009. *Asia & Spanish America:
Trans-Pacific Artistic and Cultural Exchange, 1500–1850: Papers from the 2006
Mayer Center Symposium at the Denver Art Museum*. Denver: Denver Art
Museum. Print.

Pimentel, Juan. 2000. "The Iberian Vision: Science and Empire in the Framework
of a Universal Monarchy, 1500–1800." *Osiris* 15 (1): 17–30. Print.

Ponsonby-Fane, Richard Arthur Brabazon. 1956. *Kyoto: The Old Capital of Japan,
794–1869*. Kyoto: Ponsonby Memorial Society. Print.

Portuondo, María. 2009. *Secret Science: Spanish Cosmography and the New World*.
Chicago: University of Chicago Press. Print.

Rafael, Vicente L. 1988. *Contracting Colonialism: Translation and Christian
Conversion in Tagalog Society under Early Spanish Rule*. Ithaca, NY: Cornell
University Press. Print.

Ramachandran, Ayesha. 2015. *The Worldmakers: Global Imagining in Early Modern Europe*. Chicago, London: University of Chicago Press. Print.

Randles, W. G. L. 2000. "Classical Models of World Geography and Their Transformation Following the Discovery of America." In *Geography, Cartography
and Nautical Science in the Renaissance: The Impact of the Great Discoveries*,
5–76. Aldershot, Hampshire; Burlington, VT: Ashgate / Variorum. Print.

Reyes, Raquel A. G., and William G. Clarence-Smith. 2012. *Sexual Diversity in
Asia, c. 600–1950*. London: Routledge. Print.

Rodrigues Lourenço, Miguel. 2010. "De São Lázaro às Filipinas: imagens de um
arquipélago na cartografia náutica ibérica doséculo XVI." In *Mapas de metade
do mundo: A cartografia e a construção territorial dos espaços americanos,
séculos XVI a XIX = Mapas de la mitad del mundo: la cartografía y la construc-
ción territorial de los espacios americanos, siglos XVI al XIX*, edited by Francisco

Roque de Oliveira and Héctor Mendoza Vargas, 387–422. Centro de Estudos Geográficos, Universidadede Lisboa; Instituto de Geografía, Universidad Nacional Autónoma de México. Print.

Romano, Antonella. 2013. *Impressions de Chine: L'Europe et l'englobement du monde*. Paris: Fayard. Print.

Romm, James S. 1992. *The Edges of the Earth in Ancient Thought: Geography, Exploration, and Fiction*. Princeton, NJ: Princeton University Press. Print.

———. 2001. "Biblical History and the Americas: The Legend of Solomon's Ophir, 1492–1591." In *The Jews and the Expansion of Europe to the West, 1450 to 1800*, edited by Paolo Bernardini and Norman Fiering, 27–46. European Expansion and Global Interaction 2. New York: Berghahn Books. Print.

Roque de Oliveira, Francisco M. 2003. "A construção do conhecimento europeu sobre a China, c. 1500–c. 1630. Impressos e manuscritos que revelaram o mundo chinês à Europa culta." PhD Dissertation, Universitat Autònoma de Barcelona, Barcelona. Print.

Ross, Andrew C. 1994. *A Vision Betrayed: The Jesuits in Japan and China 1542–1742*. Edinburgh: Edinburgh University Press.

Rubial García, Antonio. 1996. *La hermana pobreza: El franciscanismo de la edad media a la evangelización novohispana*. Universidad Nacional Autónoma de México, Facultad de Filosofía y Letras. Print.

Rubiés, Joan-Pau. 2003. "The Spanish Contribution to the Ethnology of Asia in the Sixteenth and Seventeenth Centuries." In *Asian Travel in the* Renaissance, edited by Daniel Carey, 93–123. Print.

———. 2005. "Oriental Despotism and European Orientalism: Botero to Montesquieu." *Journal of Early Modern History* 9 (1/2): 109–80. Print.

———. 2006. "Theology, Ethnography, and the Historicization of Idolatry." *Journal of the History of Ideas* 67 (4): 571–96. Print.

Ruiz, Enrique Martínez. 1999. *Felipe II, la ciencia y la técnica*. Madrid: Editorial Actas. Print.

Safier, Neil. 2014. "The Tenacious Travels of the Torrid Zone and the Global Dimensions of Geographical Knowledge in the Eighteenth Century." *Journal of Early Modern History* 18 (1–2): 141–72. Print.

Said, Edward. 1978. *Orientalism*. New York: Random House. Print.

Sánchez, Antonio. 2013. *La espada, la cruz y el Padrón: soberanía, fe y representación cartográfica en el mundo ibérico bajo la Monarquía Hispánica, 1503–1598*. Madrid: Consejo Superior de Investigaciones Científicas. Print.

Sandman, Alison. 2001. "Cosmographers vs. Pilots: Navigation, Cosmography, and the State in Early Modern Spain." PhD Dissertation, University of Wisconsin. Print.

———. 2007. "Spanish Nautical Cartography in the Renaissance." In *Cartography in the European Renaissance*, edited by David Woodward, 1095–1449. Print.

Schmidt, Benjamin. 1997. "Mapping an Empire: Cartographic and Colonial Rivalry in Seventeenth-Century Dutch and English North America." *William and Mary Quarterly* 55, no. 3: 549–78. Print.

———. 2015. *Inventing Exoticism: Geography, Globalism, and Europe's Early Modern World*. Philadelphia: University of Pennsylvania Press. Print.

Schurz, William Lytle. 1939. *The Manila Galleon; Illustrated with Maps*. New York: E. P. Dutton & Co. Print.

Seed, Patricia. 1995. *Ceremonies of Possession: Europe's Conquest of the New World, 1492–1640*. Cambridge: Cambridge University Press. Print.

Seijas, Tatiana. 2014. *Asian Slaves in Colonial Mexico: From Chinos to Indians*. Cambridge: Cambridge University Press. Print.

Shalev, Zur. 2003. "Sacred Geography, Antiquarianism and Visual Erudition: Benito Arias Montano and the Maps in the Antwerp Polyglot Bible." *Imago Mundi: The International Journal for the History of Cartography* 55 (1): 56. Print.

Sheehan, Kevin Joseph. 2008. "Iberian Asia: The Strategies of Spanish and Portuguese Empire Building, 1540–1700." PhD Dissertation, University of California, Berkeley. Print.

Sheppard, Warren W., and Charles Carroll Soule. 1922. *Practical Navigation*. Jersey City, NJ: World Technical Institute. Print.

Shirley, Rodney. 2001. *Mapping of the World: Early Printed World Maps 1472–1700*. Riverside, CT: Early World Press. Print.

Slack Jr., Edward R. 2009. "The Chinos in New Spain: A Corrective Lens for a Distorted Image." *Journal of World History* 20 (1): 35–67. Print.

Sola, Emilio. 2012. *Historia de un desencuentro: España y Japón, 1580–1614*. Madrid: Fugaz Ediciones. Available online at Archivo de la Frontera, http://www.archivodelafrontera.com/wp-content/uploads/2012/05/Espana-y-Japon-XVI-XVII-Desencuentro.pdf. Accessed July 31, 2018. Web.

Spate, Oskar K. 1977. "'South Sea' to 'Pacific Ocean': A Note on Nomenclature." *Journal of Pacific History* 12 (4): 205–11. Print.

———. 1979. *The Spanish Lake*. Volume 1 of *The Pacific since Magellan*. Minneapolis: University of Minnesota Press. Print.

———. 1983. *Monopolists and Freebooters*. Volume 2 of *The Pacific since Magellan*. Minneapolis: University of Minnesota Press. Print.

———. 1988. *Paradise Found and Lost*. Volume 3 of *The Pacific since Magellan*. Minneapolis: University of Minnesota Press. Print.

Spence, Jonathan D. 1999. *The Chan's Great Continent: China in Western Minds*. New York: W. W. Norton. Print.

Steinberg, Philip E. 2001. *The Social Construction of the Ocean*. New York and London: Cambridge University Press. Print.

Stevenson, Edward Luther. 1971. *Terrestrial and Celestial Globes, Volume 1: Their History and Construction Including a Consideration of Their Value as Aids in the Study of Geography and Astronomy*. New Haven: Yale University Press for the Hispanic Society of America. Google Books. Web.

Suárez, Thomas. 1999. *Early Mapping of Southeast Asia*. Singapore: Periplus. Print.

———. 2004. *Early Mapping of the Pacific: The Epic Story of Seafarers, Adventur-

ers, and Cartographers Who Mapped the Earth's Greatest Ocean. Singapore: Periplus. Print.

Swecker, Zoe. 1960. "The Early Iberian Accounts of the Far East, 1550–1600." Ph.D. Dissertation, University of Chicago, Chicago. Print.

Thomas, Nicholas. 2014. "The Age of Empire in the Pacific." In Armitage and Bashford, *Pacific Histories: Ocean, Land, People,* 75–96. Print.

Trakulhun, Sven. 2004. "The Widening of the World and the Realm of History: Early European Approaches to the Beginning of Siamese History, c. 1500–1700." In Daniel Carey, ed., *Asian Travel in the Renaissance,* 67–92. Print.

Tremml-Werner, Birgit. 2015. *Spain, China, and Japan in Manila, 1571–1644: Local Comparisons and Global Connections.* Amsterdam: Amsterdam University Press. Print.

Üçerler, M. Antoni J. 2003. "Alessandro Valignano: Man, Missionary, and Writer." In Daniel Carey, ed., *Asian Travel in the Renaissance,* 12–41. Print.

Unger, Richard. 2010. *Ships on Maps: Pictures of Power in Renaissance Europe.* New York: Palgrave Macmillan. Print.

Unno, Kazutaka. 1994. "Cartography in Japan." In *The History of Cartography,* vol. 2, bk. 2. Harley and Woodward, eds., *Cartography in the Traditional African, American, Arctic, Australian, and Pacific Societies,* 346–477. Chicago: University of Chicago Press.

Valensi, Lucette. 1993. *The Birth of the Despot: Venice and the Sublime Porte.* Ithaca, NY: Cornell University Press. Print.

Valladares, Rafael. 2001. *Castilla y Portugal en Asia: (1580–1680): Declive imperial y adaptación.* Leuven: Leuven University Press. Print.

Vega y de Luque, Carlos Luís de la. 1979. "Un proyecto utópico: La conquista de China por España." *Boletín de la asociación española de orientalistas,* no. 15: 45–69. Print.

———. 1980. "Un proyecto utópico: La conquista de China por España." *Boletín de la asociación española de orientalistas,* no. 16: 33–56. Print.

———. 1981. "Un proyecto utópico: La conquista de China por España." *Boletín de la asociación española de orientalistas,* no. 17: 3–38. Print.

———. 1982. "Un proyecto utópico: La conquista de China por España." *Boletín de la asociación española de orientalistas,* no. 18: 3–46. Print.

Vilà, Lara. 2013. "La *Historia del gran reino de la China* de Juan González de Mendoza. Hacia un estudio de las crónicas de oriente en la España del siglo de oro." *Boletín hispánico helvético* 21 (Primavera): 71–97. Print.

Villiers, John. 2003. "'A Truthful Pen and an Impartial Spirit': Bartolomé Leonardo de Argensola and the *Conquista de las islas Malucas.*" In Carey, ed., *Asian Travel in the Renaisssance,* 124–49. Print.

Vogeley, Nancy. 1997. "China and the American Indies: A Sixteenth-Century 'History'." *Colonial Latin American Review* 6 (December): 165–84. Print.

Wey Gómez, Nicolás. 2008. *The Tropics of Empire: Why Columbus Sailed South to the Indies.* Cambridge, MA: MIT Press. Print.

———. 2013. "Memorias de la zona tórrida: El naturalismo clásico y la «tropicalidad» americana en el *Sumario de la natural historia de las Indias* de Gonzalo Fernández de Oviedo (1526)." *Revista de Indias* 73 (259): 609–32. Print.

White, Hayden. 1988. *The Content of the Form.* Baltimore: Johns Hopkins University Press. Print.

Wiesner-Hanks, Professor Merry E. 2015. *Mapping Gendered Routes and Spaces in the Early Modern World.* London: Taylor & Francis. Print.

Wigen, Kären. 2006. "Introduction." *American Historical Review* 111 (3): 717–21. Print.

Willeke, B. H. 1990. "Biographical Data on Early Franciscans in Japan (1582 to 1640)." *Archivum Franciscanum Historicum* 83 (1–2): 162–223. Print

Winius, George Davison. 1985. *The Black Legend of Portuguese India: Diogo Do Couto, His Contemporaries, and the "Soldado Prático": A Contribution to the Study of Political Corruption in the Empires of Early Modern Europe.* New Delhi: Concept Publications. Print.

Wood, Denis, and John Fels. 1992. *The Power of Maps.* London: Guilford Press. Print.

Woodward, David. 1987. "Medieval Mappaemundi." In *Cartography in Prehistoric, Ancient, and Medieval Europe and the Mediterranean,* edited by J. Brian Harley and David Woodward, 286–370. Print.

———, ed. 2007. *Cartography in the European Renaissance. The History of Cartography,* vol. 3. Chicago: University of Chicago Press.

Wright, Ione Stuessy. 1939. "The First American Voyage across the Pacific, 1527–1528: The Voyage of Alvaro de Saavedra Cerón." *Geographical Review* 29 (3): 472–82. Print.

Wroth, Lawrence C. 1944. "The Early Cartography of the Pacific." *Bibliographical Society of America, Papers* 38: 87–268. Print

Zerubavel, Eviatar. 1992. *Terra Cognita: The Mental Discovery of America.* New Brunswick, NJ: Rutgers University Press. Print

———. 1993. *The Fine Line.* Chicago: University of Chicago Press. Print.

———. 1996. "Lumping and Splitting: Notes on Social Classification." *Sociological Forum* 11 (3): 421–33. Print.

Zhang, Qiong. 2015. *Making the New World Their Own: Chinese Encounters with Jesuit Science in the Age of Discovery.* Leiden: Brill. Print.

Index